TEUBNER-TEXTE zur Mathematik · Band 132

Herausgeber/Editors:

Herbert Kurke, Berlin
Joseph Mecke, Jena
Rüdiger Thiele, Leipzig
Hans Triebel, Jena
Gerd Wechsung, Jena

Beratende Herausgeber/Advisory Editors:

Ruben Ambartzumian, Jerevan
David E. Edmunds, Brighton
Alois Kufner, Prag
Burkhard Monien, Paderborn
Rolf J. Nessel, Aachen
Claudio Procesi, Rom
Kenji Ueno, Kyoto

Gennadij A. Leonov · Volker Reitmann · Vera B. Smirnova

Non-Local Methods for Pendulum-Like Feedback Systems

Springer Fachmedien Wiesbaden GmbH 1992

ISBN 978-3-663-12262-3   ISBN 978-3-663-12261-6 (eBook)
DOI 10.1007/978-3-663-12261-6

This book is related to the non-local qualitative theory of systems with cylindrical phase space, generated, in particular, by phase synchronization problems. For the investigation of these systems (ordinary and functional differential equations, discrete systems) is used a synthesis of qualitative methods of two-dimensional systems (existence of cycles, homoclinic orbits, global bifurcations) and methods of the mathematical stability theory (Lyapunov functions, Popov functionals, frequency-domain criteria).

Dieses Buch ist der nicht-lokalen qualitativen Theorie von Systemen mit zylindrischem Phasenraum gewidmet, die u.a. in der Phasensynchronisation entstehen. Für die Untersuchung dieser Systeme (gewöhnliche und funktionale Differentialgleichungen, diskrete Systeme) wird eine Synthese von qualitativen Methoden für zwei-dimensionale Systeme (Existenz von Zyklen, homokline Orbits, globale Bifurkationen) und Methoden der mathematischen Stabilitätstheorie (Lyapunow-Funktionen, Popov-Funktionale, Frequenzgang-Bedingungen) benutzt.

Le livre est consacré à la théorie qualitative globale des systèmes de l'espace de phase cylindrique. On obtient ces systèmes si on regard, en particulier, des problèmes de synchronisation de phase en Mécanique et Electrotechnique. Nous considérons des systèmes dynamiques décrivés par des équations différentielles ordinaires, des équations integro-différentielles de Volterra et des équations et des systèmes discrètes et nous les étudions à l'aide des méthodes qualitatives développées pour les systèmes de dimension deux qui ont eté hybridé aux méthodes de la théorie de stabilité (fonctions de Liapounov fonctionnelles de Popov, etc.). En cadre de la théorie moderne de contrôle nonlinéaire nous obtenons à l'aide de ces méthodes hybridées quelques critères pour la convergence globale et l'existence des cycles.

Книга посвящена проблемам нелокальной качественной теории систем с цилиндрическим фазовым пространством. Такие проблемы возникают, в частности, при изучении фазовой синхронизации механических и радиотехнических систем. Рассматриваются системы, описываемые дифференциальными, интегро-дифференциальными и разностными уравнениями. Систематически излагаются методы нелинейного анализа таких уравнений, основанные на синтезе качественной теории двумерных систем (циклы,петли сепаратрис,глобальные бифуркации) с методами теории устойчивости (функции Ляпунова, функционалы Попова, частотные критерии).

Preface

The investigation of global behavior of dissipative pendulum-type equations is one of the fundamental problems in applied qualitative theory of differential equations. The typical motions in
such systems, i.e. damped oscillatory motions and rotatory motions under external forces, are
well-known in mechanics. However, even for the forced pendulum equation

$$\ddot{\vartheta} + \alpha\dot{\vartheta} + \sin\vartheta = \gamma \tag{0.1}$$

with a constant torque $\gamma \geq 0$ and a viscous resistance $\alpha \geq 0$ it is non-trivial to give a complete
decomposition of the parameter space according to the various qualitative types of motions and
to estimate the global bifurcation values.

Experimental results concerning the pendulum equation in the case $\gamma = 0, \alpha > 0$ are already
described by I. Newton [116]. However it was 250 years later that F. Tricomi [147] carried out
the first non-local qualitative investigation of equation (0.1) with arbitrary $\alpha \geq 0$ and $\gamma \geq 0$.

It was proved by F. Tricomi that any solution of (0.1) with $\alpha > 0$ corresponds either to a
rotatory motion or to a damped oscillatory motion. Moreover, he showed that in the non-trivial
case $\gamma \leq 1$ there exists a bifurcation value $\alpha_{cr}(\gamma)$ corresponding to a separatrix-loop, i.e. to a
double-asymptotic to a saddle-point trajectory. For $\alpha < \alpha_{cr}(\gamma)$ equation (0.1) admits damped
oscillations as well as rotatory motions. For $\alpha > \alpha_{cr}(\gamma)$ global asymptotic stability takes place,
i.e. every motion is a damped oscillation. The papers of F. Tricomi became familiar immediately.
This was stimulated by the fact that already in 1937 Andronov, Witt and Chaikin included these
results in their book [5]. But more than 15 years separate Tricomi's papers from the ones of
L. Amerio [3, 4]. However, 20 years after the papers of F. Tricomi there is a long series of
publications in which the estimates of the bifurcation value α_{cr} were improved. To mention but
a few we refer to the papers of G. Seifert [133, 134, 135], C. Böhm [37], W.D. Hayes [58] and
A. Giger [50].

These results were extended by L.N. Belyustina [29], J. Barbalat and A. Halanay [16] and
V.A. Tabueva [144] on some classes of more complicated two-dimensional systems with cylindrical phase space. As a final step of this evolution we have to consider the monographs by
E.A. Barbashin and V.A. Tabueva [20] and G. Sansone and R. Conti [132], where the mentioned
results are thoroughly described.

The further investigation of pendulum-type equations was stimulated by other models of
systems with cylindrical phase space - the models of a synchronous machine [8, 56, 74, 115],
of phase synchronization systems [11, 27, 34, 38, 48, 67, 140] and of Josephson junctions [27,
7]. It turned out that a certain class of physical and technical phenomena in such models
may be described by ordinary differential equations, integro-differential equations and difference
equations, which possess similiar properties to (0.1).

Above all this concerns the possibility of introducing for such equations, a cylindrical phase
space, and to prove the existence of a bounded global attractor in such a phase space.

Since many engineering problems of non-local analysis of PLL's and synchronous machines
allow a strong mathematical formulation [107, 48, 140] it is necessary to develop a substantial
mathematical theory, which concentrates on solving these problems. The book under consideration reflects the actual state of such a mathematical theory. Note that in the 60-70-s the
investigation of two-dimensional systems on the cylinder was intensively continued. One of the

most significant results in this direction was the fact, that the transition from a global convergence behavior to the existence of circular solutions does not necessarily go through the creation of a separatrix–loop. A second possibility is the creation of a semi-stable circular limit cycle. It turned out that such a global bifurcation, which is impossible in equation (0.1), may occur in particular PLL's with a proportional-integrating filter. A further possibility of loosing the global convergence property in planar systems is a local bifurcation of a limit cycle; the so-called Andronov-Hopf bifurcation [5, 110]. In this situation, however, almost all trajectories are attracted by limit cycles. This phenomenon is often acceptable for certain working states of electrical systems of phase-synchronization [14]. For that reason, in the stability theory of pendulum-type systems, the new term "stable in the sense of Bakaev" or "synchronism of the second kind" was introduced.

One of the most fruitful ways of stability investigation of higher-dimensional systems seems to be the use of comparison principles. Its basic idea is as follows. In case one knows certain properties of an associated with the given one comparison system, (which is as a rule more simple for investigation than the original system), one tries to get the analogous properties for the original system with the help of coupling conditions between original and comparison systems. There are various ways of realizing this concept.

First of all we note one of the methods, which is described by A.M. Lyapunov [108]. The comparison system may be given by a linearization of the original system in an equilibrium. It is well-known that the Lyapunov function constructed for the linear system can be used in non-critical cases for the investigation of the original system. A further development of this technique can be found in N.N. Krasovskii [76].

A significant direction in the framework of the comparison principle consists of the exploitation of differential inequalities. Fundamental results in differential inequalities were obtained by E. Kamke [63] and S.A. Chaplygin [45]. These results were extended to systems of differential inequalities by T. Wazewski [152].

The ways of exploitation of differential inequalities in the qualitative theory of differential equations are rather different.

First of all, let us distinguish here an approach which is simply called the comparison method. According to this approach the classical conditions put on the derivative of a Lyapunov function along the solutions of the system, are considered as differential inequalities. These inequalities are replaced by more complex ones, namely differential inequalities built on the basis of certain comparison differential equations, which have well-known stability properties. According to the comparison method the derivative of a positive definite function V with respect to the given system has not necessarily to be negative definite and thereby the closed suface $\{x \; : \; V(x) = const\}$ is in general not a surface which is transversal to the vector field. Essentially in the comparison method is used the fact that Lyapunov functions considered along the solutions of the original system are bounded by certain solutions of the comparison system.

The idea of generalization of Lyapunov's direct method with the aid of differential inequalities, which was formulated for the first time in the papers [46, 111, 24], envoked a flood of results, reflected in a number of survey articles [54, 131, 111] and monographs [77, 151]. A new branch in the comparison method, which uses vector-valued Lyapunov functions, was founded by V.M. Matrosov [111] and R. Bellman [23].

Nowadays the comparison method, which is based on vector-valued Lyapunov functions, is applied not only in stability theory but also in system theory and the theory of artificial intelligence.

Another method, which proceeds from the Kamke–Chaplygin principle, is the so-called method of two-dimensional comparison systems or, which is the same, the method of monotone rotation of the vector field. This method, which essentially differs from the comparison method, described above, goes back to the paper of E.A. Barbashin and V.S. Serebryakova [136], devo-

devoted to oscillations of coupled pendulums. Then the method was generalized by V.N. Belykh for higher-dimensional systems [26, 27]. Note that this method is not based on Lyapunov's direct method. The transition from the n-dimensional original system to the two-dimensional system is carried-out by means of splitting the n-dimensional right-hand side of the differential equations into a $(n-2)$-dimensional dissipative part, whose components are considered as non-autonomous bounded perturbations of the remaining two-dimensional system. Using only the upper and lower bounds of these components one gets autonomous two-dimensional systems, the singular integral curves (separatrices) which generate surfaces, transversal to the vector field of the original system. Note that the ideas of this method can be realized for comparion systems with dimensions greater than two [25].

Another realization of the comparison technique is the non-local reduction priniciple. It is based on two ideas: On the idea of Lyapunov functions, which generate in the phase space a family of surfaces transversal to the vector field, and on the idea of comparison systems which have the desirable qualitative properties of the original system, but have a lower dimension than the original system. As a rule the information about the comparison system is used in Lyapunov functions for the original system by including particular solutions of the comparison system. Initially the non-local reduction method was developed for the stability investigation of systems with angular coordinates [80, 83, 81]. Later on it was applied to general systems of automatic control [88, 102, 96]. These applications of the non-local reduction method led to the necessity for receiving new theorems generalizing the basic theorem of T. Yoshizawa [159] and E.A. Barbashin and N.N. Krasovskij [18, 19].

In the present book differential equations are interpreted in general as feedback systems. It is obvious that for such systems Lyapunov functions can be constructed by using the customary techniques, in particular the famous frequency-domain theorem of V.A. Yakubovich and R.E. Kalman [153, 62] As a consequence, the sufficient conditions in this book for global convergence, existence of limit cycles, boundedness e.t.c. are formulated in the terms of transfer functions and have the form of frequency-domain inequalities. In the present book the development of the mentioned techniques in the last decade is reported and new classes of systems with angular coordinates, as integro-differential equations, vector fields on manifolds and discrete systems are included. The authors assume that the reader has a knowledge of advanced calculus and ordinary differential equations. Basic concepts from stability theory, system theory, integro-differential equations and the geometry of Riemannian manifolds are introduced as needed.

Acknowledgements

We are especially indebted to Mrs. Chr. Hertzschuch who wrote accurately and tirelessly several rough drafts of the manuscript, and to our collegues J. Lembcke, N. Koksch and T. Jerofsky for their accurate typing of the final manuscript in LaTeX. Authors are grateful to Miss. Sander-Williams for her careful reviewing of the English version of the book. For his interest and encouragement we thank J. Weiss from the Teubner Verlag.

Dresden and Saint Petersburg, January 1992

Gennadij A. Leonov
Volker Reitmann
Vera B. Smirnova

Contents

Chapter 1

Systems with Multiple Equilibria

In this part we introduce basic definitions and concepts necessary for understanding global aspects of ordinary differential equations with multiple equilibria. We briefly outline here some essential facts from the direct method of Lyapunov, the attractor concept, the theory of feedback control equations and frequency-domain methods in absolute stability theory.

1.1 Global Properties

In this part we consider the ODE

$$\dot{x} = f(t, x), \tag{1.1.1}$$

where $f : \mathbf{R}_+ \times \mathbf{R}^n \to \mathbf{R}^n$ is continuous and locally Lipschitz continuous in the second argument. As usual, the symbol $\mathbf{R}_+$ stands for the set of all non-negative real numbers. For $t_0 \geq 0$ and $x_0 \in \mathbf{R}^n$ we denote by $x(\cdot, t_0, x_0)$ the solution of (1.1.1) having $x(t_0, t_0, x_0) = x_0$. We suppose that every solution $x(\cdot, t_0, x_0)$ of (1.1.1) may be continued on $[t_0, +\infty)$.

It is important to mention that we consider for simplicity systems (1.1.1) under the assumption that f in (1.1.1) is Lipschitz continuous with respect to the second argument. This guarantees that there exists a unique solution of (1.1.1) for every initial condition within the considered region. However, the approaches developed in this book, as far as the existence of solutions is guaranteed, are applicable even if multiple solutions are possible for a given initial condition. Such non-lipschitzian systems can be studied under the "contingent equation" formulation of dynamical systems, see [49, 126].

In this book we are interested in asymptotic properties of solutions of (1.1.1) for $t \to +\infty$ which cannot be deduced from local information. A basic global property is the boundedness of a solution of (1.1.1) on $[t_0, +\infty)$. Recall that a solution $x(\cdot, t_0, x_0)$ of (1.1.1) is *bounded* if there exists a $c > 0$ such that $|x(t, t_0, x_0)| \leq c$ for all $t \geq t_0$, where c may depend on $x(\cdot, t_0, x_0)$. Here $|p| = (p, p)^{\frac{1}{2}}$ is given by the standard scalar product in $\mathbf{R}^n$, i.e., $(p, q) = \sum_{i=1}^{n} p_i q_i$ for $p, \ q \in \mathbf{R}^n$. In the case when all solutions $x(\cdot, t_0, x_0)$ are bounded one says that equation (1.1.1) is *Lagrange stable*.

The assumptions on f guarantee that the solutions of (1.1.1) are continuous with respect to the initial conditions. This means that for a given solution $x(\cdot, t_0, x_0)$ and arbitrary $\varepsilon > 0, t_1 \geq t_0$ and $T \geq 0$ there exist a number $\delta = \delta(\varepsilon, t_1, T)$ such that $|x(t_1, t_0, x_0) - x_1| < \delta$ and $t \in [t_1, t_1 + T]$ imply $|x(t, t_0, x_0) - x(t, t_1, x_1)| < \varepsilon$. Thus, if two solutions of (1.1.1) start in neighbouring points they remain neighbouring on a certain time interval. Recall that in contrast to this a solution $x(\cdot, t_0, x_0)$ of (1.1.1) is *Lyapunov stable* if for each $t_1 \geq t_0$ and each $\varepsilon > 0$ there exists $\delta = \delta(t_1, \varepsilon)$ such that $|x(t_1, t_0, x_0) - x_1| < \delta$ and $t \geq t_1$ imply $|x(t, t_0, x_0) - x(t, t_1, x_1)| < \varepsilon$.

The solution $x = x(\cdot, t_0, x_0)$ is *asymptotically stable* if x is Lyapunov stable and if for each $t_1 \geq t_0$ there exists $\eta > 0$ such that $|x(t_1, t_0, x_0) - x_1| < \eta$ implies

$$\lim_{t \to \infty} |x(t, t_0, x_0) - x(t, t_1, x_1)| = 0.$$

If $x(\cdot, t_0, x_0)$ is Lyapunov stable and if for each $t_1 \geq t_0$ and each $x_1 \in \mathbf{R}^n$ we have

$$\lim_{t \to \infty} |x(t, t_0, x_0) - x(t, t_1, x_1)| = 0$$

then $x(\cdot, t_0, x_0)$ is said to be *globally asymptotically stable*. If there is a point $p \in \mathbf{R}^n$ with $f(t, p) = 0$ for all $t \geq t_0$ we have for (1.1.1) a *stationary solution* or an *equilibrium point* $x(t) = p$ on $[t_0, \infty)$.

In the present book the systems under consideration are systems with multiple equilibria. Clearly, the Lyapunov stability or asymptotic stability of a fixed equilibrium point of (1.1.1) can be investigated by standard local methods. Note, that in the simplest case of a linear system (1.1.1) with a constant $n \times n$ matrix A

$$\dot{x} = Ax \tag{1.1.2}$$

the necessary and sufficient condition for the asymptotic stability of $x = 0$ is given by $\mathrm{Re}\,\lambda_j < 0$, where λ_j are the eigenvalues of A. The solution $x = 0$ of (1.1.2) is Lyapunov stable if and only if $\mathrm{Re}\,\lambda_j \leq 0$ for the eigenvalues of A and, if λ is an eigenvalue with $\mathrm{Re}\,\lambda = 0$ and λ has multiplicity $m > 1$, then there are m linearly independent eigenvectors belonging to λ. It will be shown in the sequel that the concept of global asymptotic stability is not applicable for systems with multiple equilibria. In order to investigate the global behavior of (1.1.1) with respect to the whole set of equilibria we introduce the following definitions. Notice that for a solution $x(\cdot, t_0, x_0)$ the *positive semi-orbit* is the set $\{x(t, t_0, x_0) \ : \ t \geq t_0\}$.

Definition 1.1.1 A solution $x(\cdot)$ or the positive semi-orbit of $x(\cdot)$ is said to be *convergent* if $x(t) \to p$ for $t \to +\infty$, where p is an equilibrium point of (1.1.1). We say that the solution $x(\cdot)$ is *quasi-convergent* if $\mathrm{dist}(x(t), \mathcal{E}) \to 0$ as $t \to +\infty$, where $\mathcal{E}$ is the set of equilibria of (1.1.1) and $\mathrm{dist}(\cdot, \mathcal{E})$ denotes the distance from a point to the set $\mathcal{E}$.

Definition 1.1.2 Equation (1.1.1) is said to be *monostable* if every bounded solution of (1.1.1) is convergent. It is called *quasi-monostable* if every bounded solution is quasi-convergent.

Definition 1.1.3 Equation (1.1.1) is said to be *gradient-like* if every solution is convergent. It is called *quasi-gradient-like* if every solution is quasi-convergent.

Remark 1.1.1 The definitions 1.1.1, 1.1.2 are due to M.W. Hirsch [59] and R.E. Kalman [61], respectively. Definition 1.1.3 goes back in spirit to J.K. Hale [57].

The global stability investigation of equation (1.1.1) can be carried out by various methods. One of the most effective is the direct or second method of A.M. Lyapunov [108]. This method is an analytical realization of certain geometrical approaches of H. Poincaré [121, 122]. A central element is the surface without contact with the vector field (1.1.1). The smooth surface $M \subset \mathbf{R}^n$ is a *surface without contact with the vector field* (1.1.1) if for every $t \geq 0$ the vector $f(t, p)$ does not belong to the tangential space $T_p M$ in p. It follows, that such surfaces may be intersected by the orbits of (1.1.1) only in one direction.   In the Lyapunov direct method one considers continuous functions $V : \mathbf{R}_+ \times \mathbf{R}^n \to \mathbf{R}$ which are non-increasing along the solutions of system (1.1.1). We call them *Lyapunov functions*. If such a function V belongs to C^1 one can compute for a solution $x(\cdot)$ of (1.1.1) the derivative

$$\frac{d}{dt}\, V(t, x(t)) = D_1 V(t, x(t)) + D_2 V(t, x(t)) f(t, x(t))$$

which is called *derivative of V along the solution x of* (1.1.1) (D_j denotes here the partial derivative with respect to the j-th argument). In general, for a C^1 function $V : \mathbf{R}_+ \times \mathbf{R}^n \to \mathbf{R}$ one defines the *derivative of V with respect to system* (1.1.1) by

$$\dot{V}_{(1.1.1)}(t,x) := D_1 V(t,x) + D_2 V(t,x) f(t,x)$$

for arbitrary $t \geq 0$ and $x \in \mathbf{R}^n$. It is easy to see that level surfaces of autonomous Lyapunov functions produce surfaces without contact with the vector field (1.1.1). Let us state now some well-known results concerning the global behavior of (1.1.1) which are derived by the second method of Lyapunov. The theorems 1.1.1, 1.1.2 are due to T. Yoshizawa [159] and E.A. Barbashin, N.N. Krasovskij [18], respectively.

Theorem 1.1.1 (Yoshizawa) *Suppose that there exist continuous functions $V : \mathbf{R}_+ \times \mathbf{R}^n \to \mathbf{R}$ and $W : \mathbf{R}^n \to \mathbf{R}$ with the following properties:*

(i) $V(t,x) \geq W(x)$ *for all* $(t,x) \in \mathbf{R}_+ \times \mathbf{R}^n$ *and* $W(x) \to +\infty$ *as* $|x| \to +\infty$;

(ii) V *is non-increasing along the solutions of* (1.1.1).

Then system (1.1.1) *is Lagrange stable.*

In the following theorem it is assumed that $x = 0$ is an equilibrium of (1.1.1).

Theorem 1.1.2 (Barbashin-Krasovskij) *Suppose there exist a continuously differentiable function $V : \mathbf{R}_+ \times \mathbf{R}^n \to \mathbf{R}$ and continuous functions $W_k : \mathbf{R}^n \to \mathbf{R}_+$ ($k = 1,2,3$) such that the following conditions are true:*

(i) $W_k(0) = 0, \quad W_k(x) > 0$ *for* $x \neq 0$ $(k = 1,2,3)$;

(ii) $W_1(x) \leq V(t,x) \leq W_2(x)$ *for all* $t \in \mathbf{R}_+$, $x \in \mathbf{R}^n$;

(iii) $\dot{V}_{(1.1.1)}(t,x) \leq -W_3(x)$ *for all* $t \in \mathbf{R}_+$, $x \in \mathbf{R}^n$;

(iv) $W_1(x) \to +\infty$ *as* $|x| \to +\infty$.

Then the solution $x = 0$ of (1.1.1) *is globally asymptotically stable.*

Let us now consider the autonomous case of equation (1.1.1)

$$\dot{x} = f(x), \quad x \in \mathbf{R}^n, \tag{1.1.3}$$

where f satisfies the conditions of system (1.1.1) with respect to x. We denote by $x(\cdot, x_0)$ the solution of (1.1.3) with $x(0, x_0) = x_0$ and assume that any solution is defined on $\mathbf{R}$. For a solution $x(\cdot, x_0)$ the set $\{x(t, x_0) : t \in \mathbf{R}\}$ is called the *orbit* of $x(\cdot, x_0)$ and denoted by $\gamma(x_0)$. Recall that a point $p \in \mathbf{R}^n$ is called *ω-limit point of the solution* $x(\cdot, x_0)$ (or of its orbit) if there exists an increasing sequence $\{t_n\}$, $t_n \to +\infty$, such that $x(t_n, x_0) \to p$ as $n \to +\infty$. In the same way we define an *α-limit point* using a decreasing sequence of numbers. The set of all ω-limit points of the solution $x(\cdot, x_0)$ is called *ω-limit set* and is denoted by $\omega(x_0)$. The set of all α-limit points of $x(\cdot, x_0)$ is called *α-limit set* and is denoted by $\alpha(x_0)$. The set $M \subset \mathbf{R}^n$ is called *invariant for* (1.1.3) if $p \in M$ implies $x(t,p) \in M$ for $t \in \mathbf{R}$. M is said to be *positively invariant* for (1.1.3), if $p \in M$ implies $x(t,p) \in M$ for $t \geq 0$. It is well known that for a given $x_0 \in \mathbf{R}^n$ the ω-limit set $\omega(x_0)$ is closed and invariant for (1.1.3). If, furthermore, the solution $x(\cdot, x_0)$ is bounded on $\mathbf{R}_+$ then $\omega(x_0)$ is non-empty, compact and connected (see for instance [2]). In the case of autonomous systems the following well-known results are available (see for instance [49]). The proof of these results for vector fields on manifolds is given in Chapter 4.

Theorem 1.1.3 *Suppose there exists a continuous function $V : \mathbf{R}^n \to \mathbf{R}$ such that:*

(i) *for any solution x of (1.1.3) $V(x(t))$ is not increasing;*

(ii) *if x is a bounded solution of (1.1.3) on $[0, +\infty)$ and there exists a $\tau > 0$ with $V(x(\tau)) = V(x(0))$ then x is a stationary solution.*

Then system (1.1.3) is quasi-monostable.

Theorem 1.1.4 *Suppose there exists a function V with the properties (i) and (ii) of Theorem 1.1.3 and in addition to this $V(x) \to +\infty$ as $|x| \to \infty$. Then system (1.1.3) is quasi-gradient-like.*

Theorem 1.1.5 *Suppose that the conditions of Theorem 1.1.4 are satisfied and the set $\mathcal{E}$ of equilibria of (1.1.3) is discrete. Then system (1.1.3) is gradient-like.*

Remark 1.1.2 $\mathcal{E}$ is discrete means that every equilibrium of (1.1.3) is isolated. If $f(p) = 0$, $f \in C^1$ and $f'(p)$ is non-singular then p is an isolated equilibrium of (1.1.3).

Let us interpret now the gradient-like or quasi-gradient-like behavior of (1.1.3) from the point of view of the attractor concept given, for instance, in [2]. One says that the set $M \subset \mathbf{R}^n$ *attracts* the point $p \in \mathbf{R}^n$ under the action of (1.1.3) if $\text{dist}\,(x(t, p), M) \to 0$ as $t \to +\infty$. The set $\mathcal{A}(M) := \{p \in \mathbf{R}^n \ : \ M \text{ attracts } p\}$ is the *domain of attraction of M* under (1.1.3). The set M is called an *attractor* for (1.1.3) if the domain of attraction $\mathcal{A}(M)$ under (1.1.3) is a neighborhood of M. One says that the attractor M is a *global attractor* of (1.1.3) if $\mathcal{A}(M) = \mathbf{R}^n$. We state the following well-known properties of an arbitrary set $M \subset \mathbf{R}^n$ which is considered under the flow of (1.1.3) (cf [2]):

1) $\mathcal{A}(M)$ is invariant for (1.1.3);

2) If M is an attractor for (1.1.3) then $\mathcal{A}(M)$ is open;

3) For any $p \in \mathcal{A}(M)$ we have $\omega(p) \subset \overline{M}$.

The connection between gradient-like behavior and local stability in (1.1.3) is given in the next assertion.

Proposition 1.1.1 *Suppose that (1.1.3) has at least two isolated equilibria and (1.1.3) is quasi-gradient-like. Then there exists at least one equilibrium of (1.1.3) which is not an attractor for (1.1.3) and which, consequently, is not asymptotically stable.*

Proof Since the system (1.1.3) is quasi-gradient-like the set $\mathcal{E}$ of equilibria of (1.1.3) is a global attractor for this system. It follows that $\mathbf{R}^n = \bigcup_{p \in \mathcal{E}} \mathcal{A}(p)$, where $\mathcal{A}(p) \cap \mathcal{A}(q) = \emptyset$ for $p \neq q$ and $\mathcal{A}(p)$ is open for all $p \in \mathcal{E}$ supposed that any $p \in \mathcal{E}$ is an attractor. But this contradicts the fact that $\mathbf{R}^n$ is connected. $\blacksquare$

Suppose now that f in (1.1.3) is C^1 and consider an equilibrium p of (1.1.3). This equilibrium is said to be *hyperbolic* if for any eigenvalue λ_j of $f'(p)$ yields $\text{Re}\,\lambda_j \neq 0$. For a hyperbolic equilibrium p the *stable manifold* $W^s(p)$ and the *unstable manifold* $W^u(p)$ are defined by

$$W^s(p) := \{x_0 \in \mathbf{R}^n \ : \ \omega(x_0) = p\} \qquad \text{and} \qquad W^u(p) := \{x_0 \in \mathbf{R}^n \ : \ \alpha(x_0) = p\}\,.$$

Sometimes $W^s(p)$ is called by us *ω-separatrix* and $W^u(p)$ *α-separatrix*.

Let p be a hyperbolic equilibrium and let E^s and E^u denote the projection into $\mathbf{R}^n$ of the direct sum of the eigenspaces of $f'(p)$ in $\mathbf{C}^n$ corresponding to those eigenvalues of negative and positive real parts, respectively. They are called the *stable* and *unstable subspace* of $\dot{x} = f'(p)x$, respectively. The Hadamard-Perron theorem [2] says that for f belonging to C^k and for a hyperbolic equilibrium p of (1.1.3) with $f'(p)$ having m eigenvalues in the left-hand side and $n - m$ eigenvalues in the right-hand side of the complex plane the sets $W^s(p)$ and $W^u(p)$ are m-dimensional and $(n - m)$-dimensional invariant surfaces of class C^k, respectively, and for the tangential spaces in p we have

$$T_p W^s(p) = p + E^s, \qquad T_p W^u(p) = p + E^u.$$

Every orbit $\gamma(x_0)$ of (1.1.3) with $x_0 \in W^s(p) \bigcap W^u(q)$ is called *heteroclinic* if $p \neq q$ and *homoclinic* if $p = q$.

The main question under consideration in this book is to give conditions which guarantee that every solution of system (1.1.3) with a set $\mathcal{E}$ of (isolated) hyperbolic equilibria converges to an equilibrium. It follows that for such a gradient-like system (1.1.3) of class C^1 there exists a (maximal) global attractor $\mathcal{M}$ which is the union of $\mathcal{E}$ and the heteroclinic orbits joining one point of $\mathcal{E}$ to another point of $\mathcal{E}$: $\mathcal{M} = \bigcup_{p \in \mathcal{E}} W^u(p)$. Chaos and strange attractors in this case cannot occur in system (1.1.3).

1.2 Feedback Control Equations

In this section we shall consider the differential equation

$$\dot{x} = Px + q\varphi(t, r^*x), \tag{1.2.1}$$

where P, q, r are constant real matrices of order $n \times n$, $n \times m$, and $n \times l$, respectively, and $\varphi : \mathbf{R}_+ \times \mathbf{R}^l \to \mathbf{R}^m$ is continuous and locally Lipschitz in the second argument. The asterisk denotes Hermitian conjugate (in particular, transposition in the case of real vectors and matrices and complex conjugate in the case of numbers). Such an equation in the system theory is called *feedback control equation*. Equation (1.1.1) can always be rewritten in the form (1.2.1) with arbitrary non-singular $n \times n$ matrices q and r and an $n \times n$ matrix P such that

$$\varphi(t, \sigma) = q^{-1}[f(t, (r^*)^{-1}\sigma) - P(r^*)^{-1}\sigma] \quad (t \in \mathbf{R}_+, \ \sigma \in \mathbf{R}^n).$$

We may consider (1.2.1) as an interconnection of a *linear part*, (with "input" ξ and "output" σ), and a *nonlinear part*:

$$\dot{x} = Px + q\xi, \quad \sigma = r^*x, \quad \xi = \varphi(t, \sigma). \tag{1.2.2}$$

In the following we give some basic definitions and results from system theory, where the constant matrices and vectors may also be complex.

Definition 1.2.1 The linear part of (1.2.2) [resp. the pair (P, q)] is called *controllable* if rank $[q, P, \ldots, P^{n-1}q] = n$. The linear part of (1.2.2) [resp. the pair (P, r)] is called *observable* if rank$[r, P^*r, \ldots, (P^*)^{n-1}r] = n$.

We now give a characterization of controllability which is useful in applications. For the proof of the following theorems 1.2.1, 1.2.2 the reader may consult [49, 124].

Theorem 1.2.1 *The following conditions are equivalent:*

(i) *the pair (P, q) is controllable;*

(ii) *there is no non-singular $n \times n$ matrix Q such that*

$$Q^{-1}PQ = \begin{bmatrix} S_{11} & S_{12} \\ 0 & S_{22} \end{bmatrix} \quad and \quad Q^{-1}q = \begin{bmatrix} s_1 \\ 0 \end{bmatrix},$$

where S_{11} is a $k \times k$ matrix and s_1 is a k-vector $(k < n)$;

(iii) *for arbitrary $t_1, t_2 \in \mathbf{R}$ $(t_1 < t_2)$ and arbitrary $x_1, x_2 \in \mathbf{C}^n$ there exists a continuous function $\xi : (t_1, t_2) \mapsto \mathbf{C}^m$ such that the equation $\dot{x} = Px + q\xi$ has a solution x with $x(t_1) = x_1$ and $x(t_2) = x_2$;*

(iv) $\mathrm{rank}\,[Q_1, Q_2, \dots, Q_n] = n$, *where the $n \times m$ matrices Q_i are given by*

$$(sI - P)^{-1}q \det(sI - P) = Q_1 s^{n-1} + \dots + Q_{n-1}s + Q_n;$$

(v) $\mathrm{rank}\,[P - sI, q] = n$ *for any $s \in \mathbf{C}$.*

In some cases we replace the assumption about controllability of the linear part of (1.2.2) with the weaker assumption of stabilizability.

Definition 1.2.2 We say that the linear part of (1.2.2) [resp. the pair (P, q)] is *stabilizable* if there exists an $n \times m$ matrix h such that $P + qh^*$ is Hurwitzian.

Theorem 1.2.2 *Suppose that the pair (P, q) is controllable. Then it holds:*

(i) *the pair (P, q) is stabilizable;*

(ii) *for any polynomial ψ of degree n there exists an $n \times m$ matrix h (real if P, q and ψ are real) such that $\psi(s) = \det(sI - P - qh^*)$ for all $s \in \mathbf{C}$.*

We now consider linear feedback systems. The following auxiliary result will be needed throughout the book. For the proof see [49, 156].

Lemma 1.2.1 *Suppose P, $H = H^*$ and r are matrices of order $n \times n$, $n \times n$ and $n \times m$, respectively. Suppose also that the pair (P, r) is observable and $x^*(HP + P^*H)x \leq -|r^*x|^2$ for all $x \in \mathbf{R}^n$. Then the matrix P does not have purely imaginary eigenvalues, it holds $\det H \neq 0$ and the number of negative eigenvalues of H coincides with the number of eigenvalues of P with positive real part.*

1.3 The Transfer Function

In order to characterize the linear part of a feedback control equation it is custumary to determine the transfer function. Let us recall first some facts on Laplace transforms. A function $h : \mathbf{R}_+ \to \mathbf{R}$ is said to be of *exponential order* if there are positive constants K and α with

$$|h(t)| \leq Ke^{\alpha t} \quad \text{on } \mathbf{R}_+. \tag{1.3.1}$$

In the following definition and the subsequent list of properties it is assumed that all functions of t under consideration are integrable and of exponential order (see also [43, 113]).

Definition 1.3.1 If $h : \mathbf{R}_+ \to \mathbf{R}$, then the *Laplace transform* of h (the *original*) is $L(h)$ or $\widetilde{h}$ defined by

$$\widetilde{h}(s) = \int_0^{+\infty} h(t)e^{-st}\,dt.$$

Note that for $s \in \mathbf{C}$ with $\mathrm{Re}\, s > \alpha$, where α is defined by (1.3.1), the integral converges absolutely and $\widetilde{h}$ is analytic in the domain $\mathrm{Re}\, s > \alpha$. If $D(\cdot) = (d_{ij}(\cdot)) : \mathbf{R}_+ \to \mathcal{L}(\mathbf{R}^m, \mathbf{R}^n)$ is an $n \times m$ matrix function then one puts $\widetilde{D} := (\widetilde{d}_{ij})$. The following properties will be needed in the sequel:

(i) if c is a constant and $h_1, h_2 : \mathbf{R}_+ \to \mathbf{R}$ then $L(ch_1 + h_2) = cL(h_1) + L(h_2)$;

(ii) if $D(\cdot) : \mathbf{R}_+ \to \mathcal{L}(\mathbf{R}^n, \mathbf{R}^n)$ is an $n \times n$ matrix function and $h : \mathbf{R}_+ \to \mathbf{R}^n$, then for the function $F(t) := \int\limits_0^t D(t - \tau)h(\tau)\,d\tau$ it follows that $\widetilde{F} = \widetilde{D} \cdot \widetilde{h}$;

(iii) if $h : \mathbf{R}_+ \to \mathbf{R}$ is C^1 then $\widetilde{h'}(s) = s\widetilde{h}(s) - h(0)$;

(iv) if $h_1, h_2 : \mathbf{R}_+ \to \mathbf{R}$ are continuous, then $\widetilde{h}_1 = \widetilde{h}_2$ implies that $h_1 = h_2$ on $\mathbf{R}_+$;

(v) if $h : [-\alpha, \infty) \to \mathbf{R}$ and $h(t) = 0$ for $t \in [-\alpha, 0]$ then $L[h((\cdot) - \alpha)](s) = e^{-s\alpha}L[h](s)$;

Let us now consider the linear part of equation (1.2.2) and suppose that ξ is of exponential order and integrable. If we assume for the time being that $x(0) = 0$ and if we take the Laplace transform of both sides for the first two equations in (1.2.2) (the linear part) we receive for the Laplace transforms $\widetilde{\sigma}$ and $\widetilde{\xi}$ the relation $\widetilde{\sigma}(s) = -\chi(s)\widetilde{\xi}(s)$, with $\chi(s) = r^*(P - sI)^{-1}q$, χ is called the *transfer function* of the linear part of (1.2.2). It is defined for all complex s with $\det(P - sI) \neq 0$. Note that the definition of χ is coordinate free, i.e. after any change of coordinates $y = Sx$ (S a non-singular $n \times n$ matrix), χ remains the same function. The function $\omega \mapsto \chi(i\omega)$ $(\omega \in \mathbf{R})$ is the *frequency response* of the linear part of (1.2.2).

Let us now consider the scalar case $m = l = 1$ in (1.2.2). In this situation the transfer function χ is a rational function

$$\chi(s) = \alpha(s)/\delta(s), \tag{1.3.2}$$

where $\delta(s) = \det(sI - P)$ and α is a polynomial of degree less than n. The following notion is useful for characterizing linear parts which are controllable and observable.

Definition 1.3.2 The transfer function (1.3.2) of (1.2.2) with $m = l = 1$ is called *non-degenerate* if α and δ are coprime polynomials.

Theorem 1.3.1 *If in* (1.2.2) $m = l = 1$ *then the linear part is controllable and observable if, and only if, the transfer function* (1.3.2) *is non-degenerate.*

1.4 The Yakubovich-Kalman Theorem

In closing the first chapter we offer some basic facts from the absolute stability theory, in particular a theorem of V.A. Yakubovich and R.E. Kalman [153, 62] which gives neccessary and sufficient conditions for the solvability of certain matrix inequalities. Let us consider the system (1.2.1) with a scalar continuous nonlinearity φ and suppose that for some $\mu_1, \mu_2 \in \mathbf{R}$ with $\mu_1 \leq \mu_2$ it holds

$$\mu_1 \leq \varphi(t, \sigma)/\sigma \leq \mu_2 \qquad \text{for all} \qquad t \in \mathbf{R}_+,\ \sigma \neq 0. \tag{1.4.1}$$

In this case we say that φ *belongs to the class* $M[\mu_1, \mu_2]$. Since φ is continuous it follows from (1.4.1) that $\varphi(t, 0) = 0$ on $\mathbf{R}_+$. Thus, $x = 0$ is an equilibrium of (1.2.1). Note, that in (1.4.1) it is possible that $\mu_1 = -\infty$ (but then $\mu_2 \neq +\infty$) or $\mu_2 = +\infty$ (but then $\mu_1 \neq -\infty$).

Definition 1.4.1 The system (1.2.1) is said to be *absolutely stable with respect to* $M[\mu_1, \mu_2]$ if for any nonlinearity φ satisfying (1.4.1) the equilibrium $x = 0$ of (1.2.1) is globally asymptotically stable.

The problem is to find conditions on the linear part of (1.2.1) which ensure that system (1.2.1) is absolutely stable with respect to $M[\mu_1, \mu_2]$. For fixed nonlinearity φ in (1.2.1) we want to apply the Barbashin-Krasovskij theorem (Theorem 1.1.2) and choose as a Lyapunov function the quadratic form $V(x) = x^* H x$, where $H = H^*$ is an unknown positive definite constant $n \times n$ matrix. For the derivative of V with respect to (1.2.1) we have

$$\dot{V}_{(1.2.1)}(t, x) = 2x^* H[Px + q\varphi(t, r^* x)] \qquad \text{for} \qquad t \geq 0, \ x \in \mathbf{R}^n.$$

According to Theorem 1.1.2 the function V guarantees global asymptotic stability of $x = 0$ (for fixed φ) if $\dot{V}_{(1.2.1)}(t, x) \leq -W(x)$, where $W(x) > 0$ for $x \neq 0$ and $W(0) = 0$. Let us introduce the quadratic forms G and F by

$$G(x, \xi) := 2x^* H(Px + q\xi) \qquad \text{for} \qquad x \in \mathbf{R}^n \text{ and } \xi \in \mathbf{R}$$

and

$$F(x, \xi) := (\xi - \mu_1 r^* x)(\mu_2 r^* x - \xi) \qquad \text{for} \qquad x \in \mathbf{R}^n \text{ and } \xi \in \mathbf{R}.$$

Here it is assumeed that μ_1 and μ_2 are finite. Note, that $F(x, \varphi(t, r^* x)) \geq 0$ for all $t \geq 0$ and $x \in \mathbf{R}^n$ if φ belongs to the class $M[\mu_1, \mu_2]$. With the help of these quadratic forms the absolute stability problem can now be formulated as follows:

Problem A *Find a positive definite matrix $H = H^*$ such that*

$$G(x, \xi) < 0 \text{ for all } x \in \mathbf{R}^n, \ \xi \in \mathbf{R} \text{ with } F(x, \xi) \geq 0 \text{ and } |x|^2 + \xi^2 \neq 0.$$

Let us now define the quadratic form S by

$$S(x, \xi) := G(x, \xi) + \tau F(x, \xi) \qquad (x \in \mathbf{R}^n, \ \xi \in \mathbf{R}),$$

where τ is some real parameter. It was shown by V.A. Yakubovich [154] (see also [49]) that Problem A is equivalent to

Problem B *Find for some $\tau \geq 0$ a matrix $H = H^* > 0$ such that*

$$S(x, \xi) < 0 \text{ for all } x \in \mathbf{R}^n, \ \xi \in \mathbf{R} \text{ with } |x|^2 + \xi^2 \neq 0.$$

Note that Problem B is much more convenient for investigation than Problem A. Another pair of equivalent problems important in the absolute stability theory appears when the strict inequalities $G(x, \xi) < 0$ and $S(x, \xi) < 0$ are replaced by non-strict ones. We shall denote them as Problem A' and Problem B', respectively.

In the following we often use the fact that for every quadratic form F in $\mathbf{R}^n$ defined by $F(u) = u^* H u$, with a real symmetric $n \times n$ matrix $H = H^*$ we can construct a Hermitian form $F_{\mathbf{C}}$ by

$$F_{\mathbf{C}}(u + iv) := F(u) + F(v) \qquad \text{for any} \qquad u, v \in \mathbf{R}^n.$$

This procedure is called the *extension of a quadratic form to a Hermitian one*. In coordinates $(u_1, u_2, \ldots, u_n)$ this extension means that any term $u_j u_k$ in F is replaced by $\operatorname{Re} u_j^* u_k$. It is obvious that for real values $u \in \mathbf{R}^n$ it holds $F_{\mathbf{C}}(u) = F(u)$.

In the next Theorems 1.4.1 and 1.4.2 let P and q be complex matrices of orders $n \times n$ and $n \times m$, respectively and $F(x, \xi) = x^* A x + 2\operatorname{Re}(x^* B \xi) + \xi^* \Gamma \xi$ be a Hermitian form of $x \in \mathbf{C}^n$ and $\xi \in \mathbf{C}^m$. The complex matrices $A = A^*$, $\Gamma = \Gamma^*$ and B have the orders $n \times n$, $m \times m$ and $n \times m$, respectively. The following theorem, which is crucial for the further development, goes back to V.A. Yakubovich [153] and R. Kalman [62] (see also [78, 130, 49, 124, 69, 68, 113]).

Theorem 1.4.1 (Yakubovich-Kalman) *Suppose that the pair (P, q) is controllable. Then there exists a matrix $H = H^*$ satisfying the inequality*

$$2\operatorname{Re}\left[x^* H(Px + q\xi)\right] + F(x,\xi) \le 0 \qquad (x \in \mathbf{C}^n,\ \xi \in \mathbf{C}^m) \tag{1.4.2}$$

if and only if

$$F[(i\omega I - P)^{-1} q\xi, \xi] \le 0$$

for all $\xi \in \mathbf{C}^m$ and all $\omega \in \mathbf{R}$ with $\det(P - i\omega I) \ne 0$. In case the matrices P, q and the coefficients of F are real the matrix H may also be chosen as being real. If the matrix P is Hurwitzian, $A \ge 0$ and the pair (P, B) is observable then any matrix H, satisfying (1.4.2) is positive definite.

In the next chapters we need the following assertion which can be derived from theorem 1.4.1.

Corollary 1.4.1 *Suppose that P, q and r are matrices of order $n \times n$, $n \times m$, $n \times m$ respectively and (P, q) is controllable. Then there exists a matrix $H = H^*$ (which is real in case P, q, r are real) such that*

$$HP + PH^* \le 0, \qquad Hq + r = 0 \tag{1.4.3}$$

if and only if

$$\operatorname{Re} r^*(P - i\omega I)^{-1} q \ge 0$$

for all $\omega \in \mathbf{R}$ with $\det(P - i\omega I) \ne 0$. If P is Hurwitzian and the pair (P, r) is observable, then any matrix H satisfying (1.4.3) is positive definite.

Remark 1.4.1 In order to prove this corollary it is sufficient to use Theorem 1.4.1 with the Hermitian form

$$F(x, \xi) = 2\operatorname{Re}(\xi^* r^* x).$$

In the next theorem, which comes from [156], one considers strong inequalities and the pair (P, q) is not assumed to be controllable.

Theorem 1.4.2 *Suppose that the pair (P, q) is stabilizable and $\det(P - i\omega I) \ne 0$ $(\omega \in \mathbf{R})$. Then there exists a matrix $H = H^*$ (which is real in case P, q and the coefficients of F are real) satisfying the inequality*

$$\operatorname{Re} x^* H(Px + q\xi) + F(x,\xi) < 0 \qquad (x \in \mathbf{C}^n,\ \xi \in \mathbf{C}^m,\ |x| + |\xi| \ne 0)$$

if and only if

$$F[(i\omega I - P)^{-1} q\xi, \xi] < 0 \qquad (\omega \in \mathbf{R},\ \xi \in \mathbf{C}^m,\ |\xi| \ne 0). \tag{1.4.4}$$

Now, on the basis of Theorem 1.4.2 a frequency-domain criterion for absolute stability can be easily obtained. With respect to the class $M[\mu_1, \mu_2]$ we define the Hermitian form

$$F(x, \xi) = \operatorname{Re}\left[(\xi - \mu_1 r^* x)^*(\mu_2 r^* x - \xi)\right] \qquad (x \in \mathbf{C}^n,\ \xi \in \mathbf{C}).$$

Suppose P has no eigenvalues with real part zero. Notice that for all $\omega \in \mathbf{R}$ and $\xi \in \mathbf{C}$

$$F[(i\omega I - P)^{-1} q\xi, \xi] = -\operatorname{Re}\left[(\mu_1 \chi(i\omega) + 1)^*(\mu_2 \chi(i\omega) + 1)\right]|\xi|^2.$$

So the inequality (1.4.4) is satisfied if

$$\operatorname{Re}\left[(\mu_1 \chi(i\omega) + 1)^*(\mu_2 \chi(i\omega) + 1)\right] > 0 \qquad (\omega \in \mathbf{R}). \tag{1.4.5}$$

Suppose (1.4.5) is fulfilled and for a certain number $\mu \in [\mu_1, \mu_2]$ the zero solution of the linear system (1.2.2) with $\varphi(t, \sigma) = \mu\sigma$ is globally asymptotically stable. Then by Theorem 1.4.2 and Lemma 1.2.1 there exists a Lyapunov function $V(x) = x^* H x$ with $H > 0$. Theorem 1.1.2 guarantees that the zero solution of (1.2.2) is globally asymptotically stable for any $\varphi \in M[\mu_1, \mu_2]$. It follows that system (1.2.1) is absolutely stable with respect to $M[\mu_1, \mu_2]$. Thus we have proved the following result [39, 49, 151].

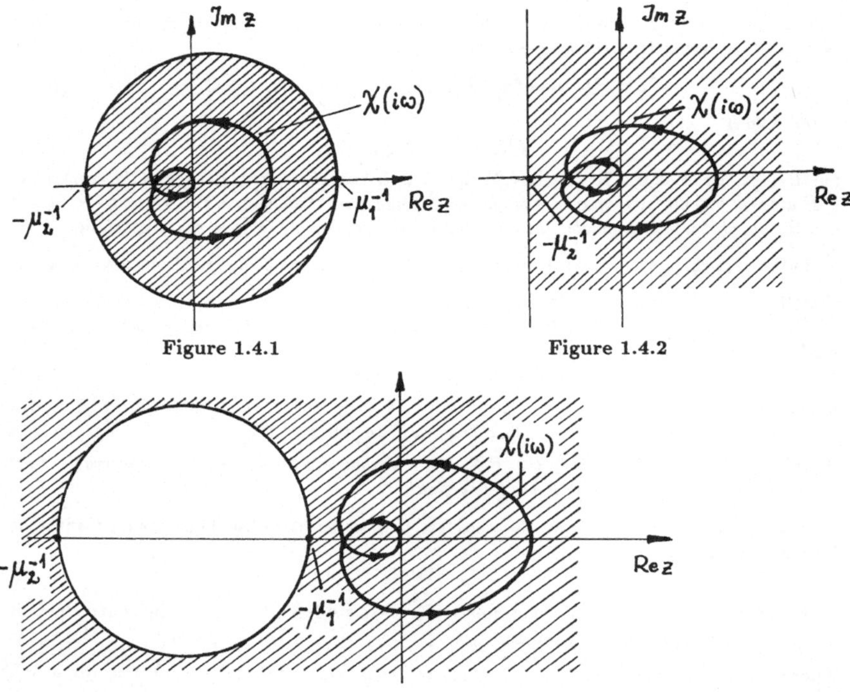

Figure 1.4.1   Figure 1.4.2

Figure 1.4.3

Theorem 1.4.3 (Circle Criterion) *Suppose that (P, q) in equation (1.2.2) is controllable, the matrix P has no eigenvalues with real part zero and there exists a $\mu \in [\mu_1, \mu_2]$ such that system (1.2.2) with $\xi = \mu\sigma$ has a globally stable zero solution. Suppose also that the frequency-domain condition (1.4.5) is satisfied. Then system (1.2.2) is absolutely stable with respect to $M[\mu_1, \mu_2]$.*

Remark 1.4.2 Theorem 1.4.3 admits a useful well-known geometric interpretation. If we plot in the complex plane $\operatorname{Re}\chi(i\omega)$ and $\operatorname{Im}\chi(i\omega)$, the condition (1.4.5) states that $\chi(i\omega)$ lies in the domain

$$\operatorname{Re}\left[(1 + \mu_1 z)^*(1 + \mu_2 z)\right] > 0, \quad z \in \mathbf{C}. \tag{1.4.6}$$

The boundary of this domain is a circle which passes through the points $z_1 = -\mu_1^{-1}$ and $z_2 = -\mu_2^{-1}$ and has the center on the real axis. In case $\mu_1 = 0$ the circle transforms into a straight line. So (1.4.6) defines for $\mu_1 < 0$ and $\mu_2 > 0$ an open circular disc (Figure 1.4.1), for $\mu_1 = 0$ a complex halfplane (Figure 1.4.2) and for $\mu_1 > 0$, $\mu_2 > 0$ the exterior of a circular disk (Figure 1.4.3).

In some cases the following theorem is applicable [49, 130, 78].

Theorem 1.4.4 *Suppose that the pair (P, q) is stabilizable, $\operatorname{rank} q = m$ and P has no eigenvalues z with $\operatorname{Re} z = 0$. Then there exists a matrix $H = H^*$ (real in case P, q and r are real) satisfying the relations*

$$HP + P^*H < 0 \quad and \quad Hq + r = 0$$

if and only if

$$\operatorname{Re} r^*(P - i\omega I)^{-1} q > 0 \quad \text{for all} \qquad \omega \in \mathbf{R}$$

and

$$\lim_{\omega \to \infty} \omega^2 \operatorname{Re} r^*(P - i\omega I)^{-1} q > 0.$$

Remark 1.4.3 In the autonomous case of system (1.2.2) we can choose the Lyapunov function V in the so called *Lur'e-Postnikov form*

$$V(x) = x^* H x + \vartheta \int_0^{r^*x} \varphi(\sigma)\, d\sigma, \tag{1.4.7}$$

where the matrix $H = H^*$ and the parameter ϑ are to be defined. The derivative of V along the solutions of (1.2.2) is

$$\dot{V}_{(1.2.2)}(x) = 2x^* H[Px + q\varphi(r^*x)] + \vartheta\varphi(r^*x)r^*[Px + q\varphi(r^*x)]. \tag{1.4.8}$$

Choosing a matrix $H = H^*$ on the basis of Theorem 1.4.2 such that $\dot{V}_{(1.2.2)}(x) \leq 0$ for all $x \in \mathbf{R}^n$, we can derive the Popov criterion [123, 49, 78].

Remark 1.4.4 The results on global behavior of feedback systems with multiple equilibria presented in this book based upon a combination of Lyapunov-type arguments and frequency-domain methods of absolute stability theory. There are two parallel directions in the theory of absolute stability. One goes back to V.M. Popov and the other to V.A. Yakubovich and R. Kalman. The techniques that are used to produce frequency-domain stability results are different. The Popov approach is functional-analytic and the strategy is to use the Parseval equality to certain functionals to get upper estimates. In [123] V.M. Popov has formulated conditions for absolute stability of nonlinear systems in terms of the frequency response of the linear part with the help of this a priori integral estimates. In 1962 V.A. Yakubovich [153] proved this Popov criterion using the solvability of certain matrix inequalities of Lur'e type for the construction of appropriate Lyapunov functions. In 1963 this result was independently obtained by R. Kalman [62].

Chapter 2

Pendulum-Like Systems

In the present chapter we develop the global bifurcation theory of two-dimensional systems with a periodic nonlinearity and the general concept of higher-dimensional pendulum-like systems and their canonical forms, which are of considerable use in studying, by frequency-domain methods, the global behavior of such systems. We show that Lyapunov functions constructed as quadratic forms are not applicable in a standard way for investigating the global behavior of pendulum-like systems. In contrast to this fact we prove global convergence results for a certain class of such systems which use the invariance principle and the theorem of Yakubovich-Kalman.

2.1 The Two Canonical Forms

Let us consider the ordinary differential equation

$$\dot{x} = f(t, x) \tag{2.1.1}$$

where $f : \mathbf{R}_+ \times \mathbf{R}^n \to \mathbf{R}^n$ is continuous and locally Lipschitz continuous in the second argument. Suppose that any solution $x(\cdot, t_0, x_0)$ with $t_0 \geq 0$ and $x_0 \in \mathbf{R}^n$ exists on $[t_0, +\infty)$. Many problems in the theory of phase synchronization [11, 27, 34, 38, 48, 67, 140], in the theory of Josephson multipoint junctions in solid physics [27, 7], in the space discretization of some boundary value problems for the sine-Gordon equation [104], and in theory of coupled pendulums [9, 136], can be described by equation (2.1.1) with solutions having an equivariance property with respect to a discrete subgroup of $\mathbf{R}^n$, defined by $\Gamma = \left\{ \sum_{j=1}^{m} k_j d_j \; : \; k_j \in \mathbf{Z}, 1 \leq j \leq m \right\}$, where the vectors $d_j \in \mathbf{R}^n$ are supposed to be linearly independent ($m \leq n$).

A typical model in the theory of phase synchronization is as follows. An input phase $\vartheta_i(t) = \omega_{i0} t + \nu_i(t)$ with actual frequency $\omega_i(t) := \omega_{i0} + \frac{d}{dt}\nu_i(t)$, where ω_{i0} is the so-called center frequency, is compared with the loop estimate of the phase $\vartheta_e(t) = \omega_{e0} t + \nu_e(t)$ and the actual frequency $\omega_e(t) := \omega_{e0} + \frac{d}{dt}\nu_e(t)$ in order to guarantee by nonlinear feedback according to the actual phase difference (phase error) $\vartheta(t) := \vartheta_i(t) - \vartheta_e(t)$ the coincidence of $\omega_i(t)$ and $\omega_e(t)$ for $t \to +\infty$ and the limit $\vartheta(t) \to const$ for $t \to +\infty$ (Locking-in). For this purpose the phase difference $\vartheta(t)$ produces in the phase detector (PD; see Figure 2.1.1) an output voltage $e_{PD}(t) := E\varphi(\vartheta(t))$, where φ is a Δ-periodical function (the phase detector characteristic) and E is a constant.

This output e_{PD} passes through the low pass filter (LPF) with a transfer function $K(s)$. Writing K as $K(s) = \int_0^{+\infty} e^{-st}\gamma(t)dt$, $\gamma : \mathbf{R}_+ \to \mathbf{R}$ an integrable function, the output of the filter

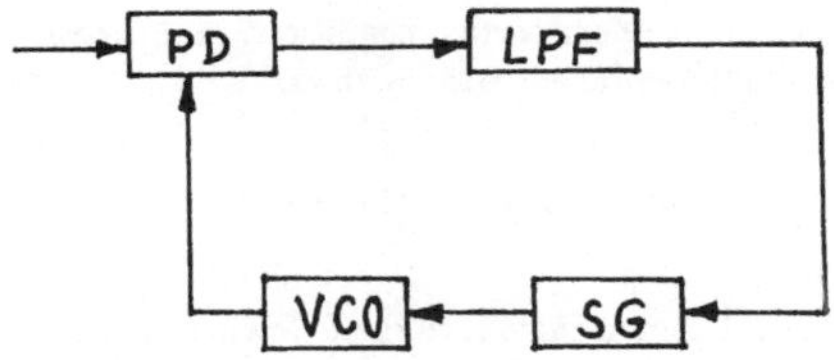

Figure 2.1.1

is the convolution term

$$e_f(t) := \int_0^t \gamma(t - \tau) E\varphi(\vartheta(\tau))\, d\tau.$$

This control signal is given on the steering gear (SG), where the new frequency is generated by

$$\omega_e(t) - \omega_{e0} = L_{SG} e_f(t),$$

where L_{SG} is a constant.

In the voltage controlled oscillator (VCO) the new phase ϑ_e is generated. Clearly,

$$\frac{d}{dt}\vartheta(t) = \frac{d}{dt}\vartheta_i(t) - \frac{d}{dt}\vartheta_e(t) = \omega_i(t) - \omega_{e0} - L_{SG}\int_0^t \gamma(t - \tau) E\varphi(\vartheta(\tau))\, d\tau.$$

Setting $\vartheta_1(t) := \vartheta_i(t) - \omega_{e0} t$ and $\Omega := L_{SG} E$ we get the basic equation of a PLL in the form

$$\dot{\vartheta}(t) = \dot{\vartheta}_1 - \Omega \int_0^t \gamma(t - s)\varphi(\vartheta(\tau))\, d\tau.$$

If we suppose now that there exist the Laplace transforms $\widetilde{\vartheta}, \widetilde{\vartheta}_1$ and $\widetilde{\varphi}$ of $\vartheta(t), \vartheta_1(t)$ and $\varphi(\vartheta(t))$, respectively, we can write the last equation as

$$s\widetilde{\vartheta} = s\widetilde{\vartheta}_1 - \Omega K(s)\widetilde{\varphi}.$$

The various types of PLL systems result from the choice of $K(s)$. If one takes, for instance, $K(s) = \frac{1}{1+Ts}$ where $T = RC$ is a constant of the filter, and $\vartheta_1(t) \equiv \gamma t$, γ a constant, one gets the equation for the Laplace originals

$$\frac{T}{\Omega}\ddot{\vartheta}(t) + \frac{1}{\Omega}\dot{\vartheta}(t) + \varphi(\vartheta(t)) = \gamma,$$

which is transformed in equation (0.1) of the Preface by the change of time $\tau := t\sqrt{\Omega/T}$ and by the new constant $\alpha := 1/\sqrt{\Omega T}$. Let us suppose that the filter LPF is, with delay $h > 0$, expressed by the transfer function

$$K(s) = -\varrho e^{-sh} + \int_0^{+\infty} e^{-st}\gamma(t)dt,$$

13

where ϱ is a constant. One can show that under certain conditions in this case from the operator equation above, results the integro-differential equation with delay

$$\dot{\vartheta}(t) = \dot{\vartheta}_1(t) + \varrho\varphi(\vartheta(t-h)) - \int\limits_0^t \gamma(t-\tau)\varphi(\vartheta(\tau))\,d\tau.$$

Let us return now to equation (2.1.1) and introduce the following notion.

Definition 2.1.1 We will say that the solutions $x(\cdot, t_0, x_0)$ of (2.1.1) satisfy the *equivariance property* with respect to Γ or that (2.1.1) is *pendulum-like* with respect to Γ if for any solution $x(\cdot, t_0, x_0)$ of (2.1.1) we have

$$x(t, t_0, x_0 + d) = x(t, t_0, x_0) + d \tag{2.1.2}$$

for all $t \geq t_0$ and all $d \in \Gamma$. A differential equation of the n-th order which can be expressed as system (2.1.1) with an equivariance property is also called by us *pendulum-like equation*.

It is easy to establish under which conditions a system (2.1.1) is pendulum-like.

Proposition 2.1.1 *System* (2.1.1) *is pendulum-like with respect to* Γ *if and only if*

$$f(t, x + d) = f(t, x) \tag{2.1.3}$$

for all $t \geq 0$, $x \in \mathbf{R}^n$ *and* $d \in \Gamma$.

Proof Suppose (2.1.3) is satisfied. Consider an arbitrary solution $x(\cdot, t_0, x_0)$ of (2.1.1) and define for a $d \in \Gamma$ the function $y(t) = x(t, t_0, x_0) + d$ for $t \geq t_0$, satisfying $y(t_0) = x_0 + d$. We have $\dot{y}(t) = \dot{x}(t, t_0, x_0) = f(t, x(t, t_0, x_0)) = f(t, x(t, t_0, x_0) + d) = f(t, y(t))$, thus $y(\cdot)$ is a solution of (2.1.1) and, by uniqueness, $y(t) \equiv x(t, t_0, x_0 + d)$. To prove the converse, consider for arbitrary $(t_0, x_0) \in \mathbf{R}_+ \times \mathbf{R}^n$ and $d \in \Gamma$ the solution $x(\cdot, t_0, x_0 + d)$ of (2.1.1). It follows that $\dot{x}(t, t_0, x_0) = f(t, x(t, t_0, x_0 + d)) = f(t, x(t, t_0, x_0))$. Setting $t = t_0$ we receive from the last equality $f(t_0, x_0 + d) = f(t_0, x_0)$. ∎

Consider now the autonomous case of system (2.1.1)

$$\dot{x} = f(x) \tag{2.1.4}$$

and assume that it is pendulum-like with respect to

$$\Gamma = \left\{ \sum_{j=1}^m k_j d_j \; : \; k_j \in \mathbf{Z}, 1 \leq j \leq m \right\}.$$

It is clear that in case system (2.1.4) has an equilibrum, the set of equilibria is infinite. The main question of this book is to deduce conditions under which all solutions of (2.1.4) approach an equilibrium, i.e. to show that (2.1.4) is gradient-like. In cases when system (2.1.4) is not gradient-like we want to ask about the existence of periodic solutions of various kinds. Recall that a non-trivial solution $x(\cdot)$ of (2.1.4) is *periodic* if there exists a time $\tau > 0$ such that $x(0) = x(\tau)$. In connection with the equivariance property of (2.1.4) such a solution is also called *cycle of the first kind*.

Suppose that $\Gamma \subset \mathbf{R}^n$ is an arbitrary discrete subgroup and define in $\mathbf{R}^n$ an equivalence relation by

$$u \sim v \Longleftrightarrow u = v + g \qquad \text{for some} \qquad g \in \Gamma.$$

If we consider now a pendulum-like system (2.1.4), (with respect to this group Γ), we can look at the solutions of (2.1.4) as curves in $\mathbf{R}^n/\Gamma$, where $\mathbf{R}^n/\Gamma$ denotes the quotient space defined by the introduced equivalence relation.

Definition 2.1.2 A solution $x(\cdot)$ of (2.1.4) is called *cycle of the second kind* if $x(\cdot)$ is a closed curve in $\mathbf{R}^n/\Gamma$ and $x(\cdot)$ is not closed in $\mathbf{R}^n$.

Let us now assume (2.1.1) as a feedback control system

$$\dot{x} = Px + q\varphi(t,\sigma), \quad \sigma = r^*x, \tag{2.1.5}$$

where P is a constant $n \times n$-matrix, q and r are constant n-vectors and $\varphi : \mathbf{R}_+ \times \mathbf{R} \to \mathbf{R}$ is continuous and locally Lipschitz continuous in the second argument. Suppose further that the transfer function of the linear part of (2.1.5)

$$\chi(s) = r^*(P - sI)^{-1}q$$

is non-degenerate. Suppose also that system (2.1.5) is pendulum-like with respect to $\Gamma = \{jd, j \in \mathbf{Z}\}$, where d is some n-vector. In the communication theory such a system is called phase-controlled [138]. In the sequel an arbitrary feedback control system (1.2.1), which is pendulum-like, is also called phase-controlled. Under our assumptions the following theorem is true.

Theorem 2.1.1 *For the pendulum-like with respect to* $\Gamma = \{jd : j \in \mathbf{Z}\}$ *system* (2.1.5), *it can be assumed w.l.o.g. that* $Pd = 0$ *and* $\varphi(t,\cdot)$ *is* r^*d-*periodic.*

Proof By Proposition 2.1.1 we have

$$Pd + q\varphi(t,r^*x + r^*d) = q\varphi(t,r^*x)$$

for all $(t,x) \in \mathbf{R}_+ \times \mathbf{R}^n$. This equality is equivalent to

$$Pd + q\varphi(t,\sigma + r^*d) = q\varphi(t,\sigma) \tag{2.1.6}$$

for all $(t,\sigma) \in \mathbf{R}_+ \times \mathbf{R}$. By the controllability of (P,q) it follows that $q \neq 0$. Let us prove also that $r^*d \neq 0$. Suppose to the contrary that $r^*d = 0$. Then from (2.1.6) it follows that

$$r^*P^kd = 0 \qquad \text{for} \qquad k = 0,1\ldots,n-1. \tag{2.1.7}$$

Thus the vector d is a solution of the linear system $Td = 0$ given by (2.1.7). Because of the observability of (P,r) the matrix T has the rank n. It follows then that $d = 0$ which contradicts the fact that (2.1.5) is pendulum-like. Thus, $r^*d \neq 0$. Let us rewrite system (2.1.5) in the form

$$\dot{x} = (P - \alpha qr^*)x + q\varphi_1(t,r^*x) \tag{2.1.8}$$

where

$$\alpha = [|\,q\,|^2\, r^*d]^{-1}q^*Pd \quad \text{and} \quad \varphi_1(t,\sigma) = \varphi(t,\sigma) + \alpha\sigma$$

for all $(t,\sigma) \in \mathbf{R}_+ \times \mathbf{R}$. Equality (2.1.6) can be transformed as

$$\alpha(\sigma + r^*d) + \varphi(t,\sigma + r^*d) = \varphi(t,\sigma) + \alpha\sigma$$

for all $(t,\sigma) \in \mathbf{R}_+ \times \mathbf{R}$. Hence for fixed t the function φ_1 is periodic in σ with the period r^*d. Using now the equivariance property (2.1.2) with respect to (2.1.8) we see that

$$(P - \alpha qr^*)d = 0.$$

∎

In the sequel system (2.1.5) with $\det P = 0$ and $\varphi(t,\cdot)$ Δ-periodic is called *pendulum-like feedback system in the first canonical form*. In order to obtain another canonical form of (2.1.5) we prove the following auxiliary result.

Proposition 2.1.2 *Suppose d is an eigenvector of P from (2.1.5) which corresponds with the eigenvalue zero. Then $r^*d \neq 0$.*

Proof Because (P, r) is observable Theorem 1.2.1, p. 5, says that the matrix $T := [P^*, r]$ has the rank n. Then the system $T^*x = 0$ has trivial solutions only. On the other hand assuming that $r^*d = 0$ it follows that $T^*d = 0$, a contradiction. ∎

Now let S be a non-singular matrix of the form $S = [S_1, d]$, where S_1 is a certain $n \times (n-1)$ matrix and $d \neq 0$ satisfies $Pd = 0$. We take in (2.1.5) the change of variables $x = Sy$. Using the notations

$$
y = \left[\begin{array}{c} z \\ w \end{array} \right], \quad S^{-1}q = \left[\begin{array}{c} b \\ \beta \end{array} \right], \quad S^*r = \left[\begin{array}{c} g \\ \gamma \end{array} \right], \quad \text{and} \quad S^{-1}PS = \left[\begin{array}{cc} A & 0 \\ a^* & 0 \end{array} \right],
$$

where z, b, g and a are $(n-1)$-vectors and w, α, β and γ are scalars, we get the system

$$
\begin{array}{rcl}
\dot{z} & = & Az + b\varphi(t, \sigma) \\
\dot{w} & = & a^*z + \beta\varphi(t, \sigma) \\
\sigma & = & g^*z + \gamma w.
\end{array}
\tag{2.1.9}
$$

From the non-degeneracy of $\chi(\cdot)$ we have $\gamma \neq 0$. It follows that we can rewrite (2.1.9) in the form

$$
\begin{array}{rcl}
\dot{z} & = & Az + b\varphi(t, \sigma) \\
\dot{\sigma} & = & c^*z + \varrho\varphi(t, \sigma)
\end{array}
\tag{2.1.10}
$$

with $c = A^*g + \gamma a$ and $\varrho = g^*b + \beta\gamma$. We call (2.1.10) with a Δ-periodic in σ nonlinearity φ the *second canonical form of a pendulum-like feedback system*. Applying to (2.1.10) the formal Laplace transform we obtain for the Laplace transform $\widetilde{\xi}$ and $\widetilde{\sigma}$ of ξ and σ, respectively, ($\xi(\cdot)$ is an arbitrary "input"-function)

$$
\widetilde{\sigma}(s) = \frac{1}{s}[c^*(sI - A)^{-1}b + \varrho]\widetilde{\xi}(s).
$$

From this we define for all $s \in \mathbf{C}$ with $\det(A - sI) \neq 0$ the function $K(s) = c^*(A - sI)^{-1}b - \varrho$. It is clear that $K(s) = s\chi(s)$. In connection with the considerations of cycles of various types for the pendulum-like system (2.1.10) it is useful to investigate circular or running solutions for this system.

Definition 2.1.3 The solution $(z(\cdot), \sigma(\cdot))$ of (2.1.10) is called *circular* if there exist an $\varepsilon > 0$ and a time $\tau \geq 0$ such that $\dot{\sigma}(t) \geq \varepsilon$ for all $t \geq \tau$.

Circular solutions will be used to construct Poincaré maps for the existence of cycles. Our main tool is Brouwer's fixed point theorem, which we cite here for completeness.

Theorem 2.1.2 (Brouwer) *Let B be a convex compact set in $\mathbf{R}^n$, and let $f : B \to \mathbf{R}^n$ be a continuous function such that $f(B) \subset B$. Then f has at least one fixed point.*

Another result needed in the sequel is Barbalat's lemma (see [124]).

Theorem 2.1.3 (Barbalat) *If $\varphi : \mathbf{R}_+ \to \mathbf{R}$ is uniformly continuous and there exists a finite limit $\lim\limits_{t\to\infty} \int_0^t \varphi(\tau)\,d\tau$, then*

$$
\lim_{t\to\infty} \varphi(t) = 0.
\tag{2.1.11}
$$

Proof Indeed, if (2.1.11) is not satisfied then there exists a positive number a such that for every positive T one can find a $t(T) \geq T$ with $\mid \varphi(t(T)) \mid \geq a$. Since φ is uniformly continuous there exists a positive number c such that for every $s > 0$ and every τ in the interval $0 \leq \tau < c$ one has $\mid \varphi(s) - \varphi(s+\tau) \mid \leq a/2$. Hence, the inequality $\mid \varphi(t) \mid - \mid \varphi(t(T)) \mid \geq -a/2$ is satisfied for $t \in [t(T), t(T) + b]$, where b is a certain number. Adding this inequality and the previous inequality $\mid \varphi(t(T)) \mid > a$ we get $\mid \varphi(t) \mid > a/2$ for every t in the mentioned interval. We thus obtain

$$\left| \int\limits_{t(T)}^{t(T)+b} \varphi(t)\,dt \right| = \int\limits_{t(T)}^{t(T)+b} \mid \varphi(t) \mid \, dt \geq \frac{1}{2}ab,$$

where the first equality holds since φ retains the same sign for $t \in [t(T), t(T)+b]$. Thus the integral $\int\limits_0^t \varphi(\tau)\,d\tau$ cannot tend to a finite limit when time tends to infinity which gives a contradiction. $\blacksquare$

Corollary 2.1.1 *If $\varphi : \mathbf{R}_+ \to \mathbf{R}$ is uniformly continuous and $\varphi(\cdot) \in \mathrm{L}^2(0, \infty)$ then $\lim\limits_{t \to \infty} \varphi(t) = 0$.*

2.2 Second-order Pendulum-Like Systems

In this section, we consider the two-dimensional system

$$\begin{aligned}
\dot{\sigma} &= \eta - \varrho\varphi(\sigma) \\
\dot{\eta} &= -a\eta - \varphi(\sigma),
\end{aligned} \tag{2.2.1}$$

where a and ϱ are parameters and $\varphi : \mathbf{R} \to \mathbf{R}$ is Δ-periodic and of class $\mathbf{C}^1$. It follows from (2.1.10) that (2.2.1) is the general case of an autonomous two-dimensional pendulum-like feedback system in the second canonical form. We suppose that $a > 0$ and $a\varrho + 1 > 0$, which implies that $K(0) > 0$ ($K(s)$ is here the transfer-function of (2.2.1)).

In case $\varrho = 0$ the system (2.2.1) with $\varphi(\sigma) = \sin \sigma - \gamma$ describes, as was already mentioned, the oscillations of a damped pendulum under the action of a constant torque. It is also the simplest model of a PLL-system and of a synchronous machine. In case $\varrho \neq 0$ system (2.2.1) is a model of a PLL-system with proportional-integrating filter and the simplest model of PLL with feedback delay [140, 107].

System (2.2.1) is a mathematical object of extreme interest since different global bifurcations are possible. There is a great number of papers in which qualitative analysis of (2.2.1) is done. The first series of results was denoted to the case of $\varrho = 0$. Its analysis was started by F. Tricomi [147, 148] and continued by L. Amerio, C. Böhm, W.D. Hayes, L.N. Belyustina, A. Giger, G. Seifert [3, 4, 29, 133, 134, 37, 58, 50]. All the results obtained in these papers were referred in the monographs of E.A. Barbashin and V.A. Tabueva [20] and of G. Sansone and R. Cont. [132]. The case of $\varrho \neq 0$ was considered in a series of papers by N.N. Bautin, L.N. Belyustina and V.N. Belykh, V.N. Belykh and V.I. Nekorkin, N.A. Gubar', M.V. Kapranov, N.P. Vlasov [21, 32, 28, 53, 64, 25, 150, 33].

In this section the main results of these papers are described. Our representation is based on the qualitative theory of differential equations which was developed in [5, 6, 22, 110]. We use of course certain ideas and approaches described in the papers, which are given above. Let us note however that the scheme of analysis which is realized in the present section differs essentially from the usual scheme of investigationof (2.2.1). We assume w.l.o.g. that the mean value of φ is non-positive, i.e.

$$\overline{\varphi} := \frac{1}{\Delta} \int\limits_0^\Delta \varphi(\sigma)d\sigma \leq 0. \tag{2.2.2}$$

Clearly that system (2.2.1) is a pendulum-like one with respect to the group $\Gamma := \{jd \ : \ j \in \mathbf{Z}\}$, where $d = (0, \Delta)$. In connection with (2.2.1) we also consider the associated first order equation

$$\frac{dF(\sigma)}{d\sigma} = \frac{-aF(\sigma) - \varphi(\sigma)}{F(\sigma) - \varrho\varphi(\sigma)}. \tag{2.2.3}$$

In the following we give a detailed qualitative study of the global behavior of (2.2.1) leading to the description of the domain of attraction of equilibrium points, to the understanding of the role of cycles of the first and second kind and to a discussion of the bifurcation values of parameters separating different types of behavior.

a) General Properties

Let us introduce the values

$$m_0 := -\frac{1}{a} \max_{\sigma \in [0,\Delta)} \varphi(\sigma) \quad \text{and} \quad M_0 := -\frac{1}{a} \min_{\sigma \in [0,\Delta)} \varphi(\sigma). \tag{2.2.4}$$

It is clear that $m_0 \leq M_0$.

Proposition 2.2.1 *Let m and M be arbitrary numbers with $m < m_0$ and $M_0 < M$. Then for any solution (σ, η) of (2.2.1) there exists a time τ such that*

$$\eta(t) \in [m, M] \quad \text{for all} \quad t \geq \tau. \tag{2.2.5}$$

Proof Suppose at first that for the solution (σ, η) we have $\eta(0) > M$. It follows from (2.2.1) that for all $t \geq 0$ with $\eta(t) \geq M$ we have

$$\dot{\eta}(t) \leq -aM - \varphi(\sigma(t)) \leq -a(M - M_0).$$

Thus the corresponding trajectory must intersect the line $\{\sigma, \eta \ : \ \eta = M\}$ in a certain time. In an analogous way we show that the trajectory of (σ, η) with $\eta(0) < m$ intersects the line $\{\sigma, \eta \ : \ \eta = m\}$. It is clear that the set $\{\sigma, \eta \ : \ m \leq \eta \leq M\}$ is positively invariant for (2.2.1). $\blacksquare$

Proposition 2.2.2 *Suppose that $F_1(\sigma)$ and $F_2(\sigma)$ are two solutions of (2.2.3) defined on $[\overline{\sigma}, +\infty)$. Suppose that the functions $y_k(\sigma) := F_k(\sigma) - \varrho\varphi(\sigma)$, $(k = 1, 2)$ are positive on $[\overline{\sigma}, +\infty)$. Suppose also that one of the following conditions holds:*

(i) *φ has no zeros on $\mathbf{R}$;*

(ii) *$a >\mid \varrho \mid r$, where $r := \max_{\sigma \in [0,\Delta)} \mid \varphi'(\sigma) \mid$;*

(iii) *$\varrho \in (-1/a, 0)$.*

Then

$$\lim_{\sigma \to +\infty} [F_1(\sigma) - F_2(\sigma)] = 0. \tag{2.2.6}$$

Proof Let $F_1(\overline{\sigma}) > F_2(\overline{\sigma})$. Then it follows from the condition of the present proposition that $F_1(\sigma) > F_2(\sigma)$ on $[\overline{\sigma}, +\infty)$. Suppose now that condition (i) holds. From (2.2.3) we have

$$\frac{d}{d\sigma}[F_1(\sigma) - F_2(\sigma)] = \frac{(a\varrho + 1)\varphi(\sigma)[F_1(\sigma) - F_2(\sigma)]}{[F_1(\sigma) - \varrho\varphi(\sigma)][F_2(\sigma) - \varrho\varphi(\sigma)]}. \tag{2.2.7}$$

18

Since φ has no zeros it is negative on $\mathbf{R}$ in virtue of (2.2.2). Consequently

$$\frac{d}{d\sigma}[F_1(\sigma) - F_2(\sigma)] < 0 \tag{2.2.8}$$

for all $\sigma > \overline{\sigma}$. Using the last inequality and the positiveness of $(F_1 - F_2)$ we get the existence of a finite limit

$$\lim_{\sigma \to +\infty}[F_1(\sigma) - F_2(\sigma)] =: A.$$

Suppose that $A > 0$. Then there exists a number $\overline{\overline{\sigma}} > \overline{\sigma}$ such that

$$F_1(\sigma) - F_2(\sigma) \geq \frac{A}{2}. \tag{2.2.9}$$

for all $\sigma > \overline{\overline{\sigma}}$. From the fact that $[F_1(\sigma) - \varrho\varphi(\sigma)][F_2(\sigma) - \varrho\varphi(\sigma)]$ is bounded and the function φ is bounded from above by a negative number, it follows from (2.2.9) that

$$\frac{d}{d\sigma}[F_1(\sigma) - F_2(\sigma)] < -C \tag{2.2.10}$$

for all $\sigma > \overline{\overline{\sigma}}$, where $C > 0$ is a constant. Hence

$$F_1(\sigma) - F_2(\sigma) \leq F_1(\overline{\sigma}) - F_2(\overline{\sigma}) - C(\sigma - \overline{\sigma})$$

for all $\sigma \geq \overline{\overline{\sigma}}$, which contradicts the positiveness of $[F_1(\sigma) - F_2(\sigma)]$ for $\sigma > \overline{\sigma}$. Thus $A = 0$. Suppose now that condition (ii) holds. Let us rewrite (2.2.3) in the form

$$F'F - F'\varrho\varphi = -aF - \varphi(\sigma) \tag{2.2.11}$$

or in the form

$$(F - \varrho\varphi)'(F - \varrho\varphi) = -aF - \varrho\varphi'F - \varphi(\sigma) + \varrho^2\varphi'\varphi. \tag{2.2.12}$$

From (2.2.12) we get

$$\frac{1}{2}\frac{d}{d\sigma}[y_1(\sigma)^2 - y_2(\sigma)^2] = -(a + \varrho\varphi'(\sigma))(y_1(\sigma) - y_2(\sigma)). \tag{2.2.13}$$

If follows from condition (ii) and (2.2.13) that

$$\frac{d}{d\sigma}[y_1(\sigma)^2 - y_2(\sigma)^2] < 0$$

for all $\sigma > \overline{\sigma}$. By this inequality and the positiveness of $y_1(\sigma)^2 - y_2(\sigma)^2$ we have that there exists a finite limit

$$\lim_{\sigma \to +\infty}[y_1(\sigma)^2 - y_2(\sigma)^2] =: B.$$

Suppose that $B > 0$. Then there exists a number $\widetilde{\sigma} > \overline{\sigma}$ such that

$$y_1(\sigma)^2 - y_2(\sigma)^2 \geq \frac{B}{2}$$

for $\sigma > \widetilde{\sigma}$. Since y_1 and y_2 are bounded and positive on $[\overline{\sigma}, +\infty)$ it follows from the last inequality that

$$y_1(\sigma) - y_2(\sigma) > C_1 > 0 \quad \text{for all} \quad \sigma > \widetilde{\sigma}.$$

It follows now from (2.2.13) that

$$y_1(\sigma)^2 - y_2(\sigma)^2 \leq y_1(\widetilde{\sigma})^2 - y_2(\widetilde{\sigma})^2 - (a- \mid \varrho \mid r)C_1(\sigma - \widetilde{\sigma})$$

for $\sigma > \tilde{\sigma}$, which contradicts the inequality $F_1(\sigma) > F_2(\sigma)$ for $\sigma > \overline{\sigma}$. Thus $B = 0$. Since $y_1(\sigma) - y_2(\sigma) > 0$, $y_k(\sigma) > 0$ $(k = 1, 2)$ and $(y_1 - y_2)^2 = (y_1^2 - y_2^2) - 2y_2(y_1 - y_2)$ if follows from $B = 0$ that

$$\lim_{\sigma \to +\infty} [F_1(\sigma) - F_2(\sigma)] = 0.$$

Let us now suppose that condition (iii) holds. Using (2.2.11) and (2.2.7) we get

$$\frac{1}{2}\frac{d}{d\sigma}[F_1^2 - F_2^2] - \frac{\varrho(a\varrho + 1)\varphi(\sigma)^2[F_1(\sigma) - F_2(\sigma)]}{[F_1(\sigma) - \varrho\varphi(\sigma)][F_2(\sigma) - \varrho\varphi(\sigma)]} = -a(F_1 - F_2). \tag{2.2.14}$$

Hence by virtue of condition (iii) we have

$$\frac{1}{2}\frac{d}{d\sigma}[F_1^2 - F_2^2] < -a(F_1 - F_2). \tag{2.2.15}$$

Note that in case ϱ is negative, then inequality $F - \varrho\varphi > 0$ implies $F > 0$, since

$$- aF - \varphi < -a\varrho\varphi - \varphi = -(a\varrho + 1)\varphi. \tag{2.2.16}$$

Suppose now that $F < 0$ for $\sigma \in (\sigma', \sigma'')$ but $F(\sigma') = F(\sigma'') = 0$. Then since $\varrho < 0$ we have that $\varphi(\sigma) > 0$ for $\sigma \in (\sigma', \sigma'')$. Thus by virtue of (2.2.17) $-aF(\sigma) - \varphi(\sigma) < 0$ for $\sigma \in (\sigma', \sigma'')$, i.e. $dF/d\sigma < 0$ on $[\sigma', \sigma'']$, which is impossible. So in virtue of (2.2.15) the argument which was used in item (ii) for y_1, y_2 is true here for functions F_1 and F_2. ∎

Proposition 2.2.3 *Suppose F_1 and F_2 are two solutions of (2.2.3) defined on $(-\infty, \overline{\sigma})$. Suppose that $y_k(\sigma) := F_k(\sigma) - \varrho\varphi(\sigma)$ $(k = 1, 2)$ is negative on $(-\infty, \overline{\sigma})$. Suppose also that one of the conditions (i), (ii) or (iii) of Proposition 2.2.2 is satisfied. Then*

$$\lim_{\sigma \to -\infty} [F_1(\sigma) - F_2(\sigma)] = 0. \tag{2.2.17}$$

Proof It is analogous to the proof of Proposition 2.2.2. ∎

From Propositions 2.2.2 and 2.2.3 follows

Corollary 2.2.1 *If one of the conditions (i), (ii) or (iii) of Proposition 2.2.2 holds then system (2.2.1) cannot have more than one cycle of the second kind in the domains $D_1 := \{\sigma, \eta : \eta > \varrho\varphi(\sigma)\}$ resp. $D_2 := \{\sigma, \eta : \eta < \varrho\varphi(\sigma)\}$. If a cycle of the second kind exists in D_1 resp. D_2 then this cycle attracts all the orbits which lie in D_1 resp. D_2 in the sense of (2.2.6) resp. (2.2.17).*

Let us consider now the set of zeros of φ and distinguish two cases.

Case 1: The function φ has no zeros, i.e. $\varphi(\sigma) < 0$ on R.
In this case system (2.2.1) has no equilibria and, consequently, no cycles of the first kind. Note also that in this case m_0 given by (2.2.4) is positive and the following proposition holds.

Proposition 2.2.4 *System (2.2.1) has a cycle of the second kind provided that φ has no zeros on R.*

Proof Define the values m_0 and M_0 as above and choose arbitrary positive $m < m_0$ and $M > M_0$. Consider an arbitrary solution σ, η of (2.2.1) which starts in the set $\{\sigma, \eta : m \leq \eta \leq M\}$ which is positively invariant for (2.2.1). For this solution we have by (2.2.1) $a\dot{\sigma}(t) + \dot{\eta}(t) = -(a\varrho + 1)\varphi(\sigma(t))$ and, consequently,

$$a\dot{\sigma}(t) + \dot{\eta}(t) \geq (a\varrho + 1)am \tag{2.2.18}$$

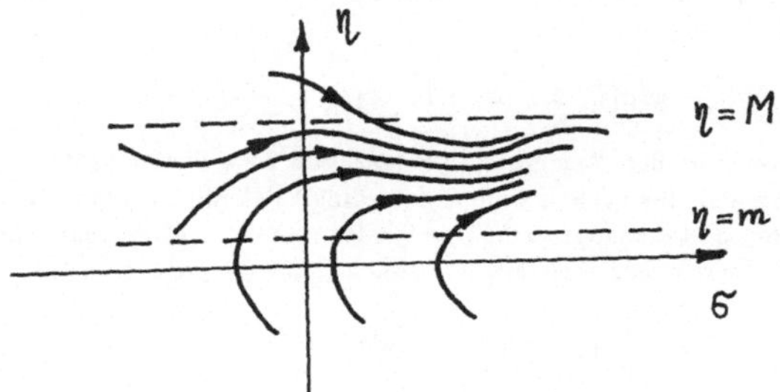

Figure 2.2.1

for all $t \geq 0$. Thus

$$a\sigma(t) + \eta(t) \to +\infty \qquad \text{as} \qquad t \to +\infty. \tag{2.2.19}$$

Let us define the sets

$$\Omega_1 := \{\sigma, \eta \; : \; \eta \in [m, M], a\sigma + \eta = 0\} \qquad \text{and} \qquad \Omega_2 := \{\sigma, \eta \; : \eta \in [m, M], a\sigma + \eta = a\Delta\}.$$

It follows from (2.2.19) that for an arbitrary $\hat{\sigma} \in \mathbf{R}$ the straight line $\{\sigma, \eta \; : \; a(\sigma - \hat{\sigma}) + \eta = a\Delta\}$ is a line without contact with the vector field of (2.2.1). The estimate (2.2.18) guarantees that for any $(\sigma_0, \eta_0) \in \Omega_1$ there exists exactly one $\tau = \tau(\sigma_0, \eta_0)$ such that

$$(\sigma_0(\tau; \sigma_0, \eta_0), \eta(\tau; \sigma_0, \eta_0)) \in \Omega_2,$$

where $(\sigma(\cdot; \sigma_0, \eta_0), \eta(\cdot; \sigma_0, \eta_0))$ denotes the solution of (2.2.1) which starts in (σ_0, η_0). Thus we have defined a mapping $T : \Omega_1 \to \Omega_2$ which is continuous since the solution depends continuously on the initial conditions. Consider also a mapping $Q : \Omega_2 \to \Omega_1$ defined by $Q(\sigma, \eta) = (\sigma - \Delta, \eta)$. It is clear that $(Q \circ T)\Omega_1 \subset \Omega_1$. Since Ω_1 is a compact convex set and $Q \circ T$ is continuous the last inclusion guarantees by Brouwer's theorem (Theorem 2.1.2) the existence of a fixed point (σ_0, η_0) of $Q \circ T$. This fact is equivalent to $\sigma(\tau(\sigma_0, \eta_0); \sigma_0, \eta_0) = \sigma_0 + \Delta$, $\eta(\tau(\sigma_0, \eta_0); \sigma_0, \eta_0) = \eta_0$. Thus system (2.2.1) has a cycle of the second kind. $\blacksquare$

From Corollary 2.2.1 and Proposition 2.2.4 follows

Corollary 2.2.2 *Suppose that φ has no zeros. Then there exists for (2.2.1) exactly one cycle of the second kind which attracts the remaining orbits of (2.2.1) in the sense of (2.2.6). All other solutions of (2.2.1) are circular solutions (Figure 2.2.1).*

Proof It is obvious that any cycle in the case considered ought to be disposed in the domain $\{\eta > \varrho\varphi(\sigma)\}$ on the plane $\{\sigma, \eta\}$. Indeed, according to Proposition 2.2.1 for an arbitrary solution $\sigma(\cdot), \eta(\cdot)$ of (2.2.1) there exists a time τ such that

$$\eta(t) > \frac{1}{a} \mid \max_{\sigma \in [0, \Delta]} \varphi(\sigma) \mid > \varrho \max_{\sigma \in [0, \Delta]} \varphi(\sigma)$$

for all $t \geq \tau$. But the cycle cannot intersect the line $\{\sigma, \eta \; : \; \eta = \varrho\varphi(\sigma)\}$ since on this line we have for a solution $\sigma(\cdot), \eta(\cdot)$ of (2.2.1)

$$\dot{\eta}(t) = -(a\varrho + 1)\varphi(\sigma(t)) > 0.$$

21

More than that, any other solution which starts in the domain $\{\sigma,\eta \ : \ \eta < \varrho\varphi(\sigma)\}$ ought to leave it. Indeed, if $\eta(t) < \varrho\varphi(\sigma(t))$ then

$$\dot{\eta}(t) = -a\eta(t) - \varphi(\sigma(t)) > -(a\varrho + 1)\varphi(\sigma(t)) > -(a\varrho + 1)m, \qquad (2.2.20)$$

where $m \in (0, m_0)$ is a certain number. According to Proposition 2.2.4 system (2.2.1) has a cycle $\overline{\sigma}(\cdot), \overline{\eta}(\cdot)$ of the second kind and this cycle is unique according to Corollary 2.2.1. On the basis of (2.2.20) all the remaining solutions approach this cycle for $t \to +\infty$. More than that for any solution $\sigma(\cdot), \eta(\cdot)$ of (2.2.1) there exists a certain $\tau > 0$ such that $\dot{\sigma}(t) \geq \varepsilon$ for all $t > \tau$, where

$$\varepsilon = \frac{1}{2} \min_{\sigma \in [0,\Delta]} [\overline{\eta}(\sigma) - \varrho\varphi(\sigma)]. \quad \blacksquare$$

Let us now consider

Case 2: The function φ has exactly two zeros on $[0, \Delta)$.
W.l.o.g. we assume that one of them is equal to zero. The other we denote by σ_1. We assume also that

$$\varphi'(0) > 0 \quad \text{and} \quad \varphi'(\sigma_1) < 0. \qquad (2.2.21)$$

In this case system (2.2.1) has infinitely many equilibrium states:

$$\mathcal{E} = \{(k\Delta, 0)\,, \ (\sigma_1 + k\Delta, 0), \ k \in \mathbf{Z}\}\,.$$

Because of the equivariance property of (2.2.1) it is sufficient for stability analysis to consider the two states $(0, 0)$ and $(0, \sigma_1)$. A stability analysis by the first approximation shows that $(0, 0)$ is either a knot or a focus (which is stable for $\varrho \geq 0$ and may be unstable for $\varrho < 0$). Indeed the characteristic equations of the linearization of (2.2.1) in $(0, 0)$ and $(\sigma_1, 0)$ is

$$\lambda^2 + (a + \varrho\varphi'(z))\lambda + (1 + \varrho a)\varphi'(z) = 0, \qquad (2.2.22)$$

where $z = 0$ or $z = \sigma_1$, respectively. The linearization of (2.2.1) in $(\sigma_1, 0)$ is characterized by the two slopes $k_1 < 0$ and $k_2 > 0$ which are roots of the equation

$$k^2 + (a - \varrho\varphi'(z))k + \varphi'(\sigma_1) = 0. \qquad (2.2.23)$$

Thus there exists by the Hadamard-Perron theorem in $(\sigma_1, 0)$ the stable manifold $W^s(\sigma_1, 0)$, given by the two ω-separatrices $(\widetilde{\sigma}_0(\cdot), \widetilde{\eta}_0(\cdot))$ and $(\widetilde{\widetilde{\sigma}}_0(\cdot), \widetilde{\widetilde{\eta}}_0(\cdot))$ which approach $(\sigma_1, 0)$ for $t \to +\infty$, and the unstable manifold $W^u(\sigma_1, 0)$, given by the α-separatices $(\overline{\sigma}_0(\cdot), \overline{\eta}_0(\cdot))$ and $(\overline{\overline{\sigma}}_0(\cdot), \overline{\overline{\eta}}_0(\cdot))$ which approach $(\sigma_1, 0)$ for $t \to -\infty$. Since

$$k_{1,2} - \varrho\varphi'(\sigma_1) = -\frac{a + \varrho\varphi'(\sigma_1)}{2} \mp \frac{1}{2}\sqrt{(a - \varrho\varphi'(\sigma_1))^2 - 4\varphi'(\sigma_1)}$$

and

$$[(a - \varrho\varphi'(\sigma_1))^2 - 4\varphi'(\sigma_1)] - (a + \varrho\varphi'(\sigma_1))^2 = -4(a\varrho + 1)\varphi'(\sigma_1) > 0$$

it follows that

$$k_1 < \varrho\varphi'(\sigma_1) < k_2. \qquad (2.2.24)$$

Thus by the mentioned theorem the phase portrait of (2.2.1) near $(\sigma_1, 0)$ has the form which is shown in Figure 2.2.2.

Furthermore, we can conclude that there exists a $T > 0$ such that $\widetilde{\eta}_0(t) > 0$ and $\widetilde{\widetilde{\eta}}_0(t) < 0$ for $t \geq T$.

22

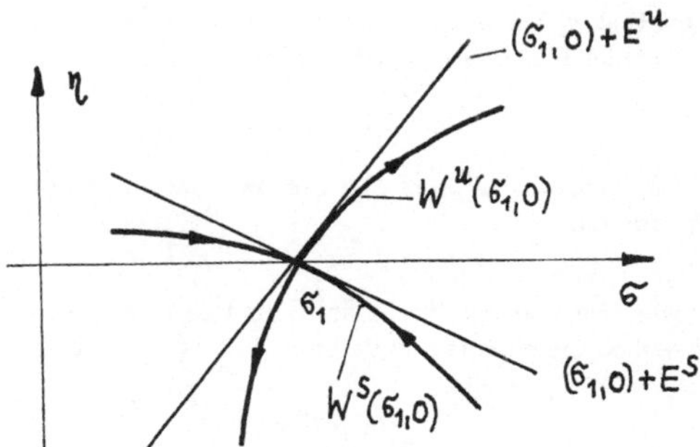

Figure 2.2.2

b) The Case $\varrho \geq 0$

Let us define the function $V : \mathbf{R} \times \mathbf{R} \to \mathbf{R}$ by

$$V(\sigma,\eta) = \frac{1}{2}\eta^2 + \int_0^{\sigma} \varphi(\vartheta)\,d\vartheta. \tag{2.2.25}$$

The derivative of V along the solutions of (2.2.1) is given by

$$\dot{V}_{(2.2.1)}(\sigma,\eta) = -a\eta^2 - \varrho\varphi^2(\sigma), \tag{2.2.26}$$

which is negative for $\eta \neq 0$.

Proposition 2.2.5 *System* (2.2.1) *with* $\varrho \geq 0$ *has no cycles of the first kind.*

Proof Suppose to the contrary that $(\sigma(\cdot),\eta(\cdot))$ defines a non-trivial τ-periodic solution of (2.2.1). Using the function $v(t) := V(\sigma(t),\eta(t))$ we get by (2.2.26) that $v(\tau) < v(0)$. But by the periodicity we have $v(\tau) = v(0)$. ∎

Let us suppose further that φ in (2.2.1) has exactly two zeros $\sigma = 0$ and $\sigma = \sigma_1$ and the condition (2.2.21) holds.

Proposition 2.2.6 *Every solution of* (2.2.1) *converges provided that* $\varrho \geq 0$ *and*

$$\int_0^{\Delta} \varphi(\vartheta)\,d\vartheta = 0. \tag{2.2.27}$$

Proof Suppose $(\sigma(\cdot),\eta(\cdot))$ is an arbitrary solution of (2.2.1) and define the functions g, v : $\mathbf{R}_+ \to \mathbf{R}$ by

$$g(t) := \int_0^{\sigma(t)} \varphi(\vartheta)\,d\vartheta \quad \text{and} \quad v(t) := V(\sigma(t),\eta(t)) = \frac{1}{2}\eta(t)^2 + g(t).$$

It follows from (2.2.26) that for any $T > 0$

$$v(T) - v(0) = -a\int_0^{T} \eta(t)^2\,dt - \varrho\int_0^{T} \varphi(\sigma(t))^2\,dt. \tag{2.2.28}$$

From (2.2.27) we get that $g(\cdot)$ is bounded on $\mathbf{R}_+$. Then it follows by Proposition 2.2.1 that function $v(t)$ is bounded on $\mathbf{R}_+$. Using (2.2.28) we see that

$$\eta \in L^2(0, +\infty). \tag{2.2.29}$$

Since η is bounded the second equation of (2.2.1) shows that $\dot{\eta}$ is also bounded on $\mathbf{R}_+$. By the Corollary 2.1.1 it follows that

$$\eta(t) \to 0 \quad \text{as} \quad t \to +\infty. \tag{2.2.30}$$

Let us demonstrate now that $\sigma(t)$ has a finite limit as $t \to +\infty$. Since v is bounded and non-increasing as $t \to +\infty$ there exists a finite limit $\lim\limits_{t \to +\infty} v(t) = c$. It follows by (2.2.30) that

$$\lim\limits_{t \to +\infty} g(t) = c. \tag{2.2.31}$$

It is obvious by (2.2.31) and (2.2.27) that $\sigma(t)$ has a finite limit $\overline{\sigma}$ as $t \to +\infty$, the point $(\overline{\sigma}, 0)$ being an equilibrium of (2.2.1). ∎

Proposition 2.2.7 *If $\varrho \geq 0$ the solution $(\sigma(\cdot), \eta(\cdot))$ of (2.2.1) converges provided that the function $g(t) := \int\limits_0^{\sigma(t)} \varphi(\vartheta)\, d\vartheta \quad (t \geq 0)$ is bounded from below on $\mathbf{R}_+$.*

Proof It is clear that the proof of Proposition 2.2.6 goes through if we change the requirement (2.2.27) by the boundedness from below of g. ∎

Proposition 2.2.8 *If $\varrho \geq 0$ then an arbitrary solution $(\sigma(\cdot), \eta(\cdot))$ of (2.2.1) either converges, or for this solution, there exists a sequence of times $\{t_k\}$ such that $\sigma(t_k) \to +\infty$ as $k \to +\infty$.*

Proof Assume that $(\sigma(\cdot), \eta(\cdot))$ is a solution of (2.2.1) with $\sigma(t)$ bounded from above on $\mathbf{R}_+$. Using (2.2.2) we see that $\int\limits_0^{\sigma(t)} \varphi(\vartheta)\, d\vartheta$ is bounded from below on $\mathbf{R}_+$. From Proposition (2.2.7) it follows that the given solution converges to an equilibrium. ∎

Let us consider now some global properties of the ω-separatrix $(\widetilde{\widetilde{\sigma}}_0(\cdot), \widetilde{\widetilde{\eta}}_0(\cdot))$ of the saddle-point $(\sigma_1, 0)$.

Proposition 2.2.9 *In case $\varrho \geq 0$ we have*

$$\widetilde{\widetilde{\eta}}_0(t) < 0 \quad \text{for all} \quad t \in \mathbf{R}. \tag{2.2.32}$$

Proof Let us consider the ω-separatrix in $(\sigma_1 - \Delta, 0)$ which is given by $(\widetilde{\widetilde{\sigma}}(t) - \Delta, \widetilde{\widetilde{\eta}}_0(t))$. In Figure 2.2.3 the direction field $(-\varrho\varphi(\sigma), -\varphi(\sigma))$ of (2.2.1) along the line $\{\sigma, \eta \ : \ \eta = \varrho\varphi(\sigma)\}$ is denoted by arrows.

It follows from the uniqueness and equivariance property of (2.2.1) that $\widetilde{\widetilde{\eta}}_0(\overline{t}) = 0$ for some $\overline{t} \in \mathbf{R}$ implies that

$$\widetilde{\widetilde{\sigma}}_0(\overline{t}) \in (0, \sigma_1). \tag{2.2.33}$$

Consider the level set

$$\Gamma := \left\{ (\sigma, \eta) \ : \ V(\sigma, \eta) = c_0 := \int\limits_0^{\sigma_1} \varphi(\vartheta)\, d\vartheta \right\}. \tag{2.2.34}$$

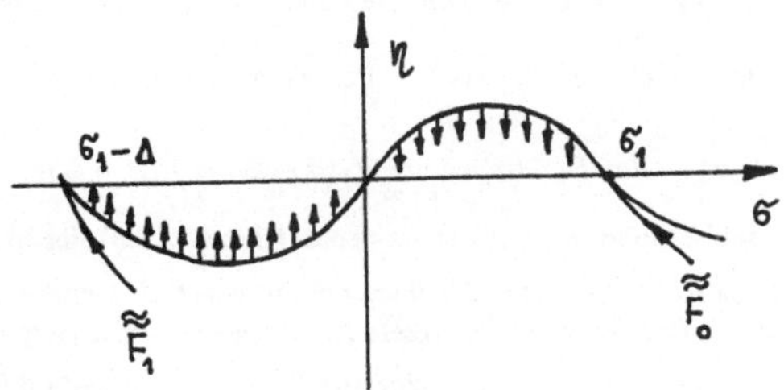

Figure 2.2.3

Note that $(\sigma_1, 0) \in \Gamma$ and Γ is a closed curve. Indeed, using (2.2.2) and (2.2.21) we see that

$$\int\limits_0^{\sigma_1 - \Delta} \varphi(\vartheta) \, d\vartheta > \int\limits_0^{\sigma_1} \varphi(\vartheta) \, d\vartheta.$$

It follows that the curve Γ also passes through a point $(\sigma_2, 0)$ with $\sigma_2 \in (\sigma_1 - \Delta, 0)$. Since Γ is symmetric with respect to the σ-axis it is a closed curve around $(0, 0)$. The curve Γ is without contact with the vector field of (2.2.1). Thus the separatrix $(\widetilde{\widetilde{\sigma}}_0(t) - \Delta, \widetilde{\widetilde{\eta}}_0(t))$ cannot intersect the curve Γ. This fact contradicts (2.2.33). Thus the equality $\widetilde{\widetilde{\eta}}_0(t) = 0$ for some t is impossible ∎

Convention 2.2.1 Denote by $\widetilde{\widetilde{F}}_0$ the solution of (2.2.3) which corresponds to $(\widetilde{\widetilde{\sigma}}_0(\cdot), \widetilde{\widetilde{\eta}}_0(\cdot))$ and which is defined on $[\sigma_1, +\infty)$.

Remark 2.2.1 It follows from Proposition 2.2.9 that the orbit of $(\widetilde{\widetilde{\sigma}}_0(\cdot), \widetilde{\widetilde{\eta}}_0(\cdot))$ cannot intersect the line $\{(\sigma, \eta) \; : \; \eta = \varphi(\sigma)\}$ and, consequently, $\widetilde{\widetilde{F}}_0(\sigma) < \varrho\varphi(\sigma)$ for all $\sigma \in (\sigma_1, +\infty)$.

Proposition 2.2.10 *In case* $\varrho > 0$ *the function* $\widetilde{\widetilde{F}}_0$, *defined by Convention 2.2.1, is not bounded as* $\sigma \to +\infty$.

Proof Let us suppose the opposite, i.e. that $\widetilde{\widetilde{F}}_0(\sigma) > -c$ for all $\sigma > \sigma_1$, where $c > 0$ is a constant. Let us show that under this assumption equation (2.2.3) has a periodic solution. Note that

$$\widetilde{\widetilde{F}}_0(\sigma + \Delta) < \widetilde{\widetilde{F}}_0(\sigma) \qquad \text{for} \qquad \sigma > \sigma_1.$$

Consider now the functions $\widetilde{\widetilde{F}}_k \quad (k = 1, 2, \ldots)$ defined by

$$\widetilde{\widetilde{F}}_k(\sigma) = \widetilde{\widetilde{F}}_0(\sigma + k\Delta) \qquad \text{for} \qquad \sigma > \sigma_1.$$

By the uniqueness theorem and the equivariance property we have that for $k = 1, 2, \ldots \widetilde{\widetilde{F}}_k(\sigma) < \widetilde{\widetilde{F}}_{k-1}(\sigma)$ and $\widetilde{\widetilde{F}}_k(\sigma - \Delta) = \widetilde{\widetilde{F}}_{k-1}(\sigma)$ in the common domain of definition. Thus for fixed $\sigma \in$

$[\sigma_1, \sigma_1 + \Delta]$ the sequence $\left\{\widetilde{\widetilde{F}}_k(\sigma)\right\}$ is monotone decreasing and bounded from below. It follows that there exists a limit function $\widetilde{\widetilde{F}}$ defined by $\widetilde{\widetilde{F}}(\sigma) = \lim\limits_{k\to\infty} \widetilde{\widetilde{F}}_k(\sigma)$. This function is Δ-periodic since

$$\widetilde{\widetilde{F}}(\sigma) = \lim\limits_{k\to\infty}\widetilde{\widetilde{F}}_k(\sigma) = \lim\limits_{k\to\infty}\widetilde{\widetilde{F}}_{k-1}(\sigma) = \lim\limits_{k\to\infty}\widetilde{\widetilde{F}}_k(\sigma - \Delta) = \widetilde{\widetilde{F}}(\sigma - \Delta).$$

Since in the right-hand neighborhood of $(\sigma_1, 0)$ the ω-separatrix of (2.2.3) lies lower than the curve $\{\sigma, \eta \ : \ \eta = \varrho\varphi(\sigma)\}$ we can conclude that $\widetilde{\widetilde{F}}_k$ does not intersect this curve on $[-\Delta(k-1), -\infty)$ (see Figure 2.2.3). Using this fact and the Arzela-Ascoli theorem it is easy to show that $\widetilde{\widetilde{F}}$ is a solution of (2.2.3). To do this it is sufficient to consider the integral equation for $\widetilde{\widetilde{F}}_k$ which follows from (2.2.3) and is given by

$$\widetilde{\widetilde{F}}_k(\sigma) = \widetilde{\widetilde{F}}_k(\sigma_1) + \int\limits_{\sigma_1}^{\sigma} \frac{-a\widetilde{\widetilde{F}}_k(\vartheta) - \varphi(\vartheta)}{\widetilde{\widetilde{F}}_k(\vartheta) - \varrho\varphi(\vartheta)}\, d\vartheta$$

for $\sigma > \sigma_1$.

It follows from Remark 2.2.1 that $\widetilde{\widetilde{F}}(\sigma) < \varrho\varphi(\sigma)$ for $\sigma > \sigma_1$ and thus on the basis of (2.2.1), $\sigma(t) \to -\infty$ along the corresponding orbit. This contradiction shows that $\widetilde{F}_0$ is not bounded as $\sigma \to +\infty$. ∎

Corollary 2.2.3 *Suppose that in (2.2.1) we have $\varrho \geq 0$ and consider the function $\widetilde{\widetilde{F}}_0$, defined by Convention 2.2.1. Then*

$$\lim\limits_{\sigma\to+\infty}\widetilde{\widetilde{F}}_0(\sigma) = -\infty.$$

Proof Since $\widetilde{\widetilde{F}}_0$ is not bounded it is possible to find a point $\overline{\sigma}$ with $\widetilde{\widetilde{F}}_0(\overline{\sigma}) < m_0$. It is clear from the proof of Proposition 2.2.1 that $\frac{d}{dt}\widetilde{\widetilde{\eta}}_0(\overline{t}) > 0$ where $\overline{t}$ is such that $\widetilde{\widetilde{\eta}}_0(\overline{t}) = \widetilde{\widetilde{F}}_0(\overline{\sigma})$. Using this and the property $\widetilde{\widetilde{\eta}}_0(t) < \varrho\varphi(\widetilde{\widetilde{\sigma}}_0(t))$ it follows that $\widetilde{\widetilde{F}}_0$ is monotone decreasing. ∎

Let us consider now the behavior of the ω-separatrix given by $(\widetilde{\sigma}_0(\cdot), \widetilde{\eta}_0(\cdot))$.

Convention 2.2.2 Denote by $\widetilde{F}_0$ the solution of (2.2.3) which corresponds to $(\widetilde{\sigma}_0(\cdot), \widetilde{\eta}_0(\cdot))$.

Let us consider the following three types of behavior:

Type (i) $\widetilde{\eta}_0(t) > 0$ for all $t \in \mathbf{R}$ and there exists a time $\overline{t}$ such that $\widetilde{\sigma}_0(\overline{t}) = \sigma_1 - \Delta$;

Type (ii) $\widetilde{\eta}_0(t) > 0$ and $\widetilde{\sigma}_0(t) \in (\sigma_1 - \Delta, \sigma_1]$ for all $t \in \mathbf{R}$;

Type (iii) there exists a $\overline{t}$ such that $\eta_0(\overline{t}) = 0$ and $\widetilde{\eta}_0(t) > 0$ for $t > \overline{t}$.

Let us investigate type (iii) first.

Proposition 2.2.11 *Suppose type (iii) occurs and $\varrho \geq 0$ in (2.2.1). Then $\widetilde{\eta}_0(t) < 0$ for all $t < \overline{t}$.*

Proof Let us again use the level curve Γ from (2.2.34). Note that the ω-separatrix given by $(\widetilde{\sigma}_0(\cdot), \widetilde{\eta}_0(\cdot))$ lies outside this closed curve. Indeed, the slope of the ω-separatrix at $(\sigma_1, 0)$ is given by

$$k_1 = k_1(a, \varrho) = -\frac{1}{2}[a - \varrho\varphi'(\sigma_1) + \sqrt{[a - \varrho\varphi'(\sigma_1)]^2 - 4\varphi'(\sigma_1)}]. \tag{2.2.35}$$

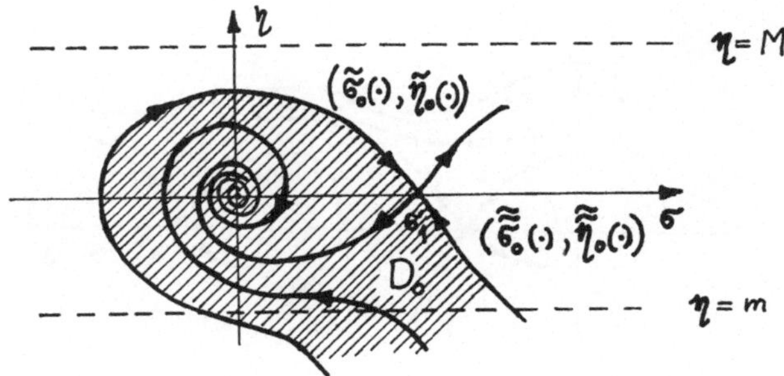

Figure 2.2.4

Then the partial derivative $D_2 k_1(a, \varrho)$ satisfies

$$D_2 k_1(a, \varrho) = \frac{1}{2}\varphi'(\sigma_1)[1 + \frac{a - \varrho\varphi'(\sigma_1)}{\sqrt{[a - \varrho\varphi'(\sigma_1)]^2 - 4\varphi'(\sigma_1)}}] < 0.$$

It follows that

$$k_1(a, \varrho) < k_1(a, 0) = \frac{-a - \sqrt{a^2 - 4\varphi'(\sigma_1)}}{2} < -\sqrt{-\varphi'(\sigma_1)} = k_1(0, 0). \tag{2.2.36}$$

On the other hand the level curve Γ in case $\varrho = 0$, $a = 0$ consists of trajectories of (2.2.1). Thus in a small neighborhood of $(\sigma_1, 0)$ the ω-separatrix lies outside Γ. Since Γ is a line without contact with the vector field of (2.2.1) the ω-separatrix cannot intersect it in the direction from outside to inside. Furthermore, the separatrix $(\widetilde{\sigma}_0(\cdot), \widetilde{\eta}_0(\cdot))$ cannot intersect the ω-separatrix $(\widetilde{\sigma}_0(\cdot) - \Delta, \widetilde{\eta}_0(\cdot))$ which approach the point $(\sigma_1 - \Delta, 0)$. So if $\widetilde{\sigma}_0(\bar{t}) = 0$ then $\widetilde{\sigma}_0(t) \in (\sigma_1 - \Delta, \sigma_2)$, where σ_2 is defined in the proof of Proposition 2.2.9. From the proof it is clear that in this case $\widetilde{\eta}_0(t)$ cannot vanish for $t < \bar{t}$.

Corollary 2.2.4 *If the type (iii) occurs there exists exactly one value $\bar{t}$ such that $\widetilde{\sigma}(\bar{t}) = 0$. More than that $\widetilde{\sigma}_0(t) \in (\sigma_1 - \Delta, \sigma_2)$, where $\sigma_2 \in (\sigma_1 - \Delta, 0)$.*

Remark 2.2.2 Repeating the argument of Proposition 2.2.10 and Corollary 2.2.3 we can prove that for the type (iii) $\widetilde{F}_0(\sigma) \to -\infty$ as $\sigma \to +\infty$.

Proposition 2.2.12 *Suppose that type (iii) occurs and $\varrho \geq 0$ in (2.2.1). Then any solution of (2.2.1) starting in the region D_0, which is bounded by the ω-separatrices $(\widetilde{\sigma}_0(\cdot), \widetilde{\eta}_0(\cdot))$ and $(\widetilde{\widetilde{\sigma}}_0(\cdot), \widetilde{\widetilde{\eta}}_0(\cdot))$ (see Figure 2.2.4) converges to $(0,0)$ as $t \to +\infty$.*

Proof By Proposition 2.1.1 any solution of (2.2.1) enters the set $H := \{(\sigma, \eta) : \eta \in [m, M]\}$ (with $m < m_0$ and $M > M_0$) and remains there for increasing t. Thus it is enough to consider solutions of (2.2.1) with initial points in $D_0 \cap H$. Let us apply to our situation Theorem 1.1.3, p. 4, using the Lyapunov function V from (2.2.25). In order to verify the conditions of the theorem, we assume, that for a solution $(\sigma(\cdot), \eta(\cdot))$ of (2.2.1) $V(\sigma(t), \eta(t)) = const$ on a time interval $[0, T]$. From (2.2.26) it follows that $\eta(t) = 0$ and $\varrho\varphi(\sigma(t)) = 0$ on $[0, T]$. Thus, $\dot{\sigma}(t) = 0$ on $[0, T]$ and $(\sigma(\cdot), \eta(\cdot))$ is a stationary solution of (2.2.1). By Theorem 1.1.3, every bounded on $\mathbf{R}_+$ solution converges to an equilibrium for $t \to +\infty$. It is obvious that the set $D_0 \cap H$ is positively invariant for (2.2.1). Thus any solution of (2.2.1) starting in D_0 is bounded and, consequently, converges to $(0,0)$. ∎

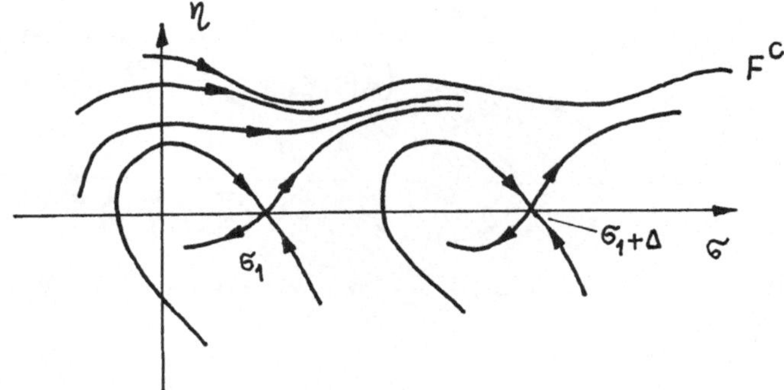

Figure 2.2.5

Convention 2.2.3 Suppose type (iii) in (2.2.1). Denote by $D_n, (n \in \mathbf{Z})$ the set

$$D_n := D_0 + (n\Delta, 0),$$

where D_0 is defined in the statement of Proposition 2.2.12.

Proposition 2.2.13 *Suppose that $\varrho \geq 0$ and type (iii) is given. Then there exists for (2.2.1) a cycle of the second kind. If the solution F^c of (2.2.3) corresponds to this cycle, then any solution F of (2.2.1) with $F(0) \notin \bigcup_n \overline{D_n}$ and $F(0) < F^c(0)$ tends to F^c as $\sigma \to +\infty$, i.e. $F(\sigma) - F^c(\sigma) \to 0$ as $\sigma \to +\infty$.*

Proof Let us consider the α-separatrix $(\overline{\sigma}_0(\cdot), \overline{\eta}_0(\cdot))$ and denote by $\overline{F}_0$ the corresponding solution of (2.2.3). Note that $\overline{F}_0(\sigma) > \varrho\varphi(\sigma)$ for all σ (see inequality (2.2.22) and Figure 2.2.3). So $\overline{F}_0$ is defined on $[\sigma_1, +\infty)$. According to Proposition 2.2.1 F_0 is bounded from above by M_0. Now we must repeat all the argument we have used while demonstrating Proposition 2.2.10. So we have to construct a sequence of functions F_k defined for $k = 1, 2, \ldots$ by $\overline{F}_k(\sigma) = \overline{F}_0(\sigma + k\Delta)$. Clearly, that $\overline{F}_k(\sigma) > \overline{F}_{k-1}(\sigma)$ and $\overline{F}_k(\sigma - \Delta) = \overline{F}_{k-1}(\sigma)$.

As a result we get the limit function F^c defined by $F^c(\sigma) = \lim_{k \to \infty} \overline{F}_k(\sigma)$, which is a Δ-periodic solution of (2.2.3).

Let us show that

$$\lim_{\sigma \to +\infty} [F^c(\sigma) - \overline{F}_0(\sigma)] = 0. \tag{2.2.37}$$

Let $\sigma \in [\sigma_1, \sigma_1 + \Delta]$. For $\hat{\sigma} = \sigma + k\Delta$ it follows that $F^c(\hat{\sigma}) = F^c(\sigma), \overline{F}_0(\hat{\sigma}) = \overline{F}_k(\sigma)$. Now for any $\varepsilon > 0$ there exists a constant K such that $| F^c(\sigma) - \overline{F}_k(\sigma) | < \varepsilon$ for all $k > K$ and $\sigma \in [\sigma_1, \sigma_1 + \Delta]$. That is why

$$|F^c(\hat{\sigma}) - \overline{F}_0(\hat{\sigma})| < \varepsilon \quad \text{if} \quad \hat{\sigma} > \sigma_1 + k\Delta.$$

Thus (2.2.37) is proved. Any solution $F_0(\sigma - k\Delta)$ tends to $F^c(\sigma)$ as $\sigma \to +\infty$. By the uniqueness theorem and the equivariance property any solution F of (2.2.3) with $F(0) < F^c(0)$ and $F(0) \notin \bigcup_n \overline{D_n}$ will tend to F^c as $\sigma \to +\infty$. $\blacksquare$

The results of Proposition 2.2.11, 2.2.12, 2.2.13 are illustrated in Figure 2.2.5.

In connection with the stability analysis of equation (2.2.3) the following definition is useful.

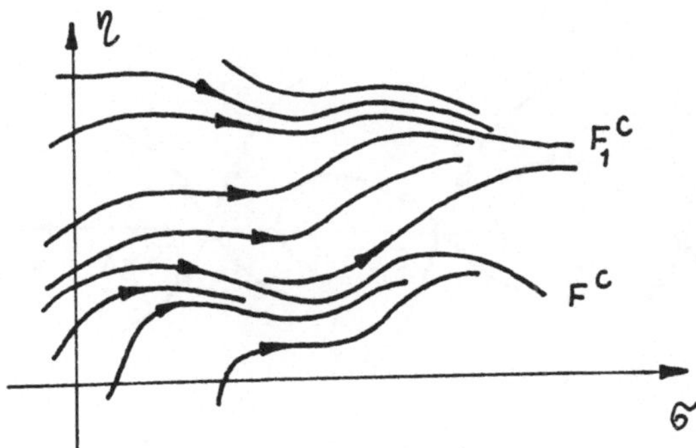

Figure 2.2.6

Definition 2.2.1 Let $F_1(\cdot)$ be a solution of (2.2.3) defined on $[\overline{\sigma}, +\infty)$. We say that F_1 is *stable from below* (resp. *from above*) for $\sigma \to +\infty$ if there exists a $\delta > 0$ such that for arbitrary $\varepsilon > 0$ and an arbitrary solution F of (2.2.3), defined at least on $[\overline{\sigma}, +\infty)$, with $F_1(\overline{\sigma}) - \delta < F(\overline{\sigma}) < F_1(\overline{\sigma})$ (resp. $F_1(\overline{\sigma}) < F(\overline{\sigma}) < F_1(\overline{\sigma}) + \delta$) there exists a $\Sigma > \overline{\sigma}$ such that $0 < F_1(\sigma) - F(\sigma) < \varepsilon$ (resp. $-\varepsilon < F_1(\sigma) - F(\sigma) < 0$) for all $\sigma < \Sigma$.

The stability from above and from below is defined in an analogous manner for $\sigma \to +\infty$.

Remark 2.2.3 It is clear that the cycle F^c in Proposition 2.2.13 is stable from below. Note that all the solutions F of (2.2.3) with $F(\overline{\sigma}) > F^c(\overline{\sigma})$ for a $\overline{\sigma} \in \mathbf{R}$ are bounded by F^c from below. Then (see the proof of Proposition 2.2.10) any such solution converges to a certain cycle of the second kind. Often there exists only one cycle to which converge other solutions (this situation takes place when $\varrho = 0$ as we shall see later on). For certain systems it is however shown that there are at least two such cycles [33, 53] (see Figure 2.2.6).

Note that the determination of the number of cycles for plane systems is connected with Hilbert's "sixteenth problem".

Let us go on to type (i) and consider the ω-separatrix in $(\sigma_1, 0)$ given by $(\widetilde{\sigma}_0, \widetilde{\eta}_0)$. The corresponding solution of (2.2.3) we denote by $\widetilde{F}_0$. It is defined on $(-\infty, \sigma_1]$.

Proposition 2.2.14 *If $\widetilde{F}_0$ is bounded from above on $(-\infty, \sigma_1]$ then (2.2.3) has a periodic solution. Any solution of (2.2.1) which starts in the domain bounded by the four ω-separatrices $(\widetilde{\sigma}_0(\cdot), \widetilde{\eta}_0(\cdot))$, $(\widetilde{\sigma}_0(\cdot) - \Delta, \widetilde{\eta}_0(\cdot))$, $(\widetilde{\widetilde{\sigma}}_0(\cdot), \widetilde{\widetilde{\eta}}_0(\cdot))$, $(\widetilde{\widetilde{\sigma}}_0(\cdot) - \Delta, \widetilde{\widetilde{\eta}}_0(\cdot))$, converges to $(0,0)$ as $t \to +\infty$.*

Proof We must repeat here the argument we have used in the Proposition 2.2.10 when proving that $\widetilde{F}_0$ generates a periodic solution of (2.2.3) if it is bounded from below. Thus the existence of a periodic solution will be proved. The proof of the second assertion is similar to the proof of Proposition 2.2.12. ∎

Proposition 2.2.15 *If $\widetilde{F}_0$ is not bounded from above then*

$$\lim_{\sigma \to -\infty} \widetilde{F}_0(\sigma) = +\infty.$$

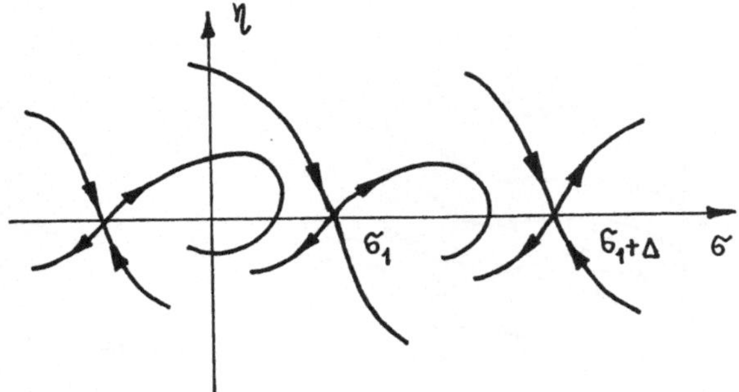

Figure 2.2.7

Proof Since $\widetilde{F}_0$ is not bounded it is possible to find a point σ with $\widetilde{F}_0(\sigma) > M_0$. It is clear from the proof of Proposition 2.2.1 that $\widetilde{F}_0(\sigma)$ is monotonic as $\sigma \to -\infty$. ∎

Proposition 2.2.16 *If $\widetilde{F}_0$ is not bounded from above then any solution of* (2.2.1) *which starts in the doamin bounded by the four ω-separatrices, adjoining the points $(\sigma_1 - \Delta, 0)$ and $(\sigma_1, 0)$, converges to $(0, 0)$. In this case the plane $\{\sigma, \eta\}$ is the union of the attraction domains for the equilibria $(k\Delta, 0)$, $k \in \mathbf{Z}$.*

Proof It repeats the argument of the proof of Proposition 2.2.12.∎

So we have distinguished the following two possibilities for type (i). One of them is described by Propositions 2.2.15 and 2.2.16. Let us call it
Type (i-a) Any solution of (2.2.1) converges to a certain equilibrium point $(k\Delta, 0)$, $k \in \mathbf{Z}$. (No cycles of the second kind are possible in this case of course.) The phase-portrait of (2.2.1) for this type is shown in Figure 2.2.7.

The second possibility is described by Proposition 2.2.14. We shall call it
Type (i-b) System (2.2.1) has an unstable cycle of the second kind.

Proposition 2.2.17 *If type (i-b) takes place then for system* (2.2.1) *there exists a stable from above cycle of the second kind.*

Proof Let us denote the unstable periodic solution of (2.2.3) by $F^c(\sigma)$. Every other solution F of (2.2.3) with $F(\overline{\sigma}) > F^c(\overline{\sigma})$ for a $\overline{\sigma} \in \mathbf{R}$ is defined on $[\overline{\sigma}, +\infty)$ and is bounded from below by F^c. Then we can prove the existence of a periodic solution F_1^c generated by F, using the scheme of the proof of Proposition 2.2.10. Furthermore, we can prove its stability from above, using the scheme described in the proof of Proposition 2.2.13. ∎

Remark 2.2.4 Of course F^c and F_1^c may coincide. This solution is then stable from above and unstable from below. We can also not exclude the possibility that there are several other cycles between F^c and F_1^c. Type (i-b) is shown in Figure 2.2.8.

In the next two subsections we shall describe which variants of phase portraits take place and indeed how they change into each other. Type (ii) will also be described here.

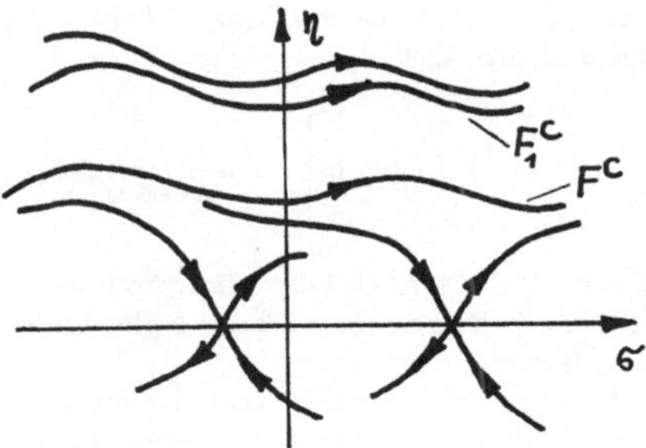

Figure 2.2.8

c)   The Case $\varrho = 0$. Existence of Separatrix-Loops

In this subsection we consider the system

$$\dot\sigma = \eta, \qquad \dot\eta = -a\eta - \varphi(\sigma), \tag{2.2.38}$$

which is equivalent to the second-order equation

$$\ddot\sigma + a\dot\sigma + \varphi(\sigma) = 0. \tag{2.2.39}$$

Let us suppose that φ has only the two zeros $\sigma = 0$ and $\sigma = \sigma_1$ such that (2.2.21) is true.

Proposition 2.2.18 *In case $\varrho = 0$ there exists for (2.2.1) a unique parameter value a_{cr} such that for any $a > a_{cr}$ the type (i-a), for any $a < a_{cr}$ type (iii) and for $a = a_{cr}$ type (ii) takes place.*

Proof The first-order equation associated with (2.2.1) for $\varrho = 0$ is

$$\frac{dF}{d\sigma}F + aF + \varphi(\sigma) = 0. \tag{2.2.40}$$

For $a = 0$ equation (2.2.40) is conservative and may be solved explicitly by

$$F(\sigma)^2 = F(\sigma_0)^2 - 2\int_{\sigma_0}^{\sigma} \varphi(\vartheta)\,d\vartheta, \tag{2.2.41}$$

where σ_0 is an arbitrary number. We are interested in a solution F satisfying $F(\sigma_1) = 0$ $(\sigma_0 = \sigma_1)$. Clearly the trajectory of F is contained in the level curve (2.2.34). Since the solutions of (2.2.40) depend continuously on the parameter a we can conclude that for sufficiently small a in (2.2.40) type (iii) takes place.

Note that the level curve (2.2.34) intersects the η-axis in the point

$$\left[2\int_{0}^{\sigma_1} \varphi(\vartheta)\,d\vartheta\right]^{\frac{1}{2}}.$$

On the other hand all the orbits of (2.2.40) with $F(\sigma_0) \in (m_0, M_0)$ are bounded from above by

$$M_0 = -\frac{1}{a}\min_{\vartheta\in[0,\Delta)} \varphi(\vartheta).$$

So the separatrix through $(\sigma_1 - \Delta, 0)$, namely $\overline{F}_1(\sigma) \equiv \overline{F}_0(\sigma - \Delta)$, is bounded by M as well.

Take now the value a so large that

$$\left[2 \int_0^{\sigma_1} \varphi(\vartheta)\, d\vartheta \right]^{\frac{1}{2}} > -\frac{1}{a} \min_{\vartheta \in [0,\Delta]} \varphi(\vartheta).$$

It is clear that $\widetilde{F}_0(0) > \overline{F}_1(0)$ and type (i) takes place. Note that for $\varrho = 0$ the condition (2.2.7) is satisfied. Then according to Proposition 2.2.2 any positive limit cycle is stable for $\sigma \to +\infty$. So type (i-b) is impossible here.

Let us show now that there exists a unique critical value $a_{cr} > 0$ such that for $a > a_{cr}$ type (i-a) takes place and for $a \in (0, a_{cr})$ type (ii) occurs. Let us consider the slope of the ω-separatrix $(\widetilde{\sigma}_0(\cdot), \widetilde{\eta}_0(\cdot))$ in $(\sigma_1, 0)$ (see formula (2.2.35)). If $\varrho = 0$ we have $k_1(a, 0) = -\frac{1}{2}\left[a + \sqrt{a^2 - 4\varphi'(\sigma_1)}\right]$ and for the partial derivative

$$D_1 k_1(a, 0) = -\frac{1}{2}\left[1 + \frac{a}{\sqrt{a^2 - 4\varphi'(\sigma_1)}}\right] < 0. \tag{2.2.42}$$

For the slope k_2 of the α-separatrix $(\overline{\sigma}_0(\cdot) - \Delta, \overline{\eta}(\cdot))$ in $(\sigma_1 - \Delta, 0)$ we have

$$k_2(a, 0) = \frac{1}{2}\left[-a + \sqrt{a^2 - 4\varphi'(\sigma_1)}\right]$$

and

$$D_1 k_2(a, 0) = -\frac{1}{2}\left[1 - \frac{a}{\sqrt{a^2 - 4\varphi'(\sigma_1)}}\right] < 0. \tag{2.2.43}$$

Let us consider the behavior of the ω-separatrix $(\widetilde{\sigma}_0(\cdot), \widetilde{\eta}_0(\cdot))$ and of the α-separatrix $(\widetilde{\sigma}_0(\cdot) - \Delta, \widetilde{\eta}_0(\cdot))$ in the band $\{\sigma, \eta \; : \; \sigma \in (\sigma_1 - \Delta, \sigma_1)\}$. In this region (2.2.40) may be rewritten as follows

$$\frac{dF}{d\sigma} = -a - \frac{\varphi(\sigma)}{F}. \tag{2.2.44}$$

In order to use the comparison principle of S.A. Chaplygin [45, 25] we also consider the solution $\widetilde{F}_0(\cdot; a)$ of (2.2.44) which corresponds to the ω-separatrix in $(\sigma_1, 0)$. According to (2.2.42) we have for σ near σ_1 $(\sigma < \sigma_1)$ the inequality $\widetilde{F}_0(\sigma; a_2) > \widetilde{F}_0(\sigma; a_1)$, where $a_2 > a_1 > 0$. On the other hand for any (σ, η) with $\sigma \in (\sigma_1 - \Delta, \sigma_1)$, $\eta \neq 0$, we get

$$-a_2 - \frac{\varphi(\sigma)}{\eta} < -a_1 - \frac{\varphi(\sigma)}{\eta}.$$

It follows by the comparison principle that $\widetilde{F}_0(0; a_2) > \widetilde{F}_0(0; a_1)$, In the same way with the help of (2.2.43) we can prove that for the solution $\overline{F}_1(\cdot; a)$ which corresponds to the α-separatrix $(\widetilde{\sigma}_0(\cdot) - \Delta, \widetilde{\eta}_0(\cdot))$ it is true that $\overline{F}_1(0; a_2) < \overline{F}_1(0; a_1)$. Since solutions of (2.2.44) continuously depend on a we can affirm that there exists a unique $a = a_{cr}$ such that $\overline{F}_1(0; a_{cr}) = \widetilde{F}_0(0; a_{cr})$.

Thus in the case $a = a_{cr}$ the α- and ω-separatrices in the halfplane $\{\sigma, \eta \; : \; \eta > 0\}$ merge and a *separatrix-loop* (or *homoclinic orbit*) is created. This curve is a periodic one, but the corresponding solution of (2.2.1) cannot, of course, be called periodic. Thus the type (ii) appears.

According to Proposition 2.2.2 the equation (2.2.44) has no other periodic solution and the created separatrix-loop is stable from above. Thus the value $a = a_{cr}$ is a single bifurcation value in the considered situation ($\varrho = 0$, φ has exactly two zeros.) The corresponding phase-portrait is illustrated in Figure 2.2.9.

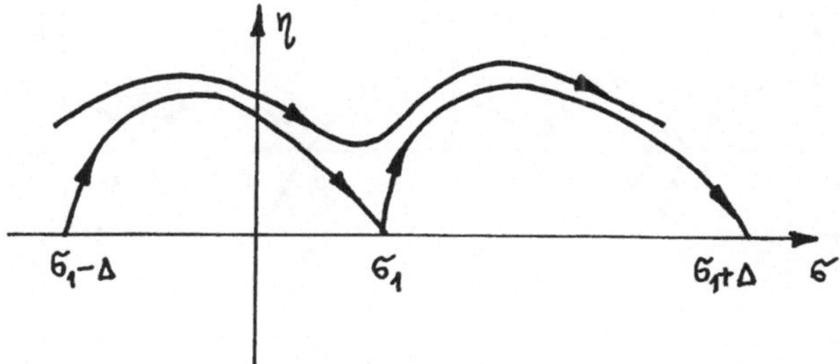

Figure 2.2.9

The calculation of a_{cr} may be done only numerically. On the other hand it is possible to use analytical estimates of a_{cr} from above and from below given in various papers [147, 148, 37, 58, 50, 133, 134, 135, 144]. Note that one of such estimates has already appeared in the proof of Proposition 2.2.18

$$a_{cr} < -\min_{\vartheta\in[0,\Delta]} \varphi(\vartheta) \left[2 \int_0^{\sigma_1} \varphi(\vartheta)\,d\vartheta \right]^{-1} . \tag{2.2.45}$$

It was obtained by evaluating the separatrices $\overline{F}_0$ and $\overline{F}_1$ and comparing their ordinates for $\sigma = 0$. By the same pattern various other estimates may be received. The following ones are taken from [20]:

$$a_{cr} < \frac{1}{\sigma_1} \left[2 \int_0^{\sigma_1-\Delta} \varphi(\sigma)\,d\sigma \right]^{\frac{1}{2}},$$

$$a_{cr}^2 < \frac{1}{\sigma_1^2} \left\{ -\int_0^1 \varphi(\sigma)\,d\sigma + \left[\left(\int_0^{\sigma_1} \varphi(\sigma)\,d\sigma \right)^2 + \sigma_1^2 \left(\min_{\sigma\in[0,\Delta]} \varphi(\sigma) \right)^2 \right]^{\frac{1}{2}} \right\},$$

$$a_{cr}^2 < -\frac{1}{\sigma_1} \min_{\sigma\in[0,\Delta]} \varphi(\sigma), \tag{2.2.46}$$

$$a_{cr} > \frac{1}{\Delta} \left[\left(2 \int_0^{\sigma_1-\Delta} \varphi(\sigma)\,d\sigma \right)^{\frac{1}{2}} - \left(2 \int_0^{\sigma_1} \varphi(\sigma)\,d\sigma \right)^{\frac{1}{2}} \right].$$

Let us now consider the system (2.2.1) for $\varrho = 0$ and $\varphi(\sigma) = \sin(\sigma + \sigma_0) - \sin\sigma_0$. Since estimate (2.2.46) takes the form

$$a_{cr}^2 < \frac{1 + \sin\sigma_0}{\pi - 2\sigma_0}$$

and coincides with the classical result of F. Tricomi [148]. Another estimate of a_{cr} for the sinusoidal nonlinearity may be received from (2.2.45) in the form

$$a_{cr} < \frac{1 + \sin\sigma_0}{[2(2\cos\sigma_0 - (\pi - 2\sigma_0)\sin\sigma_0]^{\frac{1}{2}}}$$

33

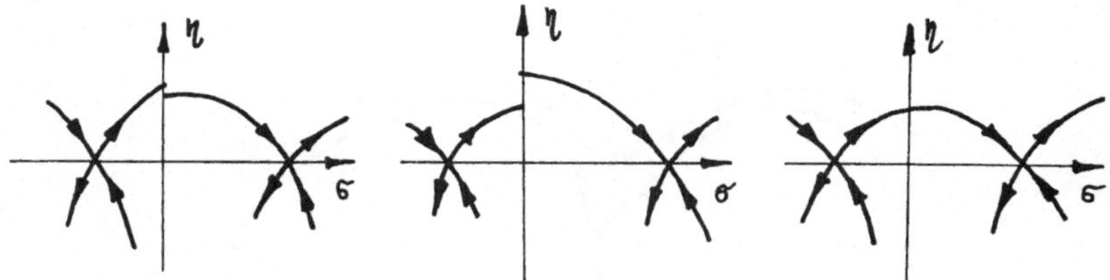

Figure 2.2.10

and gives the result of L. Amerio [3]. Of course, there are many other possible estimates of a_{cr}. Some of them we shall use in the sequel.

d) Non-Local Bifurcations in the Case $\varrho > 0$

We again consider the case when φ in (2.2.1) has exactly the two zeros $\sigma = 0$ and $\sigma = \sigma_1$ and the assumption (2.2.21) holds. Let us start our analysis with $a = 0$. Denote by $\overline{F}_1(\cdot\,;\varrho)$, $\widetilde{F}_0(\cdot\,;\varrho)$ the α- and ω-separatrices of (2.2.1) near $(\sigma_1 - \Delta, 0)$ and $(\sigma_1, 0)$, respectively, given by (2.2.3).

Proposition 2.2.19 *If $a = 0$ then there exists a single $\varrho = \varrho_0 > 0$ such that*

$$\begin{aligned}
\overline{F}_1(0;\varrho) &> \widetilde{F}_0(0;\varrho) \quad \textit{for} \quad \varrho \in (0, \varrho_0), \\
\overline{F}_1(0;\varrho) &= \widetilde{F}_0(0;\varrho) \quad \textit{for} \quad \varrho = \varrho_0, \\
\overline{F}_1(0;\varrho) &< \widetilde{F}_0(0;\varrho) \quad \textit{for} \quad \varrho > \varrho_0.
\end{aligned}$$

The three situations described in Proposition 2.2.19, are shown in Figure 2.2.10.

Proof of Proposition 2.2.19 In case $a = 0$, $\varrho = 0$ the solution of (2.2.3) is given by (2.2.41). Thus we have

$$\widetilde{F}_0(\sigma;0) = \sqrt{-2\int_{\sigma_1}^{\sigma} \varphi(\vartheta)\, d\vartheta} \qquad \text{and} \qquad \overline{F}_1(\sigma;0) = \sqrt{-2\int_{\sigma_1-\Delta}^{\sigma} \varphi(\vartheta)\, d\vartheta}$$

for $\sigma \in (\sigma_1 - \Delta, \sigma_1)$. Using these formulas and the property (2.2.2) we see that $\widetilde{F}_0(0;0) < \overline{F}_1(0;0)$.

Let us now change ϱ from 0 to $+\infty$. The vector field of (2.2.3) in case $a = 0$ is given by

$$\frac{dF}{d\sigma} = \frac{-\varphi(\sigma)}{F(\sigma) - \varrho\varphi(\sigma)} =: w(\sigma;\varrho).$$

We have for the partial derivative

$$D_2 w(\sigma;\varrho) = -\frac{\varphi(\sigma)^2}{[F(\sigma) - \varrho\varphi(\sigma)]^2} < 0.$$

Thus if ϱ increases the value of $w(\sigma;\varrho)$ decreases for $\sigma \in (\sigma_1 - \Delta, \sigma_1)$.

34

Let us keep in mind that by (2.2.35) $D_2 k_1(0,0) < 0$. It is easy to see that $D_2 k_2(0,0) < 0$ as well (k_1 resp. k_2 denotes the slope of the separatrix in $(\sigma_1, 0)$ resp. $(\sigma_1 - \Delta, 0)$). It follows that if ϱ increases the value of $\overline{F}_1(0; \varrho)$ decreases and the value of $\widetilde{F}_0(0; \varrho)$ increases monotonously. Let us note that since $\overline{F}_1 \geq 0$ we have on $(\sigma_1 - \Delta, 0)$

$$D_1 \overline{F}_1(\sigma; \varrho) < \frac{-\varphi(\sigma)}{-\varrho\varphi(\sigma)} = \frac{1}{\varrho}$$

and $k_2(a, \varrho) \to 0$ as $\varrho \to +\infty$. On the other hand the level curve (2.2.34) intersects the η-axis in the point

$$\left(2 \int\limits_0^{\sigma_1} \varphi(\vartheta)\, d\vartheta \right)^{\frac{1}{2}}$$

and $\widetilde{F}_0(\sigma; \varrho)$ is situated higher than the level curve. Thus there exists a $\varrho = \varrho_0$ such that $\widetilde{F}_0(0; \varrho_0) = \overline{F}_1(0; \varrho_0)$. The Proposition 2.2.19 proved. ∎

Convention 2.2.4 Define the value $a_0 = a_0(\varphi)$ by

$$a_0 := - \min_{\vartheta \in [0,\Delta]} \varphi(\vartheta) \left[2 \int\limits_0^{\sigma_1} \varphi(\vartheta)\, d\vartheta \right]^{\frac{-1}{2}}.$$

Proposition 2.2.20 *For any $a > a_0$ for system (2.2.1) the type (i-a) takes place.*

Proof In the same way as has been done in Proposition 2.2.18 we prove that the type (i) takes place. By assumption we have

$$\widetilde{F}_0(0) > -\frac{1}{a} \min_{\vartheta \in [0,\Delta]} \varphi(\vartheta).$$

It follows that in cases when the ω-separatrices $\widetilde{F}_k(\sigma) = \widetilde{F}_0(\sigma - k\Delta)$ are bounded above by a certain number N then for the cycle F^c the estimate

$$F^c(k\Delta) > -\frac{1}{a} \min_{\sigma \in [0,\Delta]} \varphi(\sigma)$$

for $k \in \mathbf{Z}$ would be satisfied. Thus for any solution F of (2.2.3) with $F(\sigma_0) > N$ for a $\sigma_0 \in \mathbf{R}$ it would be impossible to reach the domain

$$\left\{ \sigma, \eta \ : \ \eta \leq M_0 = -\frac{1}{a} \min_{\sigma \in [0,\Delta]} \varphi(\sigma) \right\}$$

which would contradict Proposition 2.2.1 ∎

Convention 2.2.5 Define the value $\overline{a} = \overline{a}(\varphi, \varrho) := \frac{1}{\varrho} \, | \min_{\sigma \in [0,\Delta]} \varphi(\sigma) | \left(\max_{\sigma \in [0,\Delta]} \varphi(\sigma) \right)^{-1}.$

Proposition 2.2.21 *For any $a > \overline{a}$ the type (i-a) takes place.*

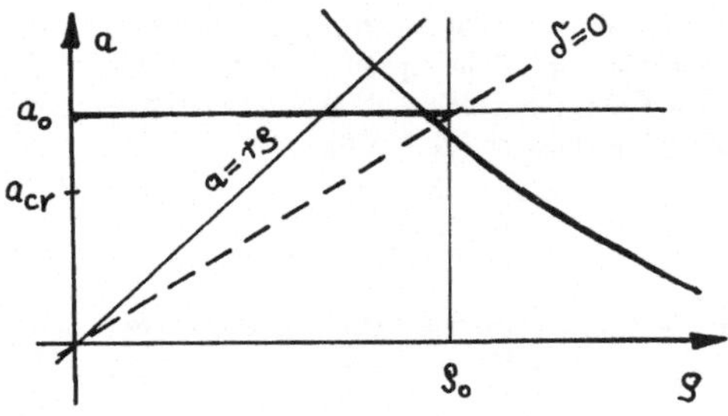

Figure 2.2.11

Proof We have already used several times the fact that $\widetilde{F}_0(\sigma) > \varrho\varphi(\sigma)$ on $[0, \sigma_1]$. Since for any solution F of (2.2.3) $\frac{d}{d\sigma}F(\sigma) < 0$ on $[0, \sigma_1]$ we see that

$$\widetilde{F}_0(0) > \varrho \max_{\vartheta \in [0, \Delta]} \varphi(\vartheta).$$

On the other hand the α-separatrix $\overline{F}_1$ in $(\sigma_1 - \Delta, 0)$ has to stay by Proposition 2.2.1 in the domain

$$\left\{ \sigma, \eta : \eta < \frac{1}{a} \mid \min_{\vartheta \in [0, \Delta]} \varphi(\vartheta) \mid < \frac{1}{\overline{a}} \mid \min_{\vartheta \in [0, \Delta]} \varphi(\vartheta) \mid = \varrho \min_{\vartheta \in [0, \Delta]} \varphi(\vartheta) \right\}.$$

Thus we have $\widetilde{F}_0(0) > \overline{F}_1(0)$ and we can repeat the argument of Proposition 2.2.20. ∎

As a result we get the parameter diagram in Figure 2.2.11. In Figure 2.2.10 the thick line characterizes an approximation from below of the boundary of the domain where the case (i-a) takes place. This approximation is obtained by means of Propositions 2.2.20 and 2.2.21. On Figure 2.2.11 the straight line $a = r\varrho$ is also shown ($r := \max_{\vartheta \in [0, \Delta]} \varphi'(\vartheta)$). According to Proposition 2.2.2 in the area $\{a, \varrho : a > r\varrho\}$ there can appear no more than one (stable) cycle of the second kind. Let us now investigate for fixed ϱ the qualitatively various phase portraits of (2.2.1) in dependence of a.

Proposition 2.2.22 *For any fixed ϱ the separatrix $\overline{F}_1(\sigma; a)$ in $(\sigma_1 - \Delta, 0)$ decreases and the separatrix $\widetilde{F}_0(\sigma, a)$ in $(\sigma_1, 0)$ increases if a increases and $\sigma \in (\sigma_1 - \Delta, \sigma_1)$ is fixed.*

Proof It is easy to verify that $D_1 k_1(a, \varrho) < 0$ and $D_1 k_2(a, \varrho) < 0$. The right-hand side of (2.2.3) is

$$w(\sigma, a) = \frac{-aF(\sigma) - \varphi(\sigma)}{F(\sigma) - \varrho\varphi(\sigma)}.$$

In the region $\{\sigma, \eta : \eta[\eta - \varrho\varphi(\sigma)] > 0\}$ we have $D_2 w(\sigma, a) < 0$. This affirmation is true for both separatrices considered. Thus the Proposition is proved. ∎

Corollary 2.2.5 *If $\varrho < \varrho_0$ there exists a single value $a = a_1(\varrho)$ for which the type (iii) changes to the type (i), i.e. for which $\widetilde{F}_0$ and $\overline{F}_1$ merge together and form a separatrix-loop.*

Let us recall now the definition of the saddle-number for a planar system [22]. Consider the system in the plane

$$\dot{\sigma} = P(\sigma, \eta), \qquad \dot{\eta} = Q(\sigma, \eta),$$

where P and Q are C^1-functions. Suppose that (σ_0, η_0) is a saddle-point of the system. Then the number

$$\delta := D_1 P(\sigma_0, \eta_0) + D_2 Q(\sigma_0, \eta_0)$$

is called *saddle-number* in (σ_0, η_0). This number characterices the stability of a separatrix-loop. According to the saddle-point $(\sigma_1, 0)$ in (2.2.1) the saddle-number is given by

$$\delta = -(a + \varrho \varphi'(\sigma_1)).$$

Proposition 2.2.23 *If for $a = a_1(\varrho)$ the saddle-number in $(\sigma_1, 0)$ is negative then the separatrix-loop is stable from above. If for $a = a_1(\varrho)$ the saddle-number is positive then the separatrix-loop is unstable from above.*

Proof Consider the solution F_0 of (2.2.3) which corresponds to the separatrix loop in $(\sigma_1, 0)$ and define for an arbitrary solution of (2.2.1) which corresponds to the solution F of (2.2.3) the functions

$$y_0(\sigma) = F_0(\sigma) - \varrho \varphi(\sigma) \qquad \text{and} \qquad y_1(\sigma) = F(\sigma) - \varrho \varphi(\sigma).$$

We shall investigate the behavior of the function

$$y_1(\sigma) = y_1(\sigma)^2 - y_0(\sigma)^2$$

on the interval $[\sigma_1 - \Delta, \sigma_1]$ provided that $y_1(\sigma_1 - \Delta) > y_0(\sigma_1 - \Delta)$. It follows from (2.2.13) that

$$\frac{1}{2} \frac{d}{d\sigma} y(\sigma) = -[a + \varrho \varphi'(\sigma)](y_1 - y_0). \tag{2.2.47}$$

Suppose at first that $\delta < 0$.

Let us demonstrate that when σ changes from $\sigma_1 - \Delta$ to σ_1 the function y decreases, i.e.

$$y(\sigma_1) < y(\sigma_1 - \Delta) \tag{2.2.48}$$

provided that $y(\sigma_1 - \Delta)$ is small enough. Let $\sigma_2 > \sigma_1 - \Delta$ and such that $y(\sigma)$ decreases on $[\sigma_1 - \Delta, \sigma_2]$. Since $-(a + \varrho \varphi'(\sigma_1 - \Delta)) = -(a + \varrho \varphi'(\sigma_1)) = \delta < 0$ such a σ_2 exists. Thus

$$y(\sigma_2) < y(\sigma_1 - \Delta). \tag{2.2.49}$$

Fix now a $\sigma_3 \in [\sigma_2, \sigma_1]$ so that

(i) $y(\sigma)$ decreases on $[\sigma_3, \sigma_1]$;

(ii) $y_0'(\sigma) < 0$ on $[\sigma_3, \sigma_1]$;

(iii) $a + \varrho \varphi'(\sigma) \geq c > 0$ on $[\sigma_3, \sigma_1]$.

Let us construct an estimate for $y(\sigma_3)$. Since $y_1(\sigma) > y_0(\sigma) > 0$ for $\sigma \in (\sigma_1 - \Delta, \sigma_1)$ it follows that

$$y_0(\sigma) + y_1(\sigma) \geq \varepsilon \quad \text{for} \quad [\sigma_2, \sigma_3],$$

where ε is a certain fixed number. It follows from (2.2.47) that

$$\frac{1}{2} \frac{y'(\sigma)}{y(\sigma)} = -\frac{a + \varrho \varphi'(\sigma)}{y_0(\sigma) + y_1(\sigma)}$$

or

$$\frac{1}{2}\ln\frac{y(\sigma_3)}{y(\sigma_2)} = -\int\limits_{\sigma_2}^{\sigma_3}\frac{a+\varrho\varphi'(\sigma)}{y_0(\sigma)+y_1(\sigma)}\,d\sigma. \qquad (2.2.50)$$

Let c_1 be such that

$$|\,a+\varrho\varphi'(\sigma)\,|\le c_1 \quad\text{for}\quad \sigma\in[\sigma_1-\Delta,\sigma_1].$$

Then from (2.2.50) we have

$$y(\sigma_3)\le y(\sigma_2)e^{\left[\frac{2c_1}{\varepsilon}(\sigma_3-\sigma_2)\right]}. \qquad (2.2.51)$$

Let us consider now the interval $[\sigma_3,\sigma_1]$. It is easy to obtain the relation

$$y_0(\sigma)+y_1(\sigma) = y_0(\sigma)+\sqrt{y_0(\sigma)^2+y(\sigma)},$$

which gives the opportunity to affirm that

$$y_0(\sigma)+y_1(\sigma)\le 2y_0(\sigma)+\sqrt{y(\sigma)}.$$

By the requirement (a) we have

$$y_0(\sigma)+y_1(\sigma)\le 2y_0(\sigma)+\sqrt{y(\sigma_3)}.$$

Thus by requirement (c)

$$-\frac{a+\varrho\varphi'(\sigma)}{y_0(\sigma)+y_1(\sigma)}\le\frac{-c}{2y_0(\sigma)+\sqrt{y(\sigma_3)}} \qquad (2.2.52)$$

for all $\sigma\in[\sigma_3,\sigma_1]$. Note that for any $\sigma\in(\sigma_3,\sigma_1)$ there exists a $\overline{\sigma}\in(\sigma,\sigma_1)$ such that $y_0(\sigma)=y_0'(\overline{\sigma})(\sigma-\sigma_1)$. Note that $y_0'(\overline{\sigma})$ is negative by the requirement (b).

Let c_2 be such that $y_0'(\sigma)\le c_2<0$ for $\sigma\in[\sigma_3,\sigma_1]$. Thus

$$-\frac{a+\varrho\varphi'(\sigma)}{y_0(\sigma)+y_1(\sigma)}\le\frac{-c}{2c_2(\sigma-\sigma_1)+\sqrt{y(\sigma_3)}} \qquad (2.2.53)$$

From (2.2.47) we have

$$\frac{1}{2}\int\limits_{\sigma_3}^{\sigma_1}\frac{y'(\sigma)}{y(\sigma)}\,d\sigma = -\int\limits_{\sigma_3}^{\sigma_1}\frac{a+\varrho\varphi'(\sigma)}{y_0(\sigma)+y_1(\sigma)}\,d\sigma.$$

So by (2.2.53) yields

$$\frac{1}{2}\ln\frac{y(\sigma_1)}{y(\sigma_3)}\le\frac{c}{2\,|\,c_2\,|}\int\limits_{\sigma_3}^{\sigma_1}\frac{d\sigma}{(\sigma-\sigma_1)-\frac{1}{2|c_2|}\sqrt{y(\sigma_3)}}$$

or

$$\ln\frac{y(\sigma_1)}{y(\sigma_3)}\le\frac{c}{|\,c_2\,|}\ln\frac{-\dfrac{1}{2\,|\,c_2\,|}\sqrt{y(\sigma_3)}}{\sigma_3-\sigma_1-\dfrac{1}{2\,|\,c_2\,|}\sqrt{y(\sigma_3)}}.$$

From this, (2.2.49) and (2.2.51) we have

$$y(\sigma_1)\le y(\sigma_1-\Delta)e^{2c_1(\sigma_3-\sigma_2)/\varepsilon}\left[\frac{\dfrac{1}{2\,|\,c_2\,|}\sqrt{y(\sigma_3)}}{\sigma_1-\sigma_3+\dfrac{1}{2\,|\,c_2\,|}\sqrt{y(\sigma_3)}}\right]^{\frac{c}{|\,c_2\,|}}.$$

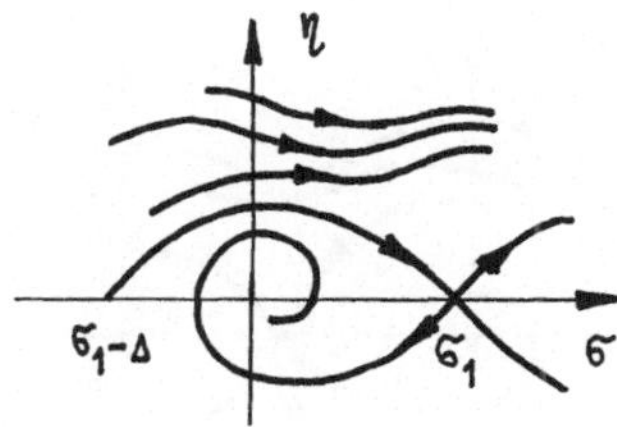

Figure 2.2.12

Now, taking into consideration (2.2.49) and (2.2.51), let us choose $y(\sigma_1 - \Delta)$ so small

$$e^{2c_1(\sigma_3 - \sigma_2)/\varepsilon} \left[\frac{\dfrac{1}{2\,|\,c_2\,|}\sqrt{y(\sigma_3)}}{\sigma_1 - \sigma_3 + \dfrac{1}{2\,|\,c_2\,|}\sqrt{y(\sigma_3)}} \right]^{\frac{c}{|\,c_2\,|}} < 1.$$

Thus (2.2.48) is fulfilled and since $y_0(\sigma_1) = y_0(\sigma_1 - \Delta) = 0$ it follows that $y_1(\sigma_1) < y_1(\sigma_1 - \Delta)$ if only $y_1(\sigma_1 - \Delta)$ is small enough.

If we suppose now that $\delta > 0$ we can show by the same pattern that

$$y_1(\sigma_1) > y_1(\sigma_1 - \Delta). \tag{2.2.54}$$

Thus Proposition 2.2.23 is proved. ∎

The straight line $\delta = 0$ is shown in Figure 2.2.10 by a dotted line.

Possible Non-Local Bifurcations in the Case $0 < \varrho < \varrho_0$

It is clear from above that $a_1(\varrho) < \min\{a_0, \overline{a}(\varrho)\}$. Suppose that $a_1(\sigma) < -\varrho\varphi'(\sigma_1)$. The separatrix-loop is unstable in this case. From the property (2.2.54) it follows that all the arguments of Proposition 2.2.13 can be applied here and there exists a stable from below cycle of the second kind (see Figure 2.2.12).

The latter is either stable from above as well, or has above itself another stable from above cycle. When a becomes greater than $a_1(\varrho)$ the separatrix-loop disappears, but the stable from above cycle does not disappear. So in this case an unstable from below cycle must appear and the situation shown in Figure 2.2.8 must take place. If a increases furthermore the stable from above and the unstable from below merge together. For the bifurcation value $a = a_2(\varrho) > a_1(\varrho)$ they disappear and form a single semi-stable cycle (Figure 2.2.13).

For $a > a_2(\varrho)$ the type (i-a) takes place. Suppose now that $r\varrho > a_1(\varrho) > -\varrho\varphi'(\sigma_1)$. The separatrix-loop is stable here and above this separatrix-loop several cycles are possible. The upper one of these cycles must be stable from below. This situation may lead only to the type (i-b). Then there must exist a value $a = a_2(\varrho)$ with $r\varrho > a_2(\varrho) > a_1(\varrho) > -\varrho\varphi'(\sigma_1)$ such that for $a > a_2(\varrho)$ the type (i-a) takes place. There may be no cycles here.

Suppose now that $\min\{a_0, \overline{a}(\varrho)\} > a_1(\varrho) > r\varrho$. Then by Proposition 2.2.2 the bifurcation is just alike that in the cas $\varrho = 0$.

39

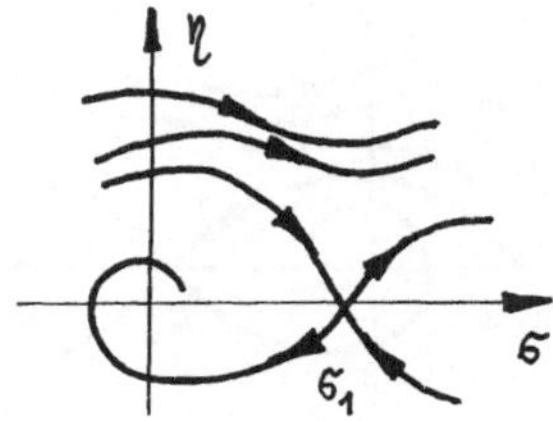

Figure 2.2.13

Possible Bifurcations in the Case $\varrho \geq \varrho_0$

If $\varrho \geq \varrho_0$ then it follows from Proposition 2.2.2 that the type (iii) is impossible for all $a > 0$. Here for $a < \overline{a}(\varrho)$ a bifurcation of the type which is shown in Figure 2.2.12 may occur. A class of systems (2.2.1) describing phase-locked loops with proportional-integrating filter is considered by L.N. Belyustina et al. in [33]. In this paper for three types of nonlinearities φ the introduced bifurcation values are calculated. In the case $\varphi(\sigma) = \sin \sigma - \gamma$, $\gamma \in (0,1)$, the bifurcation values which correspond to a separatrix-loop are obtained by means of approximate methods using the existence of such loops in certain comparison systems. Bifurcation values which correspond to double cycles are obtained in [33] with the help of numerical calculations for the solution of certain associated fixed-point problems.

e)   Several Remarks about the Case $\varrho \in (-1/a, 0)$

According to Corollary 2.2.1 in this case system (2.2.1) cannot have more than one cycle of the second kind in the domains

$$\{\sigma, \eta \,:\, \eta > \varrho\varphi(\sigma)\} \qquad \text{or} \qquad \{\sigma, \eta \,:\, \eta < \varrho\varphi(\sigma)\}.$$

A linear stability analysis of the equilibria of (2.2.1) in the case, when φ has exactly two zeros, shows that there are two essential different cases: either unstable saddle-points alternate with stable knots (or foci) or all the equilibria are unstable. The function (2.2.25) is of no use here and thus cycles of the first kind are not excluded. Cycles of the second kind are not possible in the domain $\{\sigma, \eta \,:\, \eta < \varrho\varphi(\sigma)\}$. According to the Bendixson theorem in the case $a > -\varrho \max\limits_{\sigma \in [0,\Delta]} \varphi'(\sigma)$ there are no cycles of the first kind.

In the following remarks it is assumed that the nonlinearity φ in (2.2.1) has exactly the two zeros 0 and σ_1.

A. The Equilibrium $(0,0)$ is Unstable

We can state the following possibilities:

Type (a)   For the ω-separatrix $(\widetilde{\sigma}_0(\cdot), \widetilde{\eta}_0(\cdot))$ in $(\sigma_1, 0)$ is $\widetilde{\eta}_0(t) > 0$ for all $t \in \mathbf{R}$ and there is a $\overline{t} \in \mathbf{R}$ such that $\widetilde{\sigma}_0(\overline{t}) = \sigma_1 - \Delta$;

Type (b)   For all $t \in \mathbf{R}$ is $\widetilde{\eta}_0(t) > 0$ and $\widetilde{\sigma}_0(t) \in (\sigma_1 - \Delta, \sigma_1)$;

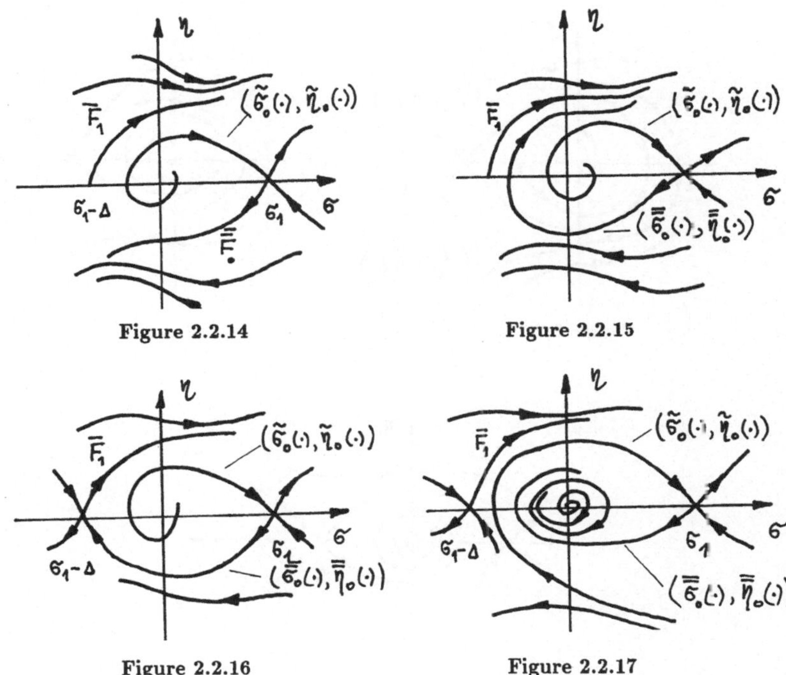

Figure 2.2.14   **Figure 2.2.15**

Figure 2.2.16   **Figure 2.2.17**

Type (c) There exists a time $\bar{t}$ and a time $\bar{\bar{t}}$ such that $\widetilde{\eta}_0(\bar{t}) = 0$, $\widetilde{\eta}_0(t) > 0$ for $t > \bar{t}$ and $\widetilde{\sigma}_0(\bar{\bar{t}}) = \sigma_1$;

Type (d) There is an unique $\bar{t}$ with $\widetilde{\eta}_0(\bar{t}) = 0$ and $\widetilde{\sigma}_0(t) \in (\sigma_1 - \Delta, \sigma_1)$ for all $t \in \mathbf{R}$;

Type (e) There is a sequence $\{t_k\}$ such that $\widetilde{\eta}_0(t_k) = 0$ and $\widetilde{\sigma}_0(t) \in (\sigma_1 - \Delta, \sigma_1)$ for all $t \in \mathbf{R}$;

Let us at first consider the type (e) in more details. In this type the α-separatrix $\overline{F}_1$ in $(\sigma_1 - \Delta, 0)$ cannot become zero for $\sigma > \sigma_1 - \Delta$ and in the same way as was done in the type (iii) above we can prove that system (2.2.1) has in the domain $\{\sigma, \eta \ : \ \eta > \varrho\varphi(\sigma)\}$ a stable cycle of the second kind. The separatrix $(\overline{\overline{\sigma}}_0(\cdot), \overline{\overline{\eta}}_0(\cdot))$ may have various types of behavior:

Type (e-i) For all t we have $\overline{\overline{\eta}}_0(t) < 0$ and there exists a time $\bar{t}$ such that $\overline{\overline{\sigma}}_0(\bar{t}) = \sigma_1 - \Delta$. According to Proposition 2.2.1 the corresponding solution $\overline{\overline{F}}_0$ of (2.2.3) is bounded. Therefore we can prove in the same way as it has been done in Proposition 2.2.13 that there exists a stable cycle of the second kind in the domain $\{\sigma, \eta \ : \ \eta < \varrho\varphi(\sigma)\}$ (Figure 2.2.14).

Type (e-ii) There exists a $\bar{t}$ such that $\overline{\overline{\eta}}_0(\bar{t}) = 0$ and $\overline{\overline{\eta}}_0(t) > 0$ for $t > \bar{t}$. The phase portrait is shown in Figure 2.2.15. Indeed, according to Proposition 2.2.2 there cannot exist any cycles of the second kind in the region $\{\sigma, \eta \ : \ \eta < \varrho\varphi(\sigma)\}$.

Type (e-iii) $\overline{\overline{\eta}}_0(t) < 0$ and $\overline{\overline{\sigma}}_0(t) \in (\sigma_1 - \Delta, \sigma_1)$ for all $t \in \mathbf{R}$. This type is a bifurcation one when a separatrix-loop appears in $\{\sigma, \eta \ : \ \eta < \varrho\varphi(\sigma)\}$ (Figure 2.2.16).

Let us now consider the type (c). In this case as well as for the type (e) the separatrix $\overline{F}_1$ cannot vanish for $\sigma > \sigma_1 - \Delta$ and thus there exists in the domain $\{\sigma, \eta \ : \ \eta > \varrho\varphi(\sigma)\}$ a limit cycle of the second kind. The separatrix $(\overline{\overline{\sigma}}_0(\cdot), \overline{\overline{\eta}}_0(\cdot))$ ought to "lie inside" the domain bounded

41

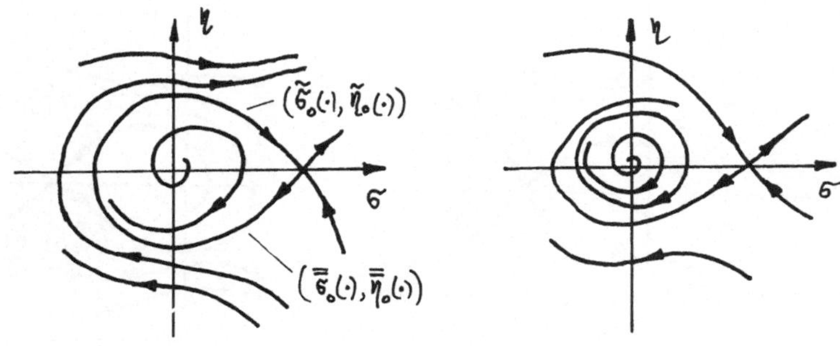

Figure 2.2.18   Figure 2.2.19

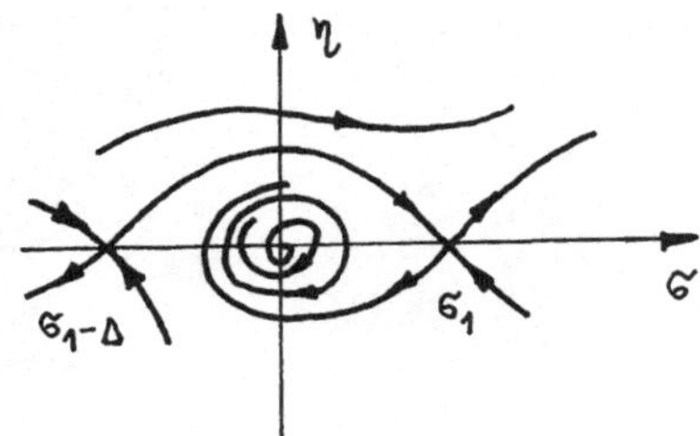

Figure 2.2.20

by $(\widetilde{\sigma}_0(\cdot),\widetilde{\eta}_0(\cdot))$. But as the equilibrium point which is situated inside this domain is unstable we may affirm that there exists a cycle of the first kind which lies inside this domain. In this case a cycle of the second kind in the domain $\{\sigma,\eta\ :\ \eta < \varrho\varphi(\sigma)\}$ cannot exist. Indeed, repeating the argument used in the case $\varrho = 0$ we may prove that if the cycle of the second kind existed it would be unstable. But this fact would contradict the Proposition 2.2.2. The type (c) behavior is shown in Figure 2.2.17.

The type (d) is a bifurcation one when type (e-ii) changes into type (c). In type (d) the separatrix $\widetilde{F}_0$ forms a loop which goes from the saddle $(\sigma_1,0)$ to the same saddle (Figure 2.2.18).

Let us consider the type (a). Analysing the disposition of the separatrices $(\overline{\sigma}_0(\cdot),\overline{\eta}_0(\cdot))$, $(\widetilde{\sigma}_0(\cdot),\widetilde{\eta}_0(\cdot))$ and $(\overline{\overline{\sigma}}_0(\cdot),\overline{\overline{\eta}}_0(\cdot))$ we come to the conclusion that the separatrix $(\overline{\overline{\sigma}}_0(\cdot),\overline{\overline{\eta}}_0(\cdot))$ "winds around" the equilibrium point $(0,0)$. But since this equilibrium point is unstable there exists a cycle of the first kind such that $(0,0)$ is situated inside this cycle and the separatrix $(\overline{\overline{\sigma}}_0(\cdot),\overline{\overline{\eta}}_0(\cdot))$ is attracted by it for $t \to +\infty$. The type (a) is illustrated in Figure 2.2.19.

A limit cycle of the second kind cannot occur in this situation. Indeed, if it appeared it would be unstable. This fact would contradict the Proposition 2.2.2.

The type (b) is a bifurcation one. It appears when the type (c) changes to the type (a). In this case a separatrix-loop appears. It is illustrated in Figure 2.2.20.

42

Remark 2.2.5 In all considered cases there may appear other cycles of the first kind surrounding the unstabel equilibrium state.

B. The Equilibrium $(0,0)$ is Stable

Various possibilities of phase portraits may be imagined here. It is important to note that, in view of Proposition 2.2.2, if for a pair a, ϱ there cannot exist any cycle of the first kind, then the three phase portraits, (and only they), realized in the case $\varrho = 0$ are possible here. The most interesting for us here is the transition from a stable equilibrium $(0,0)$ to an unstable one. This bifurcation takes place when $a + \varrho\varphi'(0) = 0$. This value separates the case of a stable focus (when $a + \varrho\varphi'(0) > 0$) and an unstable one (when $a + \varrho\varphi'(0) < 0$). It is well known [22] that when a stable focus changes to an unstable one there appears a cycle of the first kind (Androncv-Hopf bifurcation; see [110]).

2.3 The Lyapunov Direct Method in the Standard Form

Let us consider system (2.1.5). Suppose that $\varphi(\cdot, 0) = 0$ and there exist two constants μ_1 and μ_2 such that

$$\mu_1 \le \frac{\varphi(t, \sigma)}{\sigma} \le \mu_2 \tag{2.3.1}$$

for all $\sigma \ne 0$ and $t \in \mathbf{R}_+$. Because φ is periodic it is clear that $\mu_1\mu_2 \le 0$. We suppose w.l.o.g. that $\mu_1\mu_2 < 0$.

A standard Lyapunov function for investigating the global behavior of the system under consideration is the quadratic form $V(x) = x^*Hx$, where $H = H^*$ is a certain $n \times n$ matrix to be determined. In order to guarantee that the derivative of V along the solutions of (2.1.5) with nonlinearities from (2.3.1) is non-positive we have to show, (Problem B'), that for some $\tau \in \mathbf{R}_+$

$$2x^*H(Px + q\xi) + \tau(\xi - \mu_1 r^*x)(\mu_2 r^*x - \xi) \le 0 \tag{2.3.2}$$

for all $x \in \mathbf{R}^n$ and $\xi \in \mathbf{R}$.

We now prove that in our case of a phase-controlled system (2.1.5) this parameter τ may only be zero. Indeed, for $\xi = 0$ it follows from (2.3.2) that

$$2x^*HPx \le \tau\mu_1\mu_2(r^*x)^2$$

for all $x \in \mathbf{R}^n$. This inequality implies $\tau = 0$ since $r^*d \ne 0$ and $Pd = 0$ (Proposition 2.1.2). Thus (2.3.2) transforms to

$$2x^*H(Px + q\xi) \le 0 \tag{2.3.3}$$

for all $x \in \mathbf{R}^n$ and $\xi \in \mathbf{R}$.

Let us show that inequality (2.3.3) is only possible with the matrix $H = 0$. The non-degeneracy of the transfer function χ of (2.1.5) implies that the triple (P, q, r) is controllable and observable. The same is true for the triple $(-P, -q, r)$ which corresponds to the transfer function $\chi(-s)$. According to Theorem 1.2.2, p. 6, there exist two vectors r_1 and r_2 such that the matrices $P + qr_1^*$ and $-(P + qr_2^*)$ are Hurwitzian.

Putting in (2.3.3) $\xi = r_1^*x$ and $\xi = r_2^*x$ we get the inequalities

$$2x^*H(P + qr_1^*)x \le 0 \quad \text{and} \quad 2x^*H(P + qr_2^*)x \le 0 \tag{2.3.4}$$

for all $x \in \mathbf{R}^n$. Since $P + qr_1^*$ is Hurwitzian it follows from the first inequality of (2.3.4) that $H \ge 0$. (Theorem 1.4.1, p. 9) and since $-(P + qr_2^*)$ is Hurwitzian it follows from the second inequality that $H \leq 0$. Hence $H = 0$ and there is no non-trivial quadratic form as Lyapunov function for the phase-controlled system (2.1.5) with nonlinearities from the class $M[\mu_1, \mu_2]$, defined by (2.3.1).

Let us now consider the autonomous case of system (2.1.5), i.e. the case when φ does not explicitly depend on t. We want to choose a Lyapunov function in the Lur'e-Postnikov form (1.4.7). It's derivative $\dot{V}_{(2.1.5)}$ along the solutions of (2.1.5) is given by (1.4.8). In order to guarantee that $\dot{V}_{(2.1.5)}$ is non-positive in the class $M[\mu_1, \mu_2]$ we require that for a matrix $H = H^*$ and parameters ϑ and $\tau \geq 0$

$$2x^*H(Px + q\xi) + \vartheta\xi r^*(Px + q\xi) + \tau(\xi - \mu_1 r^*x)(\mu_2 r^*x - \xi) \leq 0 \tag{2.3.5}$$

for all $x \in \mathbf{R}^n$ and $\xi \in \mathbf{R}$.

As above one shows that τ has to be zero. In applying Theorem 1.4.1, p. 9, with the Hermitian form $F(x, \xi) = \mathrm{Re}\,[\vartheta\xi^* r^*(Px + q\xi)]$ $(x \in \mathbf{C}^n, \xi \in \mathbf{C})$ we see that the inequality

$$\vartheta\mathrm{Re}\,[i\omega\chi(i\omega)] \geq 0 \tag{2.3.6}$$

for all $\omega \in \mathbf{R}$ with $\det(P - i\omega) \neq 0$ is necessary and sufficient for the existence of a real matrix $H = H^*$ in (2.3.5).

Thus the frequency-domain inequality (2.3.6) guarantees that for the autonomous system (2.1.5) there exists a Lyapunov function of the type "quadratic form plus integral of the nonlinearity" which has a non-negative derivative along the solution of the system.

We consider now the second canonical form (2.1.10)

$$\dot{z} = Az + b\varphi(\sigma), \qquad \dot{\sigma} = c^*z + \varrho\varphi(\sigma). \tag{2.3.7}$$

Note that since χ is non-degenerate and $c^*(A - sI)^{-1}b \equiv s\chi(s) + \varrho$ the pairs (A, b) and (A, c) are controllable and observable, respectively.

We want to use for (2.3.7) a Lyapunov function

$$V(z, \sigma) := z^*Mz + \vartheta \int\limits_0^\sigma \varphi(u)du, \tag{2.3.8}$$

where $M = M^*$ is an $(n-1) \times (n-1)$ matrix and $\vartheta \neq 0$ is a parameter. By Theorem 1.4.1, p. 9, there exist such a matrix M and a number ϑ with

$$2z^*M(Az + b\xi) + \vartheta\xi(c^*z + \varrho\xi) \leq 0 \tag{2.3.9}$$

for all $z \in \mathbf{R}^{n-1}$ and $\xi \in \mathbf{R}$ if and only if

$$\vartheta\mathrm{Re}\,K(i\omega) \geq 0 \tag{2.3.10}$$

for all $\omega \in \mathbf{R}$ with $\det(A - i\omega I) \neq 0$. If all eigenvalues of A have a negative real part this matrix M is positive definite (cf. Theorem 1.4.1).

Consider now the case when the derivative of (2.3.8) with respect to system (2.3.7) has to be negative definite in the class $M[\mu_1, \mu_2]$:

$$2z^*M(Az + b\xi) + \vartheta\xi(c^*z + \varrho\xi) \leq -\varepsilon \mid z \mid^2 \tag{2.3.11}$$

for all $z \in \mathbf{R}^{n-1}$ and $\xi \in \mathbf{R}$, where $\varepsilon > 0$ is a number. According to Theorem 1.4.2, p. 9, (in the case $\varrho \neq 0$) and to Theorem 1.4.4 (in the case $\varrho = 0$) there exist a matrix $M = M^*$ and a number $\varepsilon > 0$ satisfying (2.3.11) if and only if

$$\vartheta\mathrm{Re}\,K(i\omega) > 0 \quad \text{for all} \quad \omega \in \mathbf{R} \quad \text{and} \quad \vartheta \lim_{\omega \to \infty}[\omega^2\mathrm{Re}\,K(i\omega)] > 0 \tag{2.3.12}$$

If A is a Hurwitz matrix it follows from (2.3.11) that $M > 0$.

2.4 Monostability and Gradient-Like Behavior of Phase-Controlled Systems with Nonlinearities having Mean Value Zero

Let us consider system (2.1.5) in the autonomous case and investigate the convergence properties of its solutions with the help of Lyapunov functions constructed in the previous section. Suppose additionally that (2.1.5) is given in the form (2.3.7) and $\det A \neq 0$. Let us at first determine the set $\mathcal{E}$ of equilibria of this system. Every equilibrium $(\overline{z}, \overline{\sigma})$ of (2.3.7) satisfies the system

$$A\overline{z} + b\varphi(\overline{\sigma}) = 0, \qquad c^*\overline{z} + \varrho\varphi(\overline{\sigma}) = 0.$$

It follows that $K(0)\varphi(\overline{\sigma}) = 0$, where K is defined in Section 2.1. Because of the non-degeneracy of χ we have $K(0) \neq 0$ and, consequently, $\varphi(\overline{\sigma}) = 0$ and $\overline{z} = 0$. Thus under our conditions the set $\mathcal{E}$ of equilibria of (2.3.7) is given by $\mathcal{E} = \{(z,\sigma) : z = 0, \varphi(\sigma) = 0\}$. In the first theorem of this section we provide frequency-domain conditions for quasi-monostability of system (2.1.5). Theorem 2.4.1 and the following Theorems 2.4.2 and 2.4.3 go back to [49].

Theorem 2.4.1 *Suppose that the transfer function χ of the linear part of (2.1.5) has $n-1$ poles with negative real part. Suppose also that*

$$\mathrm{Re}\,[i\omega\chi(i\omega)] \neq 0 \qquad \text{for all} \qquad \omega \in \mathbf{R} \qquad \text{and} \qquad \lim_{\omega \to \infty} \omega^2 \mathrm{Re}\,[i\omega\chi(i\omega)] \neq 0. \tag{2.4.1}$$

Then system (2.1.5) is quasi-monostable

Proof Assume that system (2.1.5) has the form (2.3.7). Let us choose $\vartheta = \mathrm{sgn}\,\mathrm{Re}\,[i\omega\chi(i\omega)]$. Since $K(i\omega) = i\omega\chi(i\omega)$ the frequency-domain inequalities (2.3.12) are fulfilled. So by Theorem 1.4.2, p. 9, there exist a matrix $M = M^*$ and a number $\varepsilon > 0$, such that the derivative of function (2.3.8) along a solution z, σ of (2.3.7) satisfies

$$\frac{d}{dt}W(z(t),\sigma(t)) \leq -\varepsilon \mid z(t) \mid^2, \quad t \geq 0 \tag{2.4.2}$$

Let us apply to system (2.3.7) Theorem 1.1.3, p. 4. The condition (i) of this theorem holds because of (2.4.2). Let us consider condition (ii). If for some $\tau > 0$ $W(z(\tau),\sigma(\tau)) = W(z(0),\sigma(0))$ it follows that $\dot{W} = 0$ on $(0,\tau)$ and consequently $z(t) = 0$ on $[0,\tau]$. It is easy to see that $c^*(A - sI)^{-1}b = s\chi(s) + \varrho$. So (A,b) is controllable and consequently $b \neq 0$. Using this, from the first equation of (2.3.7) we have $\varphi(\sigma(t)) = 0$ on $[0,\tau]$. From the second equation of (2.3.7)) it follows that $\sigma(t) = const$ on $[0,\tau]$. $\blacksquare$

In the next theorem we provide sufficient conditions which guarantee monostability.

Theorem 2.4.2 *Suppose at least one of the following conditions is fulfilled:*

(i) *the matrix P in (2.1.5) has $n-1$ eigenvalues with negative real part;*

(ii) *the set of equilibria $\mathcal{E}$ of (2.1.5) consists of isolated points only.*

Then if the system (2.1.5) is quasi-monostable it is monostable.

Proof Let us consider system (2.1.5) in the form (2.3.7). Because of quasi-monostability of (2.3.7) for any its bounded solution (z,σ) we have $\lim_{t \to +\infty} z(t) = 0$.

Suppose now that a certain bounded solution component does not converge for $t \to +\infty$ to an equilibrium component of (2.3.7). In this case we find at least two sequences $\{t_k\}$ and $\{s_k\}$ such that

$$\lim_{k\to\infty} \sigma(t_k) = \sigma' \quad \text{and} \quad \lim_{k\to\infty} \sigma(s_k) = \sigma'' \tag{2.4.3}$$

with $\sigma' < \sigma''$. As σ is continuous we can also find for any $\vartheta \in [\sigma', \sigma'']$ a sequence $\{u_k\}$ with

$$\lim_{k\to\infty} \sigma(u_k) = \vartheta.$$

Clearly, every point $(z, \sigma) = (0, \vartheta)$, $\vartheta \in [\sigma', \sigma'']$ is an equilibrium of (2.3.7), i.e. $\varphi(\vartheta) = 0$ on $[\sigma', \sigma'']$.

Suppose first that assumption (ii) is satisfied. Then the set of equilibria of (2.3.7) is discrete and the above construction is impossible.

Suppose now that assumption (i) is given and consider the set $\Pi := \{z = 0, \sigma \in (\sigma', \sigma'')\}$. In a neighborhood of a point $(0, \sigma_0) \in \Pi$ system (2.3.7) coincides with

$$\dot{z} = Az, \qquad \dot{\sigma} = c^* z. \tag{2.4.4}$$

Under assumption (i) the point $z = 0$, $\sigma = \sigma_0$ is Lyapunov stable. It follows that for any neighborhood U_1 of $z = 0, \sigma = \sigma_0$ there exists another neighborhood U_s such that the solutions of (2.3.7) which start in U_2 are in U_1 for all $t \geq 0$. This is a contradiction to the above construction (2.4.3). $\blacksquare$

Let us consider now system (2.3.7) with mean value zero of the nonlinearity.

The following theorem is given without proof (see [49]).

Theorem 2.4.3 *Suppose that the conditions of Theorem 2.4.1 are fulfilled and*

$$\int_0^{\Delta} \varphi(\sigma) \, d\sigma = 0,$$

where Δ is the period of φ.
Then system (2.1.5) is gradient-like.

Chapter 3

Invariant Cones

It is clear from Section 2.3 that standard Lyapunov functions are of little use in the investigation of pendulum-like systems. So such systems with multiple equilibria need new classes of Lyapunov functions to be constructed. In this chapter and in the following ones we shall demonstrate several new methods which extend the Lyapunov direct method to systems with multiple equilibria.

3.1 Extension of the Circle Criterion

Let us consider the pendulum-like feedback-system

$$\dot{x} = Px + q\xi, \quad \sigma = r^*x, \quad \xi = \varphi(t,\sigma), \tag{3.1.1}$$

where P is a constant $n \times n$-matrix, q and r are constant n-vectors and $\varphi : \mathbf{R}_+ \times \mathbf{R} \to \mathbf{R}$ is continuous and locally Lipschitz continuous in the second argument. We assume that

$$\det P = 0 \tag{3.1.2}$$

and there exists a $\Delta > 0$, such that

$$\varphi(t, \sigma + \Delta) = \varphi(t, \sigma), \ t \in \mathbf{R}_+, \ \sigma \in \mathbf{R}. \tag{3.1.3}$$

Thus system (3.1.1) to (3.1.3) is a pendulum-like system in the first canonical form. Furthermore we assume that φ belongs to the sector $M[\mu_1, \mu_2]$, e.g.

$$\mu_1 \le \frac{\varphi(t,\sigma)}{\sigma} \le \mu_2, \ t \in \mathbf{R}, \ \sigma \ne 0, \tag{3.1.4}$$

where μ_1 is either a certain number or $-\infty$ and μ_2 is either a certain number or $+\infty$. In the case $\mu_1 = -\infty$ (resp. $\mu_2 = +\infty$) we assume that $\mu_1^{-1} = 0$ (resp. μ_2^{-1}). Recall that $\chi(s) = r^*(P-sI)^{-1}q$ denotes the transfer function of the linear part of system (3.1.1).

We suppose that χ is non-degenerate. Note at first that it is impossible to use the standard Circle Criterion for system (3.1.1) - (3.1.4) to investigate the absolute stability in the class $M[\mu_1, \mu_2]$. The conditions of Theorem 1.4.3, p. 10, are not fulfilled as the matrix P has at least one eigenvalue with real part zero. And it is not surprising, because the Circle Criterion uses the existence of a quadratic form which has negative derivative along the solutions of system (3.1.1) - (3.1.4) (see Sect. 2.3).

Let us now replace the matrix P in (3.1.1) by $P + \lambda I$, where $\lambda > 0$ is a certain number. From the non-degeneracy of χ it follows that the pair $(P + \lambda I, q)$ is controllable. Suppose that the inequality

$$\text{Re } \{[\mu_1\chi(i\omega - \lambda) + 1]^*[\mu_2\chi(i\omega - \lambda) + 1]\} \ge 0 \tag{3.1.5}$$

is true for all $\omega \in \mathbf{R}$. By Theorem 1.4.1, there exists a Hermitian $n \times n$-matrix H such that the quadratic form

$$G(x, \xi) = 2x^* H[(P + \lambda I)x + q\xi] + (\mu_2^{-1}\xi - r^* x)(\mu_1^{-1}\xi - r^* x)$$

$(x \in \mathbf{R}^n, \xi \in \mathbf{R})$ is non-positive. Let us now consider the Lyapunov function $V(x) = x^* H x$. Its derivative with respect to system (3.1.1) is

$$\dot{V}_{(3.1.1)}(x, \xi) = 2x^* H(Px + q\xi).$$

Using the sector condition (3.1.4) and the fact that G is non-positive we have that

$$2x^* H(Px + q\xi) \leq -2\lambda x^* H x - (\mu_2^{-1}\xi - r^* x)(\mu_1^{-1}\xi - r^* x) \tag{3.1.6}$$

and, with $v(t) := V(x(t))$, where x is a solution of (3.1.1),

$$\frac{d}{dt}v(t) \leq -2\lambda v(t), \quad t \geq t_0. \tag{3.1.7}$$

The following assertion is useful.

Lemma 3.1.1 *Suppose that* $v : [t_0, +\infty) \to \mathbf{R}$ *is absolute continuous,* $v(t_0) \leq 0$ *(resp.* $v(t_0) < 0)$ *and there exists a number* γ *such that*

$$\dot{v}(t) \leq \gamma v(t) \quad \text{for a.e. } t \geq t_0. \tag{3.1.8}$$

Then $v(t) \leq 0$ *(resp.* $v(t) < 0)$ *for all* $t \geq t_0$.

Proof It follows from (3.1.8) that for a.e. $t \geq t_0$ we have $e^{\gamma t}\dot{v}(t) - \gamma e^{-\gamma t}v(t) \leq 0$, and, consequently, $\frac{d}{dt}[e^{-\gamma t}v(t)] \leq 0$ for a.e. $t \geq t_0$. Integrating the last inequality we get $v(t) \leq e^{+\gamma(t-t_0)}v(t_0)$ for $t \geq t_0$, and the assertion of Lemma 3.1.1 follows immediately. ∎

So if a certain Lyapunov function V for (3.1.1) has the property (3.1.7) the set $\{x : V(x) < 0\}$ is positively invariant for system (3.1.1). It is useful to note that this set is a cone in $\mathbf{R}^n$. Remember that $M \subset \mathbf{R}^n$ is a *cone with the top in* x_0 if for any $x_1 \in M$ the line $\{x_0 + s(x_1 - x_0), s \in \mathbf{R}\}$ is completely contained in M. The dimension of the maximal linear set contained in M is called the *dimension of M*. If $H = H^*$ is an $n \times n$ matrix the set $\{x : x^* H x \leq 0\}$ is obviously a cone which is called by us *quadratic*. It is clear that for H having k negative and $(n - k)$ positive eigenvalues the set $\{x : x^* H x \leq 0\}$ is a k-dimensional quadratic cone.

In the next theorem, which is taken from [49, 82] a frequency-domain condition for the boundedness of all solutions of (3.1.1) is given.

Theorem 3.1.1 *Suppose that with a* $\lambda > 0$ *the following conditions for system (3.1.1) are fulfilled:*

(i) *the matrix* $P + \lambda I$ *has* $n - 1$ *eigenvalues with negative real part;*

(ii)

$$\mu_1^{-1}\mu_2^{-1} + (\mu_1^{-1} + \mu_2^{-1})\operatorname{Re}\chi(i\omega - \lambda) + \mid \chi(i\omega - \lambda) \mid^2 \leq 0, \ \omega \in \mathbf{R}. \tag{3.1.9}$$

Then system (3.1.1) is Lagrange stable.

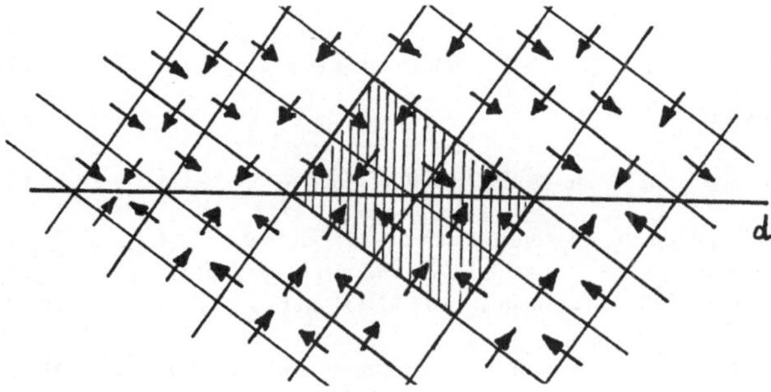

Figure 3.1.1

Proof Suppose $x(\cdot, t_0, x_0)$ is a solution of (3.1.1). Let d be an eigenvector of P, corresponding to its zero eigenvalue such that $r^*d = \Delta$. From the equivariance property (2.1.2) it follows that

$$x(t, t_0, x_0) - jd = x(t, t_0, x_0 - jd), \quad t \geq t_0, \ j \in \mathbf{Z}.$$

Hence the interior
$$\Omega_j := \{x \ : \ (x - jd)^* H(x - jd) < 0\}$$
of a quadratic cone
$$\{x \ : \ (x - jd)^* H(x - jd) \leq 0\}$$
is positively invariant if the set $\{x \ : \ x^* H x < 0\}$ is invariant for (3.1.1). Let us consider the quadratic form G from above. It follows from (3.1.9) that G is negative semi-definite. Then from the fact that $(P + \lambda I, r)$ is observable, the assumption (i) and inequality (3.1.6) are satisfied. Lemma 1.2.1, p. 6, allows us to conclude that the matrix H has one negative and $n - 1$ positive eigenvalues.

Let now $x = d$, $\xi = 0$ in (3.1.6). We see that

$$d^* H d < 0 \quad . \tag{3.1.10}$$

Let us define for an arbitrary $j \in \mathbf{Z}$ the set

$$\Gamma_j = \Omega_j \cap \Omega_{-j}.$$

The set Γ_j is positively invariant as both Ω_j and Ω_{-j} are positively invariant. The vector field of (3.1.1) with respect to the boundaries of Ω_j is shown in Figure 3.1.1 (The thick curve illustrates as an example $\Omega_{-1} \cap \Omega_1$.) Let us note that because of the spectrum of H there exists a vector $h \neq 0$ such that

$$\{x \ : \ h^* x = 0, x \neq 0\} \subset \{x \ : \ x^* H x > 0\}. \tag{3.1.11}$$

Indeed it is sufficient to reduce H to the form $\text{diag}\,(I_{n-1}, -1)$ by means of a non-singular linear transformation and to choose $h = (0, \ldots, 0, 1)$ in this case. It follows from (3.1.10) and (3.1.11) that $h^* d \neq 0$. Consider the arbitrary solution $x(\cdot, t_0, x_0)$. We can find a number j, large enough, such that the following inequalities are true

$$\mid h^* x_0 \mid < j \mid h^* d \mid, \quad x_0^* H x_0 \pm 2j x_0^* H d + j^2 d^* H d < 0. \tag{3.1.12}$$

From above it follows that

$$x(t, t_0, x_0) \in \Gamma_j \quad \text{for all} \quad t \geq t_0 \tag{3.1.13}$$

As the next we show that

$$| h^* x(t, t_0, x_0) | < j | h^* d |, \quad t \geq t_0. \tag{3.1.14}$$

Really if (3.1.14) were not satisfied there would exist a $\bar{t} > t_0$ such that $| h^* x(\bar{t}, t_0, x_0) | = j | h^* d |$. It would mean that for $\varrho = 1$ or $\varrho = -1$ we have $(x(\bar{t}, t_0, x_0) + \varrho j d)^* H(x(\bar{t}, t_0, x_0) + \varrho j d) \geq 0$ which is impossible because of (3.1.13). Inequalities (3.1.13) and (3.1.14) give the opportunity to prove the boundedness of $x(\cdot, t_0, x_0)$.

Indeed, because of (3.1.11) the matrix H can be represented in the form $H = M - \tau h h^*$, where M is a positive definite $n \times n$-matrix and τ is a positive number.

Further let ε be a positive number such that $M > \varepsilon I_n$. For a solution $x(t) = x(t, t_0, x_0)$ that satisfies (3.1.13) and (3.1.14) it is true that

$$\begin{aligned}
\varepsilon | x(t) - jd |^2 &\leq (x(t) - jd)^* M (x(t) - jd) \\
&= (x(t) - jd)^* H (x(t) - jd) + \tau | h^* (x(t) - jd) |^2 \\
&< \tau [2(h^* x)^2 + 2j^2 (h^* d)^2] \\
&< 4\tau j^2 (h^* d)^2
\end{aligned}$$

for $t \geq t_0$, which means that $x(\cdot, t_0, x_0)$ is bounded on $[t_0, +\infty)$. ∎

Remark 3.1.1 The frequency-domain inequality (3.1.9) as well as the Circle Criterion allows a simple geometrical interpretation. Consider the non-zero poles of $\chi(\cdot)$ and denote their maximal real part by $-\lambda_0 < 0$. The inequality (3.1.9) means that for a $\lambda \in (0, \lambda_0)$ the image of the frequency characteristic $\omega \mapsto \chi(i\omega - \lambda)$ lies in the complex plane inside the circle with the center on the real axis and which passes through the points $-\mu_1^{-1}$ and $-\mu_2^{-1}$.

Remark 3.1.2 Theorem 3.1.1 in connection with any monostability criterion gives sufficient conditions for a gradient-like behavior

Example 3.1.1 Let us consider system (3.1.1) to (3.1.4) with $n = 2$ and the transfer-function

$$\chi(s) = \frac{1}{s(s + \alpha)}, \tag{3.1.15}$$

where $\alpha > 0$ is a parameter. The system under consideration is equivalent to the second order equation

$$\ddot{\sigma} + \alpha \dot{\sigma} + \varphi(t, \sigma) = 0. \tag{3.1.16}$$

We assume that φ satisfies, for a certain $\mu > 0$, the condition

$$\varphi(t, \sigma) \sigma < \mu \sigma^2, \quad t \in \mathbf{R}_+, \quad \sigma \neq 0. \tag{3.1.17}$$

So the frequency-domain inequality (3.1.9) is equivalent to $-\omega^2 + \lambda^2 - \alpha\lambda + \mu \leq 0$ for $\omega \in \mathbf{R}$ and the condition (i) of Theorem 3.1.1 is given for $\lambda \in (0, \alpha)$. Thus the hypotheses of Theorem 3.1.1 are valid if and only if

$$\alpha^2 \geq 4\mu. \tag{3.1.18}$$

Example 3.1.2 Suppose $\varphi(t,\sigma) = \beta \sin \omega_0 t \sin \sigma$ where β and ω_0 are parameters and consider system (3.1.1) - (3.1.3) with the transfer function (3.1.15). This system describes the motion of a pendulum with a vibrating point of suspension [114, 44]:

$$\ddot{\sigma} + \alpha \dot{\sigma} + \frac{g - a\omega_0^2 \sin \omega_0 t}{l} \sin \sigma = 0$$

for $\beta = -a\omega_0^2/l$, $g = 0$. It describes also the hula-hoop effect [44] and the motion of a rotor which is placed on a base with sinusoidal vibration [20]. From Example 3.1.1 it follows immediately that the inequality

$$\alpha^2 \geq 4 \mid \beta \mid \tag{3.1.19}$$

is sufficient for the Lagrange stability of the given system.

Remark 3.1.3 Let us compare condition (3.1.19) with some results which have been discovered by other authors. In [114] it is shown that for

$$2\alpha l < a\omega_0 \qquad \text{and} \qquad a \ll l \tag{3.1.20}$$

in the system of Example 3.1.2 there exist circular solution. In [44] with the help of a method of approximation the condition for the existence of circular solutions in our system is given by

$$\alpha \ll 1, \quad a\omega_0^2 \ll 2l, \quad 2\alpha l < a\omega_0. \tag{3.1.21}$$

It is clear that (3.1.21) contradicts (3.1.19) when

$$2\omega_0 \sqrt{\frac{a}{l}} < \alpha < \omega_0 \frac{a}{2l}. \tag{3.1.22}$$

In this case both (3.1.19) and (3.1.21) are fulfilled which is impossible because circular solutions are not bounded. Thus when α satisfies (3.1.22) the conditions (3.1.21) are not correct

Note that (3.1.20) does not contradict (3.1.19) because $a \ll l$ and so the inequalities (3.1.22) cannot be satisfied. Note also that it would be useful to establish sufficient conditions of Lagrange stability of the type (3.1.20) by the invariant cones method. We shall try to do so in the next section.

Example 3.1.3 Let us consider the equation which describes the work of synchronous single-phase motor with pulsing vibrating moment [115]

$$\ddot{\eta} + \alpha \dot{\eta} + \beta \cos t \cos \eta + \gamma \sin 2\eta + \delta \cos 2t \sin 2\eta = 0, \tag{3.1.23}$$

where $\alpha > 0, \beta, \gamma, \delta$ are numbers. With the help of the change of variables $\sigma = \eta + \frac{\pi}{2}$ we have

$$\ddot{\sigma} + \alpha \dot{\sigma} + \varphi(t,\sigma) = 0,$$

where

$$\varphi(\sigma,t) = \beta \cos t \sin \sigma - \gamma \sin 2\sigma - \delta \cos 2t \sin 2\sigma.$$

From Example 3.1.1 we establish for (3.1.23) the following sufficient condition of Lagrange stability

$$\alpha^2 \geq 4(\mid \beta \mid + 2 \mid \gamma \mid + 2 \mid \delta \mid).$$

Example 3.1.4 Let us consider system (3.1.1) with

$$\varphi(t, \sigma) \equiv \varphi(\sigma) \tag{3.1.24}$$

(φ is a Δ-periodic function) and the transfer function (3.1.15). It is clear all the hypotheses of Theorem 2.4.1 are satisfied here. So system (3.1.1), (3.1.24) is quasi-monostable and more than that, according to Theorem 2.4.2 this system is monostable. It follows then from Theorem 3.1.1 and Example 3.1.1 that condition (3.1.18) is a sufficient condition of gradient-like behavior of system (3.1.1), (3.1.24) for any $\varphi(\cdot)$ which satisfies the sector condition

$$\frac{\varphi(\sigma)}{\sigma} < \mu, \qquad \sigma \neq 0.$$

Note that in the particular case $\varphi(\sigma) = \sin \sigma - 1$ we can choose $\mu = 3/4$. It follows that for $\alpha > \sqrt{3}$ the equation

$$\ddot{\sigma} + \alpha \dot{\sigma} + \sin \sigma = 1 \tag{3.1.25}$$

is gradient-like. This condition corrects the numerical result from [74], p. 306, where for any $\alpha > 0$ in (3.1.25) there is observed instability.

Example 3.1.5 Let us consider the system (3.1.1) - (3.1.4) with $n = 3$ and the transfer function

$$\chi(s) = \frac{1}{s(s^2 + \alpha s + \beta)}, \tag{3.1.26}$$

where α, β are positive parameters. Since

$$\det[(s - \lambda)I + P] = (s - \lambda)[s^2 + (\alpha - 2\lambda)s + \lambda^2 - \alpha\lambda + \beta]$$

condition (i) of Theorem 3.1.1 takes the form

$$\alpha - 2\lambda > 0, \quad \lambda^2 - \alpha\lambda + \beta > 0. \tag{3.1.27}$$

Condition (ii) of Theorem 3.1.1 with $\mu_1 = -\infty$ is as follows

$$(3\lambda - \alpha)\omega^2 - \lambda(\lambda^2 - \lambda\alpha + \beta) + \mu_2 \leq 0 \tag{3.1.28}$$

for all $\omega \in \mathbf{R}$. If $\alpha^2 \leq 3\beta$ then the inequalities (3.1.27), (3.1.28) are satisfied for

$$\lambda = \frac{\alpha}{3}, \quad \mu_2 \leq \frac{\alpha}{3}(\beta - \frac{2}{9}\alpha^2).$$

If $\alpha^2 \geq 3\beta$ it is evident that the inequalities (3.1.27), (3.1.28) are satisfied for

$$\lambda = \frac{\alpha}{3} - \sqrt{\frac{\alpha^2}{9} - \frac{\beta}{3}}$$

$$\mu_2 \leq \frac{\alpha}{3}(\beta - \frac{2}{9}\alpha^2) + 2(\frac{\alpha^2}{9} - \frac{\beta}{3})^{\frac{3}{2}}.$$

Thus if $\mu_1 = -\infty$ and

$$\mu_2 \leq \begin{cases} \frac{\alpha}{3}(\beta - \frac{2}{9}\alpha^2) & \text{for} \quad \alpha^2 \leq 3\beta, \\ \frac{\alpha}{3}(\beta - \frac{2}{9}\alpha^2) + 2(\frac{\alpha^2}{9} - \frac{\beta}{3})^{\frac{3}{2}} & \text{for} \quad \alpha^2 \geq 3\beta, \end{cases} \tag{3.1.29}$$

system (3.1.1) - (3.1.4), (3.1.26) is Lagrange stable.

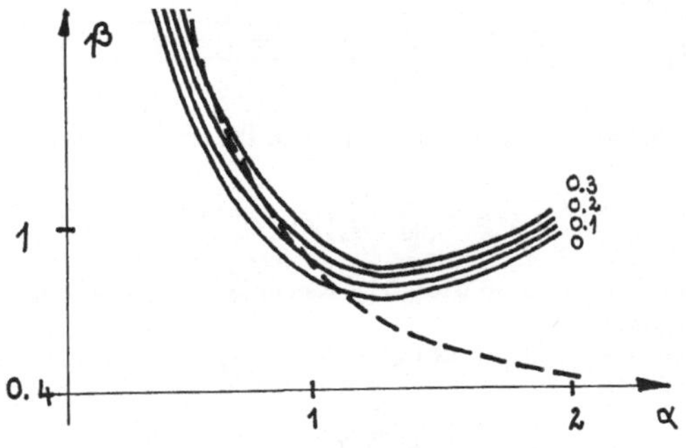

Figure 3.1.2

Let us consider the function

$$\varphi(t,\sigma) = \sin(\sigma - \sigma_0) - \sin \sigma_0$$

with $\sigma_0 \in [\frac{\pi}{2}, \pi]$, $\sin \sigma_0 \le 0.99$. It satisfies (3.1.4) with $\mu_1 = -\infty$ and

$$\mu_2 \ge \frac{1 + \sin \sigma_0}{\sqrt{2} - 1 + 0.25\pi + \sigma_0} \, .$$

In this case the inequality (3.1.29) takes the form

$$\frac{1 + \sin \sigma_0}{\sqrt{2} - 1 + 0.25\pi + \sigma_0} \le \left\{ \begin{array}{ll} \frac{\alpha}{3}(\beta - \frac{2}{9}\alpha^2) & \text{for} \quad \alpha^2 \le 3\beta, \\ \frac{\alpha}{3}(\beta - \frac{2}{9}\alpha^2) + 2(\frac{\alpha^2}{9} - \frac{\beta}{3})^{\frac{3}{2}} & \text{for} \quad \alpha^2 \ge 3\beta, \end{array} \right. \tag{3.1.30}$$

In Figure 3.1.2 the boundaries of Lagrange stability regions on the plane $\{\alpha, \beta\}$ obtained by (3.1.30) are shown. The cases $\sin \sigma_0 = 0, 0.1, 0.2, 0.3$ are considered. By the dotted line the boundary of the region where Lyapunov stable equilibria exist is drawn from the case of $\sin \sigma_0 = 0$. We see that sometimes Theorem 3.1.1 gives the opportunity to establish the Lagrange stability of an autonomous system which is unstable in the sense of Lyapunov equilibria.

Let us suppose now that $\varphi(t,\sigma) = \varphi(\sigma)$, function φ is differentiable and

$$\varphi'(\sigma) \le \nu_1 < \alpha\beta, \quad |\varphi'(\sigma)| \le \nu_2 \tag{3.1.31}$$

for all $\sigma \in \mathbf{R}$. It follows from [49], Chapter 3.2, that system (3.1.1) with such a stationary nonlinearity and the transfer function (3.1.26) under conditions (3.1.31) is monostable. Thus conditions (3.1.29) become here the conditions of gradient-like behavior of the system.

Example 3.1.6 Let in (3.1.1) - (3.1.4) $n = k + 2$, where k is a positive integer, and

$$\chi(s) = \frac{1}{s(s + \alpha)(s + \beta)^k} \tag{3.1.32}$$

with positive parameters α and β. This system describes the dynamics of PLL's with a band amplifier which is engaged between the generator and phase detector and consists of k uninetwork similarly adjucted cascades the phase detector being followed by an integrating filter [140].

We suppose that $-\mu_1 = \mu_2 = \mu$. Then Theorem 3.1.1 gives the following condition of Lagrange stability

$$\mu \le \lambda_0(\alpha - \lambda_0)(\beta - \lambda_0)^k,$$

where

$$\lambda_0 = \frac{(k+1)\alpha + 2\beta - \sqrt{k(k+2)\alpha^2 + (\alpha - 2\beta)^2}}{2(k+2)}.$$

Example 3.1.7 Let us consider (3.1.1) with $n = k + 1$ and the transfer function

$$\chi(s) = \frac{1}{s(1 + T_k s)^k},\tag{3.1.33}$$

where $T_k = T/k$, T being a positive parameter. The system (3.1.1), (3.1.33) describes the PLL of Example 3.1.6 without lowpass filter.

Suppose $\varphi(t, \sigma)$ satisfies here conditions (3.1.4) with $\mu_2 = -\mu_1 = \mu$. We apply then Theorem 3.1.1 with

$$\lambda = \frac{k}{(k+1)T}$$

and come to the conclusion that the system is Lagrange stable if

$$\mu \le \frac{1}{T}\left(\frac{k}{k+1}\right)^{k+1}.\tag{3.1.34}$$

Since

$$\lim_{k\to\infty}\left(\frac{k}{k+1}\right)^{k+1} = \frac{1}{e}$$

the inequality (3.1.34) may be transformed for great k into equality

$$\mu < (eT)^{-1}.$$

Note that in [140] the condition of Lyapunov stability in the case of stationary nonlinearity $\varphi(t, \sigma) \equiv \varphi(\sigma)$ is obtained. In the case of great k it is as follows

$$0 < \varphi'(\sigma_0) \le T^{-1}$$

where σ_0 is a certain equilibrium.

3.2 Systems with One Nonlinearity and a Bounded Forcing Term

Let us consider the pendulum-like system

$$\dot{x} = Px + q\xi, \qquad \sigma = r^*x, \qquad \xi = \varphi(t, \sigma) + g(t, x),\tag{3.2.1}$$

where P, q, r and φ are defined in system (3.1.1) - (3.1.4). We suppose that $g : \mathbf{R}_+ \times \mathbf{R}^n \to \mathbf{R}$ is continuous and locally Lipschitz continuous in the second argument. Suppose also that

$$\alpha < g(t, x) < \beta, \quad t \in \mathbf{R}_+, \quad x \in \mathbf{R}^n,\tag{3.2.2}$$

where α and β are certain numbers and that

$$g(t, x + d) = g(t, x), \quad x \in \mathbf{R}^n,\tag{3.2.3}$$

for any $d \in \mathbf{R}^n$ with $Pd = 0$.

In the following we choose such a d that $r^*d = \Delta$. Conditions (3.1.2), (3.1.3) and (3.2.3) guarantee that the system under consideration is a pendulum-like one with respect to the subgroup $\Gamma = \{jd, j \in \mathbf{Z}\}$. In analogy to the sector condition we suppose that

$$
\begin{aligned}
(\varphi(t,\sigma) + \alpha)(\sigma - \sigma_1) &\leq \mu(\sigma - \sigma_1)^2 \quad \text{and} \\
(\varphi(t,\sigma) + \beta)(\sigma - \sigma_2) &\leq \mu(\sigma - \sigma_2)^2 \quad t \in \mathbf{R}_+, \; \sigma \in \mathbf{R},
\end{aligned}
\tag{3.2.4}
$$

where $\mu > 0$, σ_1 and σ_2 are certain numbers.

In the next theorem we use the transfer function χ in the sense of Section 3.1.

Theorem 3.2.1 *Let* $\lim\limits_{s \to \infty} s\chi(s) > 0$. *Suppose there exists a* $\lambda > 0$ *such that the following conditions are true:*

(i) *the matrix* $P + \lambda I$ *has* $n - 1$ *eigenvalues with negative real part;*

(ii) $\operatorname{Re} \chi(i\omega - \lambda) + \mu \mid \chi(i\omega - \lambda) \mid^2 < 0, \quad \omega \in \mathbf{R}.$

Then system (3.2.1) to (3.2.4) is Lagrange stable.

Proof We only give a sketch of the proof which is just like that of Theorem 3.1.1. Note that conditions (i) and (ii) of Theorem 3.2.1 coincide with the analogous conditions of Theorem 3.1.1 for the case $\mu_1 = -\infty$ and $\mu_2 = \mu \neq +\infty$. So the matrix H is constructed here in the same way as in the previous section. With the help of H the two Lyapunov functions

$$
V(x) = (x - x_1)^* H(x - x_1) \qquad \text{and} \qquad W(x) = (x - x_2)^* H(x - x_2)
$$

are defined where $x_1, x_2 \in \mathbf{R}^n$ satisfy the conditions

$$
Px_1 = Px_2 = 0 \qquad \text{and} \qquad r^*x_1 = \sigma_1, \; r^*x_2 = \sigma_2.
$$

The condition

$$
r^*q = -\lim_{s \to \infty} s\chi(s) < 0
$$

does not occur in Theorem 3.1.1. It gives the opportunity to demonstrate the positive invariance and boundedness of the sets

$$
\Omega_j = \{x \; : \; V(x - jd) \leq 0, \quad r^*(x - x_1 - jd) \leq 0\}
$$

and

$$
\Phi_j = \{x \; : \; W(x - jd) \leq 0, \quad r^*(x - x_2 - jd) \leq 0\}
$$

where j is an arbitrary integer. For the proof it is sufficient to note that for any $x_0 \in \mathbf{R}^n$ there exists an integer j such that $x_0 \in \Omega_j \cap \Phi_j$. ∎

Let us consider in the next theorem the case when $\lim\limits_{s \to \infty} s\chi(s) = 0$. The proof of this theorem is analogous to the one given above.

Theorem 3.2.2 *Suppose that for a certain integer* $k \in [1, n]$ *there exists* $\lim\limits_{s \to \infty} s^k\chi(s) > 0$ *and the functions* φ *and* f *are* C^{k-1} *in* $\mathbf{R}_+ \times W_i$ *and* $\mathbf{R}_+ \times G_i$, *respectively, where* W_i, G_i $(i = 1, 2)$ *are certain neighborhoods of* σ_i *and* x_i. *If* $k \geq 2$ *we assume that* f *is* C^{k-2} *in the second argument in a neighborhood of the rays* $\{x \; : \; x = x_1 + sq, s \in [0, \infty)\}$ *and* $\{x \; : \; x = x_2 + sq, s \in (-\infty, 0]\}$. *Then if there exists a* $\lambda > 0$ *such that conditions (i) and (ii) of Theorem 3.2.1 are fulfilled then system (3.2.1) to (3.2.4) is Lagrange stable.*

Example 3.2.1 Let us consider a non-autonomous PLL with a proportionally-integrating filter [140, 107]

$$\ddot{\sigma} + \gamma\dot{\sigma} + \delta\frac{d}{dt}[\Omega(t)\cos\sigma + \Psi(t)] + \Omega(t)\cos\sigma + \Psi(t) = 0,$$

where γ, δ are positive parameters and Ω, Ψ are differentiable functions. Let us write this system in the form

$$\ddot{\sigma} + \gamma\dot{\sigma} + \delta\frac{d}{dt}[\Omega_0\cos\sigma + (\Omega(t) - \Omega_0)\cos\sigma + \Psi(t)] + \Omega_0\cos\sigma + (\Omega(t) - \Omega_0)\cos\sigma + \Psi(t) = 0, \quad (3.2.5)$$

where Ω_0 is a positive parameter, and let us suppose that

$$\mid \Omega(t) - \Omega_0 \mid + \mid \Psi(t) \mid < \Omega_0 \qquad\qquad (3.2.6)$$

for all $t \in \mathbf{R}_+$. Let us apply to (3.2.5) Theorem 3.2.1. We have to put

$$\begin{aligned}
g(t,x) &= (\Omega(t) - \Omega_0)\cos\sigma + \Psi(t) \quad (\sigma = r^*x), \\
\varphi(t,\sigma) &= \Omega_0\cos\sigma, \quad \chi(s) = \frac{\delta s + 1}{s^2 + \gamma s}, \\
\det(P - sI) &= s^2 + \gamma s, \quad \alpha = -\Omega_0, \quad \beta = \Omega_0.
\end{aligned}$$

The inequalities

$$\gamma\delta < 1, \qquad \Omega_0 \le \left(\frac{\sqrt{1 - \gamma\delta} - 1}{\delta}\right)^2$$

guarantee that all the conditions of Theorem 3.2.1 are satisfied. In this case $\mu = \left(\frac{\sqrt{1 - \gamma\delta} - 1}{\delta}\right)^2$ and we can choose

$$\lambda = \frac{-\sqrt{1 - \gamma\delta} + 1}{\delta}.$$

Example 3.2.2 Let us consider a non-autonomous PLL with integrating RC-filter [140, 107]

$$T\ddot{\sigma} + \dot{\sigma} + \Omega(t)\cos\sigma + \Psi(t) = \Omega_H,$$

where $T > 0$, Ω_H are numbers, Ω, Ψ are differentiable functions. Let us write the system in the form

$$T\ddot{\sigma} + \dot{\sigma} + \Omega_0\cos\sigma + [\Omega(t) - \Omega_0]\cos\sigma + \Psi(t) = \Omega_H, \qquad\qquad (3.2.7)$$

where $\Omega_0 > 0$ is a constant, and let us apply Theorem 3.2.2. We have

$$\begin{aligned}
\chi(s) &= \frac{1}{s(Ts + 1)} \quad (\text{with } \lim_{s\to\infty} s^2\chi(s) > 0), \\
\varphi(t,\sigma) &= \Omega_0\cos\sigma, \\
g(t,x) &= (\Omega(t) - \Omega_0)\cos\sigma + \Psi(t) - \Omega_H \quad (\sigma = r^*x).
\end{aligned}$$

Let us choose $\lambda = \frac{1}{2T}$ and introduce the function $\eta(\sigma) = \cos\sigma - \gamma$. Define the value

$$D(\gamma) := \inf\left\{\mu \; : \; \mu(\sigma - \sigma_0)^2 \ge \eta(\sigma)(\sigma - \sigma_0) \text{ for all } \sigma \in \mathbf{R}\right\}.$$

Suppose that

$$\mid \Omega(t) - \Omega_0 \mid + \mid \Psi(t) \mid + \mid \Omega_H \mid < \varepsilon$$

for all $t \ge 0$. Then we can take $\beta = -\alpha = \varepsilon$, $\varepsilon \le \Omega_0$ and $\frac{1}{4T} \ge \Omega_0 D(\varepsilon\Omega_0)$ then all conditions of Theorem 3.2.2 are fulfilled and all solutions of (3.2.7) are bounded as well as their first derivatives.

Remark 3.2.1 The results of the present section are borrowed from [82, 93, 94, 98].

3.3 Systems with Vector-Valued Nonlinearities

Let us consider in this section the system

$$\dot{x} = Px + Q\varphi(t,x), \tag{3.3.1}$$

where P and Q are $n \times n$ and $n \times m$ matrices, respectively. Suppose that the pair (P,Q) is controllable. Suppose also that $\varphi : \mathbf{R}_+ \times \mathbf{R}^n \to \mathbf{R}^m$ is continuous, locally Lipschitz in the second argument and bounded. Further we suppose that $\varphi(t,0) = 0$ on $\mathbf{R}$ and that there exists a non-zero vector $d \in \mathbf{R}^n$ with $Pd = 0$ and $\varphi(t, x + d) = \varphi(t,x)$ for $t \geq 0$, $x \in \mathbf{R}^n$. So the system (3.3.1) is a pendulum-like one.

Let us assume that there exists a quadratic form $F : \mathbf{R}^n \times \mathbf{R}^m \to \mathbf{R}$ and a n-vector such that

$$F(x, \varphi(t,x)) \geq 0 \quad \text{and} \quad F(x,0) \geq |\, r^*x \,|^2, t \in \mathbf{R}_+.\ x \in \mathbf{R}^n, \tag{3.3.2}$$

and the pair (P,r) is observable. Let us denote the extension of F to a Hermitian form by F_C.

Theorem 3.3.1 *Suppose that there exists a positive number λ such that the following conditions are valid:*

(i) *the matrix $P + \lambda I$ has $n - 1$ eigenvalues with negative real part;*

(ii) *for all $\omega \in \mathbf{R}$ and $\xi \in \mathbf{C}^m$ the inequality*

$$F_C(-\{P - (i\omega - \lambda)I\}^{-1} Q\xi\,,\ \xi) \leq 0 \tag{3.3.3}$$

is true.

Then system (3.3.1) is Lagrange stable.

Proof Since the proof of this theorem goes parallel to the proof of Theorem 3.1.1 we give only a sketch of this. Because of condition (ii) we may apply the Yakubovich-Kalman theorem (Theorem 1.4.1, p. 9) to conclude that there exists a $n \times n$ matrix $H = H^*$ such that

$$2x^*H[(P + \lambda I)x + Q\xi] + F(x,\xi) \leq 0 \tag{3.3.4}$$

for all $x \in \mathbf{R}^n$ and $\xi \in \mathbf{R}^m$. Using (3.3.2) and (3.3.4) for $\xi = 0$ we get

$$2x^*H(P + \lambda I)x \leq - |\, r^*x \,|^2, \qquad x \in \mathbf{R}^n.$$

From Lemma 1.2.1, p. 6, it follows that H has one negative and $n - 1$ positive eigenvalues. As was shown in the proof of Theorem 3.1.1 we can find an n-vector h such that

$$\{x\ :\ x^*Hx \leq 0\} \cap \{x\ :\ h^*x = 0\} = \{0\}\,.$$

We can assume that $h^*H^{-1}d < 0$. Let us introduce the function $V : \mathbf{R}^n \to \mathbf{R}$ by $V(x) = x^*Hx$. For an arbitrary $j \in \mathbf{Z}$ and a solution $x(\cdot) = x(\cdot, t_0, x_0)$ we have

$$\frac{d}{dt}V(x(t) - jd) \leq -2\lambda V(x(t) - jd) - F(x(t) - jd, \varphi(t, x(t) - jd)). \tag{3.3.5}$$

It is not difficult to show that the sets $\Omega_j = \{x\ :\ V(x - jd) \leq 0\}$ are positively invariant for (3.3.1) and for any solution $x(\cdot, t_0, x_0)$ there exists a $j = j(x_0)$ such that the solution $x(\cdot, t_0, x_0)$ belongs to the bounded positively invariant set

$$\Gamma_j = \Omega_j \cap \Omega_{-j} \cap \{x\ :\ |\, h^*x \,| \leq j \,|\, h^*d \,|\}\,. \quad \blacksquare$$

Example 3.3.1 Let us return to Example 3.1.2 and represent the equation of a pendulum with a vibrating point of suspension in the form

$$
\begin{aligned}
\dot{x}_1 &= x_2 - \frac{a\omega_0}{l}\varphi_1(t,x), \\
\dot{x}_2 &= -\alpha x_2 + \alpha\frac{a\omega_0}{l}\varphi_1(t,x) + \frac{a\omega_0}{l}\varphi_2(t,x) - \frac{g}{l}\varphi_3(t,x).
\end{aligned}
$$

Here $x_1 = \sigma$, $x_2 = \dot{\sigma} + \frac{a\omega_0}{l}\cos\omega_0 t \sin\sigma$,

$$
\begin{aligned}
x_1 &= \sigma, \quad x_2 = \dot{\sigma} + \frac{a\omega_0}{l}\cos\omega_0 t \sin\sigma, \\
\varphi_1(t,x) &= \cos\omega_0 t \sin x_1, \\
\varphi_2(t,x) &= x_2 \cos\omega_0 t \cos x_1 - \frac{a\omega_0}{l}\cos^2\omega_0 t \sin x_1 \cos x_1, \\
\varphi_3(t,x) &= \sin x_1.
\end{aligned}
$$

Let us introduce the parameters $\nu = \frac{a\omega_0}{l}$, $m = \frac{g}{l}$. Then

$$
(P - sI)^{-1} = \frac{1}{s(s+\alpha)}
\begin{bmatrix}
\nu s & -\nu & m \\
-\alpha\nu s & -\nu s & ms
\end{bmatrix}.
$$

Define the quadratic form $F : \mathbf{R}^3 \times \mathbf{R}^3 \to \mathbf{R}$ by

$$
F(x,\xi) = \tau_1(x_1^2 - \xi_3^2) + \tau_2(\xi_3^2 - \xi_1^2) + (x_2 - \nu\xi_1)^2 - \xi_2^2,
$$

where $\tau_1 > 0$, $\tau_2 > 0$ are parameters. Define also the vectors $r = \begin{bmatrix} \sqrt{\tau_1} \\ 0 \end{bmatrix}$ and $d = \begin{bmatrix} 2\pi \\ 0 \end{bmatrix}$. We apply Theorem 3.3.1. Let us choose $\lambda = \frac{\alpha}{2}$ and then condition (i) is satisfied. The second condition of the theorem can be verified by means of the Silvester criterion. Finally the sufficient condition of Lagrange stability is given by

$$
\frac{\alpha^2}{4} > \frac{g}{l} + \alpha\frac{a\omega_0}{l}.
$$

In case $g = 0$ we have

$$
\alpha > 4\frac{a\omega_0}{l}. \tag{3.3.6}
$$

Note that (3.3.6) has the same structure as (3.1.20).

Remark 3.3.1 The results of this section are taken from [93].

3.4 Bakaev Stability

In the paper [12] Yu. N. Bakaev has noticed that for pendulum-like systems (3.1.1) – (3.1.3) it is useful to guarantee that the difference $[\sigma(t_1) - \sigma(t_2)]$ of the output σ of the system in various moments t_1 and t_2 is bounded by the period of the nonlinearity φ if t_1, t_2 are sufficiently large. More precisely we give the following

Definition 3.4.1 The system (3.1.1) – (3.1.3) is called *Bakaev stable* if for any solution σ there exists a $T > 0$ such that for $t_1, t_2 > T$ we have $|\sigma(t_1) - \sigma(t_2)| < \Delta$.

Remark 3.4.1 It is important to note that the Bakaev stability domain in the system parameters space may be larger than the domain where the system is gradient-like. For two-dimensional systems this fact follows from the papers [150] and [30]. The Bakaev stability criteria which are proved in the following give the opportunity to get parameter domains for higher dimensional systems in which the system is Bakaev stable but not gradient-like.

Suppose that for a certain $\mu > 0$

$$\sigma\varphi(t,\sigma) \leq \mu\sigma^2, \quad t \in \mathbf{R}_+, \ \sigma \in \mathbf{R}. \tag{3.4.1}$$

In order to construct positively invariant cones for the system (3.1.1) - (3.1.3) we need the following purely algebraic result from [42].

Lemma 3.4.1 *Consider the k-dimensional quadratic cone $M := \{x \ : \ x^*Hx \leq 0\}$ and the hyperplane $\{x \ : \ c^*x = 0\}$ $(c \neq 0)$. Then the relation*

$$\operatorname{int} M \cap \{x \ : \ c^*x = 0\} = \emptyset \tag{3.4.2}$$

is satisfied if and only if $k = 1$ and $c^*H^{-1}c \leq 0$.

Proof Suppose at first that $k = 1$ and $c^*H^{-1}c < 0$. We choose a basis $c, c_1, \ldots, c_{n-1}$ of $\mathbf{R}^n$ with $c^*c_i = 0$, $i = 1, 2, \ldots, n-1$. Let f be a vector with $f^*c = 1$ and $Hf = \beta c$. Because of $\det H \neq 0$ and $c^*H^{-1}c < 0$ such a vector exists. The vectors $f, c_1, \ldots, c_{n-1}$ form a new basis of $\mathbf{R}^n$. Thus any $x \in \mathbf{R}^n$ can be written in the form

$$x = (c^*x)f + \sum_{i=1}^{n-1} \alpha_i c_i.$$

From this it follows that

$$x^*Hx = \frac{(c^*x)^2}{c^*H^{-1}c} + \sum_{i,j=1}^{n-1} \alpha_i\alpha_j c_i^*Hc_j = \frac{(c^*x)^2}{c^*H^{-1}c} + \sum_{j=1}^{n-1} \varepsilon_j(g_j^*x)^2, \tag{3.4.3}$$

where $c, g_1, \ldots, g_{n-1}$ are linearly independent vectors. By assumption H is regular so we have $\varepsilon_j \neq 0$, $j = 1, \ldots, n-1$. Using $c^*H^{-1}c < 0$ and the fact that $k = 1$ we see that $\varepsilon_j > 0$, $j = 1, \ldots, n-1$. It follows from (3.4.3) that for $x \neq 0$, $c^*x = 0$, we have $x^*Hx > 0$.

It remains to investigate the case $c^*H^{-1}c = 0$.

Suppose that (3.4.2) is not satisfied. Since $\operatorname{int} M$ is open there exists a vector $h \neq 0$ such that $h^*H^{-1}h < 0$ and $\operatorname{int} M \cap \{x \ : \ h^*x = 0\} \neq \emptyset$. But this is not possible.

Suppose now that (3.4.2) is given. We have to show that $k = \dim M = 1$ and $c^*H^{-1}c \leq 0$. Suppose to the contrary that $k > 1$. In this case there exists a subspace Π with $\dim \Pi \geq 2$ and $\Pi \setminus \{0\} \subset \operatorname{int} M$. Employing the fact that $\dim \{x \ : \ c^*x = 0\} = n - 1$ we get $\dim (\Pi \cap \{x \ : \ c^*x = 0\}) \geq 1$ which contradicts (3.4.2) because $\Pi \cap \{x \ : \ c^*x = 0\} \setminus \{0\} \subset \operatorname{int} M$.

In cases when $c^*H^{-1}c \neq 0$ the representation (3.4.3) is given for an arbitrary vector x. For x with $c^*x = 0$ it follows that $\sum_{j=1}^{n-1} \varepsilon_j(\cdots)^2 \geq 0$, so we have $\varepsilon_j > 0$, $j = 1, \ldots, n-1$. Because of $\operatorname{int} M \neq \emptyset$ we see from (3.4.2) that $c^*H^{-1}c < 0$. $\blacksquare$

Theorem 3.4.1 *Suppose that $r^*q \leq 0$ and that there exists a positive number λ which satisfies the following conditions:*

(i) the matrix $P + \lambda I$ has $n - 1$ eigenvalues with negative real part;

(ii)

$$\pi(\omega) := \operatorname{Re}\chi(i\omega - \lambda) + \mu \mid \chi(i\omega - \lambda) \mid^2 < 0, \quad \omega \in \mathbf{R} \quad and \quad \lim_{\omega \to +\infty} \omega^2 \pi(\omega) < 0. \qquad (3.4.4)$$

Then the system (3.1.1) *to* (3.1.4) *is Bakaev stable.*

Proof It follows by the inequalities (3.4.4) that there exists a number $\delta > 0$ such that

$$\operatorname{Re}\chi(i\omega - \lambda) + (\mu + \delta) \mid \chi(i\omega - \lambda) \mid^2 \le 0, \ \omega \in \mathbf{R}. \qquad (3.4.5)$$

This inequality coincides with the inequality (3.1.9) in case $\mu_1^{-1} = 0$, $\mu_2 = \mu + \delta$. So we can use the proof of Theorem 3.1.1. It follows from this that there exists a matrix $H = H^*$ with one negative and $n - 1$ positive eigenvalues such that

$$2x^*H[(P + \lambda I)x + q\xi] + r^*x[(\mu + \delta)r^*x - \xi] \le 0, \qquad (3.4.6)$$

$x \in \mathbf{R}^n, \xi \in \mathbf{R}$. Hence $2Hq = r$ and consequently $r^*H^{-1}r = 2r^*q \le 0$. Then from Lemma 3.4.1 we have that

$$\{x \ : \ x^*Hx \ge 0\} \supset \{x \ : \ r^*x = 0\}. \qquad (3.4.7)$$

Consider now the function $V_j(x) = (x - jd)^*H(x - jd)$, where $j \in \mathbf{Z}$ and d is an non-zero vector with $Pd = 0$ and $r^*d = \Delta$. The derivative of V_j with respect to t along a solution x of (3.1.1) can be evaluated with the help of (3.4.1) and (3.4.6) in the following way:

$$\begin{aligned}
\dot{V}_j(x(t)) + 2\lambda V_j(x(t)) &\le -r^*(x(t) - jd)[\mu r^*(x(t) - jd) - \varphi(t, r^*(x(t) - jd))] - \\
&\quad -\delta[r^*(x(t) - jd)]^2 \\
&\le -\delta[r^*(x(t) - jd)]^2.
\end{aligned} \qquad (3.4.8)$$

So the sets $\Omega_j = \{x \ : \ V_j(x) < 0\}$ are positively invariant.

Let us now put in (3.4.6) $\xi = 0$ and $x = d$. We establish that $2\lambda d^*Hd \le -(\mu + \delta)\Delta^2$. Hence for any fixed y we have $\lim_{j \to \infty} V_j(y) = -\infty$. It follows that for any solution $x(\cdot, t_0, x_0)$ there exists such a N that $V_j(x_0) < 0$ for $\mid j \mid > N$. Hence in virtue of (3.4.7) it follows that $r^*x(t, t_0, x_0) \ne j\Delta$ for $t \ge t_0, \mid j \mid > N$.

Suppose that for a certain k we have $V_k(x(t)) \ge 0$ for all $t \ge t_0$, where $x(\cdot) = x(\cdot, t_0, x_0)$. Then from (3.4.8) it follows that

$$V_k(x(t)) - V_k(x_0) \le -\delta \int_{t_0}^t [r^*(x(\tau) - kd)]^2 \, d\tau$$

and consequently for any $t \ge t_0$

$$\int_{t_0}^t [r^*(x(\tau) - kd)]^2 \, d\tau \le \frac{1}{\delta} V_k(x_0). \qquad (3.4.9)$$

As the conditions of Theorem 3.1.1 are fulfilled we can conclude that the solution $x(\cdot)$ is bounded on $[t_0, +\infty)$. Then the derivative $\frac{d}{dt}(r^*x(t))$ is also bounded. Using Barbalat's lemma (Theorem 2.1.3, p. 16) and the property (3.4.9) we establish that

$$\lim_{t \to +\infty} r^*(x(t) - kd) = 0. \qquad (3.4.10)$$

Suppose now that for any k there exists a τ_k such that $V_k(x(\tau_k)) < 0$. As the set

$$\{x \ : \ V_k(x) < 0\}$$

is positively invariant we get $V_k(x) < 0, \ t \ge \tau_k$. So either there exists a finite limit of $r^*x(t)$ for $t \to +\infty$ or $V_k(x(t)) < 0, \ t \ge T$, where $T = \max_{|k| \le N} \tau_k$. In the latter case it follows from (3.4.7) that $r^*x(t) \ne k\Delta, \ t \ge T, \mid k \mid \le N$. Thus the Bakaev stability is proved. $\blacksquare$

Remark 3.4.2 Let us return again to the Example 3.1.5. It is clear that all the conditions of Theorem 3.4.1 are satisfied for this system. It follows that condition (3.1.30) guarantees the Bakaev stability of the given system. Comparing the parameter region where the system has only Lyapunov stable equilibria (see Fig 3.1.2) with the mentioned parameter region of Bakaev stability we conclude that the inequality (3.1.30) defines some regions in the parameter space of the system in Example 3.1.5 where this system is Bakaev stable, but all its equilibria are not stable in the sense of Lyapunov.

Often it is necessary for pendulum-like systems to get asymptotic estimates which are sharper than in the Definition 3.4.1. The following definition and the theorem are due to [95, 96].

Definition 3.4.2 The system (3.1.1) to (3.1.4) is called $\varrho\Delta$-*stable* with $0 < \varrho < 1$ if for any of its solution σ there exists a T such that $\mid \sigma(t_1) - \sigma(t_2) \mid < \varrho\Delta$ for all $t_1, t_2 > T$.

Theorem 3.4.2 *Suppose that all the conditions of Theorem 3.4.1 are fulfilled. Suppose also that there exists numbers $\delta_0 > 0$ and $\delta_1 > 0$ with $\delta_0 + \delta_1 < \Delta$ and there exists a continuous Δ-periodic function φ_1 such that the following estimates are true:*

(i) $\varphi(t,\sigma) \le \varphi_1(\sigma) < \mu(\sigma - \gamma)$ *for all* $\gamma \in (0, \delta_0]$, $\sigma \in [\gamma, \Delta]$ *and* $t \ge 0$;

(ii) $\varphi(t,\sigma) \ge \varphi_1(\sigma) > \mu(\sigma - \gamma)$ *for all* $\gamma \in [-\delta_1, 0)$, $\sigma \in [-\Delta, \gamma]$ *and* $t \ge 0$.

Then for any solution $x(\cdot)$ of (3.1.1) to (3.1.4) for which $r^*x(t)$ does not converge to an $j\Delta$ (j some integer) and for any $\varepsilon_1 \in (0, \delta_0)$ and $\varepsilon_2 \in (0, \delta_1)$ there exists a positive integer k and a $T > 0$ such that

$$r^*x(t) \in [k\Delta + \delta_0 - \varepsilon_1, (k+1)\Delta - \delta_1 + \varepsilon_2], \qquad t \ge T.$$

Corollary 3.4.1 *Under the conditions of Theorem 3.4.2 the system (3.1.1) - (3.1.4) is $\varrho\Delta$-stable with*

$$\varrho = 1 - \frac{1}{\Delta}(\delta_0 + \delta_1 - \varepsilon_1 - \varepsilon_2).$$

Remark 3.4.3 The method of invariant cones was developed in the papers [1, 73, 79, 82, 93, 94, 96, 117, 118].

Chapter 4

The Bakaev-Guzh Technique

In the present chapter we show that questions concerning the global convergence of solutions of pendulum-like systems can well be formulated in terms of factor-manifolds. Our factor-manifolds are non-compact. If we want to investigate the convergence theory on these manifolds we have to give additional conditions which ensure that Lyapunov functions are bounded from below.

This is not a book on the differential geometric approach to nonlinear systems, but in this chapter there are given some connections between this approach and the classical approaches developed in [15, 49, 72, 129, 99].

4.1 Basic Tools for Vector Fields on Riemannian Manifolds

In the sequel we repeat some well-known facts and basic definitions on vector fields on manifolds, using the notation of J. Mawhin and M. Willem [112]. Suppose M is a set. An *n-dimensional chart* x for M is a bijection $x : D(x) \to R(x) \subset \mathbf{R}^n$, where $R(x)$ is open in $\mathbf{R}^n$ and $D(x) \subset M$.

An *n-dimensional atlas* of class C^k ($k \geq 0$) on M is a set A of n-dimensional charts such that:

1) $\bigcup_{x \in A} D(x) = M$;

2) $x(D(x) \cap D(y))$ is open in $\mathbf{R}^n$ for arbitrary $x, y \in A$;

3) The mapping $y \circ x^{-1} : x(D(x) \cap D(y)) \to y(D(x) \cap D(y))$ is of class C^k for each $x \in A$ and $y \in A$.

An *n-dimensional manifold of class* C^k is a pair (M, A), where M is a set and A is an n-dimensional C^k-atlas for M. In the sequel we write M instead of (M, A).

The topology on M is given by the following basis: For any $x \in A$ the set $D(x) \subset M$ is assumed to be open and x is assumed to be a homeomorphism. It follows that a set $U \subset M$ is open if and only if for any $u \in U$ there exist a chart x and a open set $W \subset R(x)$ with $u \in x^{-1}(W) \subset U$.

In order to define for a C^1-manifold the tangent space we introduce the following equivalence relation. Consider an arbitrary $u \in M$ and suppose that $u \in D(x)$ and $u \in D(y)$ for $x, y \in A$. Let be $\xi \in \mathbf{R}^n$ and $\eta \in \mathbf{R}^n$ two arbitrary vectors. We define the relation

$$(u, x, \xi) \sim (u, y, \eta) \Leftrightarrow \eta = (y \circ x^{-1})'(x(u))\xi,$$

where $(y \circ x^{-1})'(x(u))$ is the Jacobi matrix computed in the point $x(u)$. It is easy to verify that this is an equivalence relation with the equivalence classes

$$[u, x, \xi] := \{(u, y, \eta) \: : \: u \in D(y) \quad \text{and} \quad (u, y, \eta) \sim (u, x, \xi)\} \,.$$

The *tangent space of M at* $u \in M$ is then the set $T_u M$ of all equivalence classes $[u, x, \xi]$ such that $u \in D(x)$ and with a vector space structure defined by

$$[u, x, \xi] + [u, x, \eta] := [u, x, \xi + \eta]$$

and

$$\lambda[u, x, \xi] := [u, x, \lambda \xi] \qquad (\lambda \in \mathbf{R}),$$

where x is a chart at u.

It can be shown that this definition does not depend on the chart x.

The *tangent bundle TM of M* is defined by $TM := \bigcup_{u \in M} T_u M$ and the *projection τ* is given by $\pi : TM \to M$ with $[u, x, \xi] \to u$.

Suppose M and N are C^k-manifolds of dimension m and n, respectively. The function $f : M \to N$ is called *locally Lipschitz continuous* (resp. *of class C^k, $k \geq 0$*) if for any charts x at $u \in M$ and y at $f(u) \in N$ the mapping $y \circ f \circ x^{-1} : R(x) \to y(f(D(x)) \cap D(y))$ is locally Lipschitz continuous (resp. of class C^k).

Suppose $f : M \to N$ is of class C^1. The *differential of f* is the mapping $df : TM \to TN$ definded by

$$df([u, x, \xi]) := [f(u), y, (y \circ f \circ x^{-1})'(x(u))\xi]$$

where x is a chart at $u \in M$, y is a chart at $f(u)$ in $\xi \in \mathbf{R}^n$. One can easily show that this definition is independent of x and y.

If $N = \mathbf{R}^n$ we can write $T\mathbf{R}^n \approx \mathbf{R}^n \times \mathbf{R}^n$ and $df : TM \to \mathbf{R}^n \times \mathbf{R}^n$ is defined by

$$df([u, x, \xi]) := (f(u), (f \circ x^{-1})'(x(u))\xi).$$

In particular one can show that, if x is a chart on M, $dx : \pi^{-1}(D(x)) \to R(x) \times \mathbf{R}^n$ is a chart and $\{dx : x \in A\}$ is a $2n$-dimensional C^{k-1} atlas on TM, such that $dx([u, x, \xi]) = (x(u), \xi)$.

Suppose M is an n-dimensional Hausdorff manifold of class C^2. (Recall, that M is Hausdorff if for any two points $u_1, u_2 \in M$, $u_1 \neq u_2$, there exist open neighborhoods $B(u_1)$ and $B(u_2)$ with $B(u_1) \cap B(u_2) = \emptyset$.)

A vector field on M is a mapping $f : M \to TM$ such that $\pi \circ f = id$.

For a C^1-path $\alpha : J \to M$, $J = (a, b)$, the corresponding *differential* is $d\alpha : TJ \to TM$, where $TJ = \{[t, id, \xi], t \in J, \xi \in \mathbf{R}\}$.

We define $\dot{\alpha}(t)$ by

$$\dot{\alpha}(t) := d\alpha([t, id, (1)]) = [\alpha(t), x, (x \circ \alpha)'(t)].$$

The C^1-curve $\alpha : J \to M$ with $0 \in J$, J an open interval in $\mathbf{R}$, $\alpha(0) = p \in M$ and $\dot{\alpha}(t) = f(\alpha(t))$ for $t \in J$ is called an *integral curve* of the vector field f through p at $t = 0$ or *solution of the Cauchy problem*

$$\begin{aligned} \dot{\alpha}(t) &= f(\alpha(t)), \\ \alpha(0) &= p. \end{aligned} \qquad (4.1.1)$$

We assume that the vector field f is locally Lipschitz continuous.

Consider a chart x on M with $p \in D(x)$. Define in $R(x)$ the locally Lipschitz continuous function $F(\eta) := (P \circ dx \circ f \circ x^{-1})(\eta)$, with $P : \mathbf{R}^n \times \mathbf{R}^n \to \mathbf{R}^n$ given by $P(u, v) = v$. By the Picard Lindelöf theorem the Cauchy problem in $R(x) \subset \mathbf{R}^n$ defined by $\dot{\eta} = F(\eta)$, $\eta(0) = x(p)$, has on a certain interval $(-\varepsilon, \varepsilon)$ the unique solution $\eta(\cdot, x(p))$. It is clear that $\alpha(\cdot, p) := x^{-1}(\eta(\cdot, x(p)))$ will be the solution of (4.1.1) on $(-\varepsilon, \varepsilon)$.

In order to define, on the manifold, the length of a curve we introduce a Riemannian metric. We say that on the n-dimensional C^k-manifold M there is given a *Riemannian metric G of the class C^{k-1}* if there is defined a mapping which associates to each pair (u, x) with $u \in M$ and x a chart at u, a positive definite matrix $G_x(u)$ such that

1) the mapping $G_x : D(x) \to \mathcal{L}(\mathbf{R}^n, \mathbf{R}^n)$ given by $u \to G_x(u)$ is C^{k-1} for every x,

2) if x and y are two charts at $u \in M$, then

$$[(y \circ x^{-1})'(x(u))]^* G_y(u)[(y \circ x^{-1})'(x(u))] = G_x(u).$$

An *inner product* on $T_u M$ we define by

$$\langle [u, x, \xi], [u, x, \xi] \rangle := (G_x(u)\xi, \eta),$$

where $(\cdot, \cdot)$ is the standard inner product in $\mathbf{R}^n$ with $(\xi, \eta) = \sum_{i=1}^{n} \xi_i \eta_i$. It can be shown that this inner product definition is independent on x.

The corresponding *norm* on $T_u M$ is given by

$$\|[u, x, \xi]\| := \langle [u, x, \xi], [u, x, \xi] \rangle^{1/2}.$$

A connected C^k-manifold with a Riemannian metric G of class C^{k-1} is called *Riemannian manifold* of class C^k and denoted by (M, G).

Consider on a Riemannian manifold of class C^1 a piece-wise C^1-path $\alpha : [a, b] \to M$. For such a path α we define the *length* $s(\alpha) := \int_a^b |\dot{\alpha}(t)| \, dt$. It can be shown that under our conditions every two points $u, v \in M$ can be connected by a piece-wise C^1-path. So we can define on M the *geodesic distance*

$$d(u, v) := \inf \left\{ s(\alpha) \ : \ \alpha \text{ is piece-wise } C^1 \text{ and connects } u \text{ with } v \right\}.$$

One can prove that this function d defines a distance on M and M becomes a metric space whose topology is compatible with the manifold topology.

Let us consider now C^1-functions $V : M \to \mathbf{R}$ along the solutions of (4.1.1). Recall that for a C^1-function $h : U \to \mathbf{R}$, $U \subset \mathbf{R}^n$, the gradient of h is by definition $\nabla h(u) := (D_1 h(u), \ldots, D_n h(u))$, where $D_i h$ indicates the i-th partial derivate.

For an arbitrary integral curve α of (4.1.1) we can compute

$$\frac{d}{dt} V(\alpha(t)) = \frac{d}{dt}(V \circ x^{-1} \circ x \circ \alpha)(t),$$

where x is a chart at $\alpha(t)$. It follows that

$$\begin{aligned}
\frac{d}{dt} V(\alpha(t)) &= (\nabla(V \circ x^{-1})(x(\alpha(t))), \frac{d}{dt}(x \circ \alpha)(t)) \\
&= (G_x(\alpha(t)) G_x^{-1}(\alpha(t)) \nabla(V \circ x^{-1})(x(\alpha(t))), \frac{d}{dt}(x \circ \alpha(t)) \\
&= \langle [\alpha(t), x, G_x^{-1}(\alpha(t)) \nabla(V \circ x^{-1})(x(\alpha(t)))], [\alpha(t), x, \frac{d}{dt}(x \circ \alpha)(t)] \rangle.
\end{aligned}$$

Thus we can define for a function $V \in C^1(M, \mathbf{R})$ the associated vector field, called the *gradient* of V, by

$$\operatorname{grad} V(u) := [u, x, G_x^{-1}(u) \nabla(V \circ x^{-1})(x(u))],$$

where x is a chart at u. It can be shown that this definition does not depend on x.

Let us now recall some facts on the fundamental group of a manifold. Consider a continuous path $\alpha : J \to M$, $J = [0, 1]$, with $\alpha(0) = \alpha(1)$, which is called *closed*. Suppose a point $u \in M$

is given and consider the set C_u of all closed paths with $\alpha(0) = \alpha(1) = u$. The two paths α_1, $\alpha_2 \in C_u$ are *homotopic* in M ($\alpha_1 \sim \alpha_2$) if there exists a continuous function $H : J \times J \to M$ with

$$H(0,\cdot) = \alpha_1, \quad H(1,\cdot) = \alpha_2 \quad \text{and} \quad H(\cdot,0) = H(\cdot,1) = u.$$

Intuitively, α_1 is homotopic to α_2 if one can continuously deform α_1 into α_2. Moreover, it is easily shown that the homotop relation is an equivalence relation in C_u. Suppose $[\alpha]$ denotes the set of all paths from C_u homotopic to α.

Let us define the composition of two classes $[\alpha_1]$ and $[\alpha_2]$ by the class $[\alpha_1 * \alpha_2]$, where

$$(\alpha_1 * \alpha_2)(t) = \left\{ \begin{array}{ll} \alpha_1(2t), & 0 \leq t \leq \frac{1}{2}, \\ \alpha_2(2t - 1), & \frac{1}{2} \leq t \leq 1. \end{array} \right.$$

Note, that the class $[\alpha_1 * \alpha_2]$ depends only on the classes $[\alpha_1]$ and $[\alpha_2]$. This product is associative but in general not commutative. Thus we can define a group structure on the equivalent classes of C_u. As the unit we take the constant mapping $\alpha : J \mapsto u$. The inverse for a class $[\alpha]$ is given by the class $[\alpha(1 - (\cdot))]$. In this way we have defined a group which is called the *fundamental group of M in u* and which is denoted by $\pi_1(u, M)$.

It is well known, that for a connected manifold M the fundamental groups $\pi_1(u, M)$ are isomorphic for various u. Thus we can consider the *fundamental group of M* as $\pi_1(M)$.

If a closed piece-wise C^1-path γ is homotopic to a constant we say that γ is *contractible* in M. A manifold M is *simply connected* if M is connected and every closed piece-wise C^1-path is contractible.

Suppose $M = N_1 \times N_2$ with the product topology, where N_1 and N_2 are manifolds Consider $u \in N_1$ and $v \in N_2$ and define $z = (u, v)$. Then it follows that

$$\pi_1(M, z) \approx \pi_1(N_1, u) \times \pi_1(N_2, v).$$

Notice that for $M = S^1$ the group $\pi_1(S^1)$ is free and cyclic with a generator α_1.

Suppose $\gamma : [a, b] \to M$ is a piece-wise C^1-path on the Riemannian manifold M and $h : M \to TM$ is a C-vector field on M. We define the *curve integral of the second kind of h along the curve* γ as

$$\int_\gamma \langle h, du \rangle := \int_a^b \langle h(\gamma(t)), \dot{\gamma}(t) \rangle \, dt.$$

If $f : M \to \mathbf{R}$ is continuous and γ as above, the *curve integral of the first kind of f along γ* is

$$\int_\gamma f ds := \int_a^b f(\gamma(t)) |\dot{\gamma}(t)| \, dt.$$

One can show that these definitions are correct.

The vector field h of class C is called an *exact* or *gradient vector field* if there exists a C^1-function $V : M \to \mathbf{R}$ such that $h(u) = \operatorname{grad} V(u)$ on M. We say that the C-vector field h is *closed* if for an arbitrary closed piece-wise C^1-path γ an M

$$\int_\gamma \langle h, du \rangle = \int_{\overline{\gamma}} \langle h, du \rangle$$

for all $\overline{\gamma} \in [\gamma]$.

An important class of vector fields on Riemannian manifolds is given by ordinary differential equations in $\mathbf{R}^n$ with an equivariance property.

Let us consider the equation

$$\dot{u} = f(u), \qquad u \in \mathbf{R}^n, \tag{4.1.2}$$

with f locally Lipschitz continuous. Suppose that the solutions of (4.1.2) exist on $\mathbf{R}_+$. Suppose also that there exist vectors $d_i \in \mathbf{R}^n$, $i = 1, 2, \ldots, m$, with $\dim \operatorname{span} \{d_1, \ldots, d_m\} = m$ and

$$f(u + \gamma) = f(u)$$

for all $u \in \mathbf{R}^n$ and

$$\gamma \in \Gamma := \left\{ \sum_{j=1}^m k_j d_j \ : \ k_j \in \mathbf{Z}, 1 \le j \le m \right\}. \tag{4.1.3}$$

Hence (4.1.2) is pendulum-like with respect to Γ. Clearly, Γ is a discrete subgroup of $\mathbf{R}^n$. Let $\pi : \mathbf{R}^n \to \mathbf{R}^n/\Gamma$ be the canonical projection in the quotient space $\mathbf{R}^n/\Gamma$ defined by the equivalence relation in $\mathbf{R}^n$

$$u \sim v \quad \Longleftrightarrow \quad u = v + \gamma$$

with some $\gamma \in \Gamma$. Then

$$\left\{ \pi^{-1} : \pi(U) \to U, U \text{ is open in } \mathbf{R}^n, \text{ and } \pi : U \to \mathbf{R}^n/\Gamma \text{ is injective} \right\}$$

is an atlas of class C^∞ for $\mathbf{R}^n/\Gamma$.

The C^∞-manifold $\mathbf{R}^n/\Gamma$ is called *cylinder*, the special case $m = n$ produces the *torus*.

We can interpret $\mathbf{R}^n/\Gamma$ as Riemannian manifold by setting, as in the case $M = \mathbf{R}^n$, $G_x = I$ for any chart x of $\mathbf{R}^n/\Gamma$.

The cylinder $\mathbf{R}^n/\Gamma$ equipped with this metric is called *flat*. It follows that the geodesic distance in this case is given by

$$d(u, v) = \inf_{\substack{[\xi] = u \\ [\eta] = v}} |\xi - \eta|,$$

where $|\xi - \eta| = (\xi - \eta, \xi - \eta)^{\frac{1}{2}}$ is determined by the standard scalar product in $\mathbf{R}^n$.

Let us return to equation (4.1.1). We say that the solution $\alpha(\cdot, p)$ of (4.1.1) is *bounded* on $\mathbf{R}_+$, if the positive semi-orbit $\gamma^+(p) := \{\alpha(\cdot, p), \ t \ge 0\}$ is relatively compact in M.

We say that the solution $\alpha(\cdot, p)$ (or the positive semi-orbit $\gamma^+(p)$) *converges* if there exists a point $q \in M$ with $\lim_{t \to +\infty} \alpha(t, p) = q$. It follows that q is an *equilibrium point*, e. g. $\alpha(t, q) = q$ for all t.

The solution $\alpha(\cdot, p)$ is *quasi-convergent* if the set $\mathcal{E}$ of equilibria of (4.1.1) attracts the orbit of $\alpha(\cdot, p)$, i. e., $\operatorname{dist}(\alpha(t, p), \mathcal{E}) \to 0$ as $t \to +\infty$, where $\operatorname{dist}(\cdot, \mathcal{E})$ is computed by using the geodesic distance.

System (4.1.1) is called *gradient-like* (resp. *quasi-gradient-like*) if every solution is convergent (resp. quasi-convergent).

We say that system (4.1.1) is *monostable* (resp. *quasi-monostable*) if every bounded solution is convergent (resp. quasi-convergent).

In a parallel manner as for differential equations in $\mathbf{R}^n$ we define for a solution $\alpha(\cdot, p)$ of (4.1.1) the ω-limit set $\omega(p)$ as

$$\omega(p) := \left\{ y \ : \ \exists (t_n) \text{ with } t_n \to +\infty, \ \lim_{n \to \infty} \alpha(t_n, p) = y \right\}.$$

It is well-known, (for instance, [51]), that this set is closed and invariant for (4.1.1). If the solution $\alpha(\cdot, p)$ is bounded on $\mathbf{R}_+$ the limit set $\omega(p)$ is compact, connected and non-empty. Furthermore, we have $\operatorname{dist}(\alpha(t, p), \omega(p)) \to 0$ as $t \to +\infty$.

4.2 Lyapunov-Type Results for Boundedness and Convergence

Let us consider system (4.1.1) on the non-compact Riemannian manifold (M, G). Suppose that $d(\cdot, \cdot)$ is the geodesic distance introduced above. Suppose also that the set of equilibria of (4.1.1) is discrete.

The problem of establishing gradient-like behavior for (4.1.1) can be broken down into two parts: First, to show, that the solutions are bounded, and second, that bounded solutions converge. The Lyapunov-type results given in this section are slight generalizations of well-known results for dynamical systems in $\mathbf{R}^n$ presented, for instance in [49].

Lemma 4.2.1 *Suppose that there exists a continuous function $V : M \to \mathbf{R}$ such that:*

(i) *For any sequence $\{u_k\}_{k=0}^\infty \subset M$ with $d(u_k, u_0) \to +\infty$ for $k \to +\infty$ it follows that $V(u_k) \to +\infty$.*

(ii) *For any solution α of (4.1.1) the function $V \circ \alpha$ is non-increasing.*

Then every solution of (4.1.1) is bounded on $\mathbf{R}_+$.

Proof Employing (ii) we obtain that the positive semi-orbit of an arbitrary solution $\alpha(\cdot, p)$ lies in the set $\Gamma := \{u \in M : V(u) \leq V(p)\}$. From this it follows that $\Gamma \subset \{u \in M : d(u, p) \leq l\}$, where l is sufficiently large. Assuming the opposite, we get a sequence $\{u_k\} \subset \Gamma$ with $d(u_k, p) \to +\infty$. By condition (i) we have $V(u_k) \to +\infty$, which contradicts the definition of Γ. ∎

Lemma 4.2.2 *Suppose $\alpha(\cdot, p)$ is a bounded solution of (4.1.1) and there exists a positively invariant set $N \subset M$ with $\omega(p) \cup \gamma^+(p) \subset N$ and a continous function $V : N \to \mathbf{R}$ satisfying the following conditions:*

(i) *for any $q \in N$ the function V is not increasing along $\alpha(t, p)$;*

(ii) *if for some $q \in N$ there exists a $\tau > 0$ with $V(q) = V(\alpha(\tau, q))$ then it follows that $\alpha(\cdot, q)$ is constant.*

Then $\alpha(t, p)$ converges to an equilibrium for $t \to +\infty$.

Proof By (i) it follows that there exists the limit $\lim_{t \to +\infty} V(\alpha(t, p)) =: \beta$. Consider an arbitrary $q \in \omega(p)$. Clearly, $\alpha(t, q) \in \omega(p)$ for all $t \in \mathbf{R}$. Furthermore, $V(\alpha(t, q)) = \beta$ for all t. Indeed, for every t there is a sequence $t_n \to +\infty$ with $\alpha(t_n, p) \to \alpha(t, q)$. By continuity of V we get $V(\alpha(t_n, p)) \to V(\alpha(t, q)) = \beta$.

Using assumption (ii) we see that $\alpha(t, q) \equiv q$ and, consequently, $\omega(p) \subset \mathcal{E}$.

Because $\omega(p)$ is connected and the set $\mathcal{E}$ is by assumption discrete, it follows that $\omega(p)$ is single. Since $\mathrm{dist}\,(\alpha(t, p), \omega(p)) \to 0$ as $t \to +\infty$, the assertion of Lemma 4.2.2 follows. ∎

Remark 4.2.1 It easy to see that the assertions of Theorems 1.1.3, 1.1.4 and 1.1.5 follow from Lemmata 4.2.1, 4.2.2.

As an application of Lemma 4.2.2 let us deduce from this lemma a variant of the Barbashin-Krasovskij Theorem for the cylinder. Let us consider system (4.1.2), which is supposed to be equivariant with respect to the group defined by (4.1.3).

Theorem 4.2.1 *Assume that the set of equilibria of (4.1.2) consists of isolated points only and there exists a continuous function $V : \mathbf{R}^n \to \mathbf{R}$ satisfying the following properties:*

(i) $V(u+g) = V(u)$ *for all* $u \in \mathbf{R}^n$, $g \in \Gamma$;

(ii) $V(u) + \sum\limits_{i=1}^{m}(d_i^* u)^2 \to +\infty$ *as* $|u| \to +\infty$;

(iii) *for every solution* $\alpha(\cdot)$ *of* (4.1.2) *the function* $V \circ \alpha$ *is non-increasing;*

(iv) *if* $\alpha(\cdot)$ *is a solution of* (4.1.2) *and there exists a* $\tau > 0$ *with* $V(\alpha(0)) = V(\alpha(\tau))$ *then* α *is a stationary solution.*

Then system (4.1.2) *is gradient-like.*

Proof Consider the flat Riemannian cylinder $\mathbf{R}^n/\Gamma$. The properties of f allow us to interpret (4.1.2) as a continuous function on this manifold. Denote the geodesic on the flat manifold $\mathbf{R}/\Gamma$ by d. In order to use the conclusion of Lemma 4.2.1 it remains to show that for a sequence $\{u_k\} \subset \mathbf{R}^n/\Gamma$ from $d(u_k, u_0) \to +\infty$ it follows that $V(u_k) \to +\infty$ as $k \to +\infty$. For each $\overline{u}_k \in \mathbf{R}^n$ with $[\overline{u}_k] = u_k$ we have the unique representation

$$\overline{u}_k = v_k + \sum_{i=1}^{m} \varrho_i^k d_i + \sum_{i=1}^{m} l_i^k d_i$$

with $\varrho_i \in [0,1)$, $l_i^k \in \mathbf{Z}$ and $v_k \perp \mathrm{span}\,\{d_1, \ldots, d_m\}$. Because of $d(u_k, u_0) \to +\infty$ we have $\left| v_k + \sum\limits_{i=1}^{m} \varrho_i^k d_i \right| \to +\infty$ as $k \to \infty$. We conclude from (ii) that $V(v_k + \sum\limits_{i=1}^{m} \varrho_i^k d_i) \to \infty$ as $k \to \infty$. On the other hand one has $V(\overline{u}_k) = V(v_k + \sum\limits_{i=1}^{m} \varrho_i^k d_i)$ for all k. $\blacksquare$

In the next section Theorem 4.2.1 will be used to establish sufficient conditions for the gradient-like behavior of pendulum-like feedback-systems.

4.3   The Bakaev-Guzh Technique for Vector Fields

The purpose of this section is to derive sufficient conditions for gradient-like behavior of pendulum-like feedback equations using the Barbashin-Krassovskij theorem from Section 4.2 and general results on the global convergence of solutions of vector fields on Riemannian manifolds.

Let us consider equation (4.1.1) on the Riemannian manifold (M, G) and suppose that the set of equilibria of (4.1.1) is discrete.

The following Theorems 4.3.1 and 4.3.2 are borrowed from [97], [49].

Theorem 4.3.1 *Suppose that for system* (4.1.1) *there are satisfied the following assumptions:*

(i) *there exist a* C^1-*function* $V : M \to \mathbf{R}$, *a closed vector field* h *on* M *and a continuous function* $\vartheta : M \to \mathbf{R}_+$ *such that*

$$\langle \mathrm{grad}\, V(u) + h(u), f(u) \rangle \leq -\vartheta(u) \qquad \text{for all} \qquad u \in M;$$

(ii) *if* $\int\limits_0^{+\infty} \vartheta(\alpha(t,p))\, dt < \infty$ *for some bounded on* $\mathbf{R}_+$ *solution* $\alpha(\cdot, p)$ *of* (4.1.1) *then* $\alpha(t,p)$ *converges to an equilibrium for* $t \to +\infty$;

(iii) *there exists an $\varepsilon > 0$ such that for any closed piece-wise C^1-path γ in M, which is not contractible in M, it follows that*

$$\inf_{\overline{\gamma} \in [\gamma]} \int_{\overline{\gamma}} \frac{\vartheta}{|f|} ds \geq \varepsilon + \int_{\gamma} \langle h, du \rangle.$$

Then every solution of (4.1.1), bounded on $\mathbf{R}_+$, converges to an equilibrium as $t \to +\infty$.

Proof Assume that $\alpha(\cdot) = \alpha(\cdot, p)$ is an arbitrary bounded on $\mathbf{R}_+$ solution of (4.1.1). Along this solution we have for arbitrary $\overline{t} < \overline{\overline{t}}$

$$
\begin{aligned}
V(\alpha(\overline{\overline{t}})) - V(\alpha(\overline{t})) &= \int_{\overline{t}}^{\overline{\overline{t}}} \langle \operatorname{grad} V(\alpha(\tau)), f(\alpha(\tau)) \rangle \, d\tau \\
&= \int_{\overline{t}}^{\overline{\overline{t}}} \langle \operatorname{grad} V(\alpha(\tau)) + h(\alpha(\tau)), f(\alpha(\tau)) \rangle \, d\tau - \\
&\qquad - \int_{\overline{t}}^{\overline{\overline{t}}} \langle h(\alpha(\tau)), f(\alpha(\tau)) \rangle \, d\tau \\
&\leq -\int_{\overline{t}}^{\overline{\overline{t}}} \vartheta(\alpha(\tau)) \, d\tau - \int_{\overline{t}}^{\overline{\overline{t}}} \langle h(\alpha(\tau)), f(\alpha(\tau)) \rangle \, d\tau.
\end{aligned}
\tag{4.3.1}
$$

Let us choose a point $q \in \omega(p)$ and two sequences $\overline{t}_n \to +\infty$ and $\overline{\overline{t}}_n \to +\infty$ with $\alpha(\overline{t}_n) \to q$, $\alpha(\overline{\overline{t}}_n) \to q$ and $\overline{\overline{t}}_n - \overline{t}_n \to +\infty$ as $n \to \infty$. For every n we construct a closed piece-wise smooth path γ_n by adding to the orbit $\left\{ \alpha(t) : \overline{t}_n \leq t \leq \overline{\overline{t}}_n \right\}$ the smooth path $\Delta\gamma_n$ with the length $s(\Delta\gamma_n) = d(\alpha(\overline{t}_n), \alpha(\overline{\overline{t}}_n))$. This guarantees that

$$\left| \int_{\Delta\gamma_n} \langle h, du \rangle \right| < \varepsilon_n,$$

where $\varepsilon_n \to 0$ for $n \to +\infty$.

The following two situations are possible.

First, every cycle γ_n for $n \geq n_0$, n_0 sufficiently large, is contractible. Then by (4.3.1) we conclude that for all $n \geq n_0$

$$\int_{\overline{t}_n}^{\overline{\overline{t}}_n} \vartheta(\alpha(\tau)) \, d\tau \leq V(\alpha(\overline{t}_n)) - V(\alpha(\overline{\overline{t}}_n)) + \varepsilon_n.$$

Second, for any n_0 there exists a cycle γ_n with $n \geq n_0$ which is not contractible in M. Because of (4.3.1) this implies that there exists also a sequence of positive numbers $\{\delta_n\}$ satisfying $\delta_n \to 0$ and

$$
\begin{aligned}
\delta_n &\geq V(\alpha(\overline{t}_n)) - V(\alpha(\overline{\overline{t}}_n)) \\
&\geq \int_{\gamma_n} \vartheta \, d\tau - \varepsilon_n + \int_{\gamma_n} \langle h, du \rangle \\
&\geq \inf_{\overline{\gamma} \in [\gamma_n]} \int_{\overline{\gamma}} \frac{\vartheta}{|f|} ds - \varepsilon_n + \int_{\gamma_n} \langle h, du \rangle \\
&\geq \varepsilon - \varepsilon_n,
\end{aligned}
$$

where we have used assumption (iii).

It is clear, that for large n the inequality $\delta_n \geq \varepsilon - \varepsilon_n$ is impossible. Thus, the second case cannot occur.

From the above it follows that we can take a fixed $\bar{t}_{n_0}$ sufficiently large and $\bar{\bar{t}}_n \to +\infty$ such that the closed paths γ_n consisting of $\left\{\alpha(t) \; : \; \bar{t}_{n_0} \leq t \leq \bar{\bar{t}}_n\right\}$ and $\Delta\gamma_n$ as above are also contractable for n sufficiently large. It follows that we can suppose that there exists a constant c and a sequence $\bar{\bar{t}}_n \to +\infty$ with $\int\limits_{\bar{t}_{n_0}}^{\bar{\bar{t}}_n} \vartheta(\alpha(\tau))d\tau \leq c$.

Using now assumption (ii) of the theorem we get the stated assertion. $\blacksquare$

Let us consider now the pendulum-like feedback system

$$\begin{aligned}
\dot{z} &= Az + B\varphi(\sigma), \\
\dot{\sigma} &= C^*z + R\varphi(\sigma),
\end{aligned} \tag{4.3.2}$$

where A, B, C, and R are real matrices of orders $n \times n$, $n \times m$, $n \times m$ resp. $m \times m$. Suppose that the pair (A, B) is controllable, the pair (A, C) is observable, and the matrix A is Hurwitzian. Suppose also that $\varphi : \mathbf{R}^m \to \mathbf{R}^m$ is a vector-valued function having the components $\varphi_i(\sigma) = \varphi_i(\sigma_i)$ with $\sigma = (\sigma_1, \sigma_2, \ldots, \sigma_m)$. We assume that every component $\varphi_i : \mathbf{R} \to \mathbf{R}$ is Δ_i-periodic, satisfies a local Lipschitz condition and possesses a finite number of zeros on $[0, \Delta_i)$. Introduce the vector $d_i = [0, \ldots, 0, \Delta_i, 0, \ldots, 0]$, where Δ_i is the $(n+i)$-th component of d_i. Let us define the discrete group $\Gamma := \left\{\sum\limits_{j=1}^{m} k_j d_j \; : \; k_j \in \mathbf{Z}\right\}$ and the cylinder $\mathbf{R}^{n+m}/\Gamma$ equipped with the flat metric $\langle \cdot, \cdot \rangle$, which is produced by the standard metric in $\mathbf{R}^{n+m}$. Recall, that $K(s) = -R + C^*(A - sI)^{-1}B$ denotes the transfer function of system (4.3.2)

Theorem 4.3.2 *Suppose there exist diagonal $m \times m$ matrices $\kappa = \mathrm{diag}\{\kappa_1, \ldots, \kappa_m\}$, $\delta = \mathrm{diag}\{\delta_1, \ldots, \delta_m\}$ and $\varepsilon = \mathrm{diag}\{\varepsilon_1, \ldots, \varepsilon_m\}$ with $\delta > 0$ and $\varepsilon > 0$ satisfying the following conditions:*

(1) $\kappa \mathrm{Re}\, K(i\omega) - K(i\omega)^*\varepsilon K(i\omega) - \delta \geq 0$ for all $\omega \in \mathbf{R}$;

(2) $2\sqrt{\varepsilon_j \delta_j} \int\limits_0^{\Delta_j} |\varphi_j(\tau)|\, d\tau > \kappa_j \left|\int\limits_0^{\Delta_j} \varphi_j(\tau)\, d\tau\right|$ for $j = 1, 2, \ldots, m$.

Then every solution $\alpha(\cdot, p)$ of (4.3.2) converges to an equilibrium for $t \to +\infty$.

Proof In view of condition (1) we can apply the Yakubovich-Kalman theorem (Theorem 1.4.1, p. 9) to conclude that there exists a real matrix $H = H^*$ of order $n \times n$ satisfying the inequality

$$2z^*H(Az + B\xi) + \xi^*\kappa(C^*z + R\xi) + (C^*z + R\xi)^*\varepsilon(C^*z + R\xi) + \xi^*\delta\xi \leq 0 \tag{4.3.3}$$

for all $(z, \xi) \in \mathbf{R}^n \times \mathbf{R}^m$. Since A is Hurwitzian, we see by Lemma 1.2.1, p. 6, that H is positive definite. It is clear that every solution of (4.3.2), considered in the phase space $M = \mathbf{R}^{n+m}/\Gamma$, is bounded on $\mathbf{R}_+$. Introduce on M the function $V : M \to \mathbf{R}$ by $V(z, \sigma) = z^*Hz$ and the vector field $h : M \to TM$ by $h(z, \sigma) = [0, 0, \ldots, 0, \kappa\varphi(\sigma)]$ (the zero is written n-times).

Because of (4.3.3) we obtain for system (4.3.2) the relation

$$\langle \mathrm{grad}\, V(z, \sigma) + h(z, \sigma), f(z, \sigma)\rangle \leq -\vartheta(z, \sigma)$$

for all $(z, \sigma) \in M$, in which f denotes the right-hand side of (4.3.2) and

$$\vartheta(z, \sigma)) = (C^*z + R\varphi(\sigma))^*\varepsilon(C^*z + R\varphi(\sigma)) + \varphi(\sigma)^*\delta\varphi(\sigma).$$

It follows, that condition (i) of Theorem 4.3.1 is satisfied.

Suppose that $z(\cdot), \sigma(\cdot)$ is an arbitrary solution of (4.3.2) and

$$\int\limits_{0}^{+\infty} \varphi(\sigma(t))^*\delta\varphi(\sigma(t)) \, dt < +\infty.$$

By the Barbalat lemma (Theorem 2.1.3, p. 16), it follows that $\sigma(t) \to \sigma_0$ for $t \to +\infty$, where $\varphi(\sigma_0) = 0$. Since A is Hurwitzian and φ is periodic we conclude that $z(t) \to 0$ for $t \to +\infty$. Thus assumption (ii) of Theorem 4.3.1 is verified. It remains to verify condition (iii) of this theorem. It is sufficient to do this for the m generators of the fundamental group $\pi_1(\mathbf{R}^n \times S^1 \times \cdots \times S^1)$. Suppose that $\sigma_j(\cdot)$ is a parameter representation of the j-th generator defined on $[0, T_j]$. We have then

$$\int\limits_{0}^{T_j} \varepsilon_j\dot{\sigma}_j(t)^2 + \delta_j\varphi_j(\sigma_j(t))^2 \, dt = \int\limits_{0}^{\Delta_j} \varepsilon_j|\dot{\sigma}(t)| + \delta_j\frac{\varphi^2(\sigma_j)}{|\dot{\sigma}_j|} ds \geq 2\sqrt{\varepsilon_j\delta_j} \int\limits_{0}^{\Delta_j} |\varphi_j(\tau)| \, d\tau.$$

Employing the fact, that

$$\int\limits_{\sigma_j} \langle h, du \rangle \leq \kappa_j \left| \int\limits_{0}^{\Delta_j} \varphi_j(\tau) \, d\tau \right|,$$

we conclude by assumption (2) of the given theorem that the condition (iii) of Theorem 4.3.1 is satisfied. $\blacksquare$

Let us suppose now in addition to the above that the nonlinearity φ in (4.3.2) belongs to $\mathbf{C}^1$. W. l. o. g. we can assume that there are numbers $\mu_{1j} < 0 < \mu_{2j}$ $(j = 1, 2, \ldots, m)$ such that

$$\mu_{1j} \leq \frac{d}{d\sigma}\varphi_j(\sigma) \leq \mu_{2j} \tag{4.3.4}$$

for all $\sigma \in \mathbf{R}$. Suppose also that for any $j = 1, 2, \ldots, m$ we have $\varphi_j(\sigma) \not\equiv 0$ and

$$\sqrt{[1 - \mu_{1j}^{-1}\varphi_j'(\sigma)][1 - \mu_{2j}^{-1}\varphi_j'(\sigma)]} \not\equiv 0.$$

Define now for any $j = 1, 2, \ldots, m$ the values

$$\nu_j = \int\limits_{0}^{\Delta_j} \varphi_j(\sigma)d\sigma \quad / \quad \int\limits_{0}^{\Delta_j} |\varphi_j(\sigma)|d\sigma$$

and the $m \times m$ matrix $\nu = \operatorname{diag}(\nu_1, \ldots, \nu_m)$. Furthermore, let us define the values

$$\nu_{0j} = \int\limits_{0}^{\Delta_j} \varphi_j(\sigma)d\sigma \quad / \quad \int\limits_{0}^{\Delta_j} \sqrt{[1 - \mu_{1j}^{-1}\varphi_j'(\sigma)][1 - \mu_{2j}^{-1}\varphi_j'(\sigma)]}|\varphi_j(\sigma)|d\sigma$$

and the $m \times m$ matrix $\nu_0 = \operatorname{diag}(\nu_{01}, \ldots, \nu_{0m})$.

The next theorem is taken from [99].

Theorem 4.3.3 *Suppose that there exist diagonal $m \times m$ matrices $\varepsilon \geq 0$, $\delta \geq 0$, $\tau \geq 0$, κ such that*

$$\mathrm{Re}\left\{\kappa K(i\omega) - K(i\omega)^*\varepsilon K(i\omega) - [K(i\omega) + \mu_1^{-1}i\omega]^*\tau[K(i\omega) + \mu_2^{-1}i\omega]\right\} \geq \delta \qquad (4.3.5)$$

for all $\omega \in \mathbf{R}$ and at least one of the following two conditions holds:

(i) $4\varepsilon\delta > (\kappa\nu)^2$;

(ii) $4\tau\delta > (\kappa\nu_0)^2$.

Then system (4.3.2) is gradient-like.

Proof Let us introduce the designations

$$P = \begin{bmatrix} A & B \\ 0 & 0 \end{bmatrix}, \quad Q = \begin{bmatrix} 0 \\ I \end{bmatrix}, \quad D = \begin{bmatrix} C \\ R^* \end{bmatrix}, \quad Y = \begin{bmatrix} z \\ \varphi(\sigma) \end{bmatrix}$$

where P, Q, D are matrices of order $(n+m) \times (n+m)$, $(n+m) \times m$, $(n+m) \times m$, respectively and $(z,\sigma) \in \mathbf{R}^n \times \mathbf{R}^m$. It is clear that an arbitrary solution $z(\cdot)$, $\sigma(\cdot)$ of (4.3.2) satisfies for $t \geq 0$ the system

$$\dot{y}(t) = Py(t) + Q\xi(t), \qquad \dot{\sigma}(t) = D^*y(t),$$

where $\xi(t) = \frac{d}{dt}\varphi(\sigma(t))$ and $y(t) = [z(t), \varphi(\sigma(t))]^\mathsf{T}$.

Consider the quadratic form $G : \mathbf{R}^{n+m} \times \mathbf{R}^m \to \mathbf{R}$, defined by

$$G(y,\xi) = y^*Q\kappa D^*y + y^*D\varepsilon D^*y + y^*Q\delta Q^*y - (D^*y - \mu_1^{-1}\xi)^*\tau(\mu_2^{-1}\xi - D^*y).$$

From the representation

$$[Q, PQ, \ldots, P^{n+m-1}Q] = \begin{bmatrix} 0 & B & AB & \ldots & A^{n+m-2}B \\ I & 0 & 0 & \ldots & 0 \end{bmatrix}$$

it follows that $\mathrm{rank}\,[Q, PQ, \ldots, P^{n+m-1}Q] = n + m$, i. e. the pair (P,Q) is controllable. Using this fact and the frequency-domain condition (4.3.5) we get by Theorem 1.1.4, p. 4, that there exists an $(n+m) \times (n+m)$ matrix $H = H^*$ such that

$$2y^*H[Py + Q\xi] + G(y,\xi) \leq 0 \qquad (4.3.6)$$

for all $(y,\xi) \in \mathbf{R}^{n+m} \times \mathbf{R}^m$.

Indeed, it is evident that for every $s \notin \sigma(A)$

$$\begin{bmatrix} A - sI_n & B \\ 0 & -sI_n \end{bmatrix}\begin{bmatrix} (A - sI_n)^{-1}B \\ -I_m \end{bmatrix} = \begin{bmatrix} 0 \\ sI_m \end{bmatrix}$$

(For clarity the index of I_k denotes here the order of the unit matrix).

Multiplying both parts of this equation from left by the matrix

$$(P - sI_{n+m})^{-1} = \begin{bmatrix} A - sI_n & B \\ 0 & -sI_m \end{bmatrix}^{-1}$$

we obtain for $s \notin \sigma(A)$ the equation

$$s(P - sI_{n+m})^{-1}Q = \begin{bmatrix} (A - sI_n)^{-1}B \\ -I_m \end{bmatrix}.$$

72

From this it follows that

$$D^*(P - sI_{n+m})^{-1}Q = \frac{1}{s}K(s) \qquad (4.3.7)$$

and

$$Q^*(P - sI_{n+m})^{-1}Q = -\frac{1}{s}I_m. \qquad (4.3.8)$$

Using now (4.3.7) and (4.3.8) we get the identity

$$G_{\mathbf{C}}[(sI_{n+m} - P)^{-1}Q\xi, \xi] \equiv -\frac{1}{|s|^2}\mathrm{Re}\,\xi^*\{\kappa K(s) - K(s)^*\varepsilon K(s) - \delta - $$
$$-[K(s) + \mu_1^{-1}s]^*\tau[K(s) + \mu_2^{-1}s]\}\xi.$$

It follows then by (4.3.5) that the frequency-domain condition of Theorem 1.1.4, p. 4, is satisfied.

Because of (4.3.6) we have for all $z \in \mathbf{R}^n$, $\eta \in \mathbf{R}^m$, $\xi \in \mathbf{R}^m$ the inequality

$$2\begin{bmatrix} z \\ \eta \end{bmatrix}^* H \begin{bmatrix} Az + B\eta \\ \xi \end{bmatrix} + \eta^*\kappa(C^*z + R\eta) - [(C^*z + R\eta) - \mu_1^{-1}\xi]^*\tau[\mu_2^{-1}\xi - (C^*z + R\eta)]$$
$$+\eta^*\delta\eta + [C^*z + R\eta]^*\varepsilon[C^*z + R\eta] \leq 0. \qquad (4.3.9)$$

Let us put in (4.3.9) $\eta = 0$ and $\xi = 0$. It follows then that

$$2z^*H_{11}Az \leq -[C^*z]^*(\tau + \varepsilon)C^*z \qquad (4.3.10)$$

for all $z \in \mathbf{R}^n$, where it is supposed that H has the form

$$H = \begin{bmatrix} H_{11} & H_{12} \\ H_{21} & H_{22} \end{bmatrix}$$

and H_{11} is of order $n \times n$. Since A is Hurwitzian, $\tau + \varepsilon > 0$ and the pair (A, C) is observable it follows from (4.3.10) (Lemma 1.2.1, p. 6), that $H_{11} > 0$.

Let us define now the function $W : \mathbf{R}^n \times \mathbf{R}^m \to \mathbf{R}$ by $W(z, \xi) = \begin{bmatrix} z \\ \xi \end{bmatrix}^* H \begin{bmatrix} z \\ \xi \end{bmatrix}$. Note that the function $(z, \sigma) \to W(z, \varphi(\sigma))$ is periodic in σ and satisfies

$$W(z, \varphi(\sigma)) \to +\infty \quad \text{as} \quad |z| \to +\infty, \qquad (4.3.11)$$

since φ is bounded.

The derivative of W along an arbitrary solution $z(\cdot)$, $\sigma(\cdot)$ of (4.3.2) is given by

$$\frac{d}{dt}W(z(t), \varphi(\sigma(t))) = 2\begin{bmatrix} z(t) \\ \varphi(\sigma(t)) \end{bmatrix}^* H \begin{bmatrix} Az(t) + B\varphi(\sigma(t)) \\ \frac{d}{dt}\varphi(\sigma(t)) \end{bmatrix}.$$

According to (4.3.9) we get the inequality

$$\frac{d}{dt}W(z(t), \varphi(\sigma(t))) \leq -\varphi(\sigma(t))^*\kappa\dot\sigma(t) + [\dot\sigma(t) - \mu_1^{-1}\varphi'(\sigma(t))\dot\sigma(t)]^*$$
$$\tau[\mu_2^{-1}\varphi'(\sigma(t))\dot\sigma(t) - \dot\sigma(t)] - \varphi(\sigma(t))^*\delta\varphi(\sigma(t)) - \dot\sigma(t)^*\varepsilon\dot\sigma(t) \qquad (4.3.12)$$

for $t \geq 0$.

In order to consider case (i) we introduce the function $V : \mathbf{R}^n \times \mathbf{R}^m \to \mathbf{R}$ defined by

$$V(z, \sigma) = W(z, \varphi(\sigma)) + \int_0^\sigma [\varphi(\tau) - \nu|\varphi|(\tau)]^*\kappa\,d\tau$$

where

$$|\varphi|(\sigma)^* := (|\varphi_1(\sigma_1)|, |\varphi_2(\sigma_2)|, \ldots, |\varphi_m(\sigma_m)|).$$

This function is Δ-periodic in σ because of

$$\int_0^{\Delta_j} (\varphi_j(t)) - \nu_j|\varphi_j(t)|\, dt = 0 \qquad (j = 1, 2, \ldots, m)$$

and

$$V(z, \sigma) \to +\infty \quad \text{as} \quad |z| \to \infty \tag{4.3.13}$$

independently of σ, since (4.3.11) is satisfied:

According to (4.3.4) and (4.3.12) we can write now

$$\begin{aligned}
\frac{d}{dt}V(z(t), \sigma(t)) &\leq -\varphi(\sigma(t))^*\kappa\dot\sigma(t) - \varphi(\sigma(t))^*\delta\varphi(\sigma(t)) - \dot\sigma(t)^*\varepsilon\dot\sigma(t) - \\
&\quad -[\varphi(\sigma(t)) - \nu|\varphi|(\sigma(t))]^*\kappa\dot\sigma(t) \\
&= -\varphi(\sigma(t))^*\delta\varphi(\sigma(t)) - \dot\sigma(t)^*\varepsilon\dot\sigma(t) - [\nu|\varphi|(\sigma(t))]^*\kappa\dot\sigma(t).
\end{aligned}$$

From this inequality and assumption (i) it follows that there are positive constants α_1 and α_2 such that

$$\frac{d}{dt}V(z(t), \sigma(t)) \leq -\alpha_1|\varphi(\sigma(t))|^2 - \alpha_2|\dot\alpha(t)|^2 \tag{4.3.14}$$

for $t \geq 0$. Hence V cannot increase along solutions of (4.3.2). Moreover, if for some solution $z(\cdot)$, $\sigma(\cdot)$ of (4.3.2) and some $T > 0$ $V(z(T), \sigma(T)) = V(z(0), \sigma(0))$ we get immediately from (4.3.14) that $\varphi(\sigma(t)) \equiv 0$ on $[0, T]$ and consequently $[\varphi(\sigma(t)) - \nu|\varphi|(\sigma(t))] = 0$ on $[0, T]$. As zeros of φ are isolated and $\sigma(\cdot)$ is continuous it is clear that $\sigma(t) \equiv \sigma_0$ on $[0, T]$, where σ_0 is a zero of φ. It follows than that $V(z(t), \sigma_0) = W(z(t), 0) = const$ on $[0, T]$ and, consequently, $C^*z(t) \equiv 0$ on $[0, T]$. Using the fact that (A, C) is observable, we get $z(t) \equiv 0$ on $[0, T]$. Thus all the assumptions of Theorem 4.2.1 are fulfilled and Theorem 4.3.3 is proved in the case (i).

Let us pass now to the case (ii) and consider the function $V_1 : \mathbf{R}^n \times \mathbf{R}^m \to \mathbf{R}$ defined by

$$V_1(z, \sigma) = W(z, \varphi(\sigma)) + \int_0^\sigma [\varphi(\sigma) - \nu_0|\widetilde\varphi\varphi|(\sigma)]^*\kappa\, d\sigma,$$

where

$$\widetilde\varphi(\sigma) := \sqrt{[I - \mu_1^{-1}\varphi'(\sigma)]^*[I - \mu_2^{-1}\varphi'(\sigma)]} \qquad (\sigma \in \mathbf{R}^m)$$

and

$$|\widetilde\varphi\varphi|(\sigma)^* := (|(\widetilde\varphi(\sigma)\varphi(\sigma))_1|, |(\widetilde\varphi(\sigma)\varphi(\sigma))_2|, \ldots, |(\widetilde\varphi(\sigma)\varphi(\sigma))_m|).$$

The function V_1 is also periodic in σ since for $j = 1, 2, \ldots, m$

$$\int_0^{\Delta_j} \left[\varphi_j(\sigma) - \nu_{0j}\sqrt{(1 - \mu_{1j}^{-1}\varphi_j'(\sigma))(1 - \mu_{2j}^{-1}\varphi_j'(\sigma))}|\varphi_j(\sigma)|\right] d\sigma = 0.$$

The property (4.3.13) is also fulfilled. Using (4.3.12) we get along a solution $(z(\cdot), \sigma(\cdot))$ of (4.3.2) the inequality

$$\begin{aligned}
\frac{d}{dt}V_1(z(t), \sigma(t)) &= \frac{d}{dt}W(z(t), \varphi(\sigma(t)) + [\varphi(\sigma(t)) - \nu_0|\widetilde\varphi\varphi|(\sigma(t))]^*\kappa\dot\sigma(t) \\
&\leq -\varphi(\sigma(t))^*\delta\varphi(\sigma(t)) - \dot\sigma(t)^*\varepsilon\dot\sigma(t) - [\nu_0|\widetilde\varphi\varphi|(\sigma(t))]^*[\widetilde\varphi(\sigma(t))\dot\sigma(t)]^*\tau\widetilde\varphi(\sigma(t))\dot\sigma(t)
\end{aligned}$$

for $t \geq 0$.

In virtue of assumption (ii) and the last inequality there exist two positive numbers β_1 and β_2 such that

$$\frac{d}{dt}V_1(z(t),\sigma(t)) \leq -\dot{\sigma}(t)^*\varepsilon\dot{\sigma}(t) - \beta_1|\varphi(\sigma(t))|^2 - \beta_2|\widetilde{\varphi}(\sigma(t))\dot{\sigma}(t)|^2$$

for $t \geq 0$.

The remaining part of the proof is the same as in the case (i). $\blacksquare$

Let us consider now examples which show how Theorem 4.3.3 can be used to get convergence region estimates in terms of system parameters.

Example 4.3.1 Consider the scalar second-order equation

$$\ddot{\sigma} + a\dot{\sigma} + \varphi(\sigma) = 0 \tag{4.3.15}$$

with parameter $a > 0$ and the smooth and Δ-periodic function $\varphi : \mathbf{R} \to \mathbf{R}$. Suppose that $\varphi'(\sigma) \leq a/k$ for all $\sigma \in \mathbf{R}$, where $k > 0$ is constant. Suppose also that φ has a finite number of zeros on $[0,\Delta]$, $\varphi(\sigma) \not\equiv 0$ and $1 - \frac{k}{a}\varphi'(\sigma) \not\equiv 0$.

The transfer function K of (4.3.15) with respect to the nonlinearity φ is $K(s) = 1/(s+a)$. In order to apply Theorem 4.3.3 we consider the inequality (4.3.5) with $\kappa = 1$, $\mu_1^{-1} = 0$ and $\mu_2^{-1} = k/a$ in the form

$$\omega^2(\tau\frac{k}{a} - \delta) + (a - \varepsilon - \tau - \delta a^2) \geq 0 \tag{4.3.16}$$

for $\omega \in \mathbf{R}$. Let us choose $\varepsilon = 0$, $\delta = k/(ak+1)$ and $\tau = a/(ak+1)$. Then (4.3.16) is true and the only condition for gradient-like behavior of (4.3.15) is condition (ii) of Theorem 4.3.3, i. e.

$$4ka/(1+ka)^2 > \frac{\int\limits_0^\Delta \varphi(\sigma)\,d\sigma}{\int\limits_0^\Delta \sqrt{1 - \frac{k}{a}\varphi'(\sigma)}|\varphi(\sigma)|\,d\sigma}. \tag{4.3.17}$$

Example 4.3.2 Let us consider system (4.3.2) with $m = n = 1$, $\varphi \in C^1$ a Δ-periodic function with $\varphi(\sigma) \not\equiv 0$ and the transfer function

$$K(s) = T\frac{1+mTs}{1+Ts}, \tag{4.3.18}$$

where $T > 0$ and $m \in (0,1)$ are constants. This system describes an autonomous second-order PLL with proportional-integrating filter [140].

It is evident that $\chi(s) \equiv \frac{1}{s}K(s)$ is non-degenerate and thus the pair (A,B) in (4.3.2) is controllable. It is also clear that since $\det(sI - A) = s + T^{-1}$ matrix A is Hurwitzian Let us take $\kappa = 1$, $\tau = 0$ in order to verify (4.3.5). Then the latter inequality takes the form

$$T(1 + mT^2\omega^2) - \varepsilon T^2(1 + m^2T^2\omega^2) - \delta(1 + T^2\omega^2) \geq 0$$

for all $\omega \in \mathbf{R}$ or, what is the same,

$$\omega^2(-\delta T^2 + mT^3 - \varepsilon m^2T^4) - (T - \varepsilon T^2 - \delta) \geq 0 \tag{4.3.19}$$

for all $\omega \in \mathbf{R}$.

We require also that the parameters ε and δ satisfy the inequality

$$4\varepsilon\delta > \left(\int_0^\Delta \varphi(\vartheta)\,d\vartheta \; / \; \int_0^\Delta |\varphi(\vartheta)|\,d\vartheta \right)^2 =: \nu^2. \tag{4.3.20}$$

Let us choose the parameter ε and δ in such a way that the second-order polynomial in (4.3.19) is identical to zero, i. e. we take

$$\varepsilon = \frac{1}{T(m+1)} \quad\text{and}\quad \delta = T(1-\varepsilon T) = \frac{mT}{m+1}.$$

Then condition (4.3.20) takes the form

$$\frac{2\sqrt{m}}{1+m} > |\nu|. \tag{4.3.21}$$

In the special case of a sinusoidal characteristic $\varphi(\sigma) = \sin\sigma - \gamma$ with $\gamma \in (0,1)$ we have

$$|\nu| = \frac{\pi\gamma}{2\left(\gamma\arcsin\gamma + \sqrt{1-\gamma^2}\right)}. \tag{4.3.22}$$

Thus inequality (4.3.21) with $|\nu|$ from (4.3.22) guarantees the gradient-like behavior of system (4.3.2), (4.3.17) with $\varphi(\sigma) = \sin\sigma - \gamma$. In the next chapters we continue the stability investigation of system (4.3.2), (4.3.18).

Example 4.3.3 Let in (4.3.2) be $n = 2$, $m = 1$, $\varphi \in C^1$ Δ-periodic and

$$K(s) = \frac{1 + \beta_1\alpha_1 s + \beta_2\alpha_2 s^2}{1 + \alpha_1 s + \alpha_2 s^2}, \tag{4.3.23}$$

where α_1, α_2, β_1 and β_2 are positive parameters and

$$\beta_1 < \beta_2 < 1. \tag{4.3.24}$$

System (4.3.2), (4.3.23) describes the dynamics of an autonomous PLL with a second-order filter of type "2/2" [140]. By (4.3.23) the transfer-function $\chi(s) = \frac{1}{s}K(s)$ is non-degerate. Thus the pair (A, B) in (4.3.2) is controllable. Since $\det(sI - A) = s^2 + \alpha_1\alpha_2^{-1}s + \alpha_2^{-1}$, matrix A is Hurwitzian. In order to verify (4.3.5) we take $\kappa = 1$, $\tau = 0$ and receive

$$\begin{aligned} \omega^4\big(\beta_2\alpha_2^2 - \varepsilon\beta_2^2\alpha_2^2 - \delta\alpha_2^2\big) + \\ +\omega^2\big(-\beta_2\alpha_2 - \alpha_2 + \beta_1\alpha_1^2 + 2\varepsilon\beta_2\alpha_2 - \varepsilon\beta_1^2\alpha_1^2 + 2\delta\alpha_2 - \delta\alpha_1^2\big) + 1 - \delta - \varepsilon \;\geq\; 0 \end{aligned} \tag{4.3.25}$$

for $\omega \in \mathbf{R}$.

Let us choose $\varepsilon = 1 - \delta$ and $\delta = \dfrac{\alpha_1^2(1-\beta_1)\beta_1 - \alpha_2(1-\beta_2)}{\alpha_1^2(1-\beta_1^2) - 2\alpha_2(1-\beta_2)}$.

Now the free term and the coefficient of ω^2 in the left-hand side of (4.3.25) are equal to zero. Note that if

$$\alpha_1^2 > \frac{\alpha_2(1-\beta_2)}{\beta_1(1-\beta_1)} \tag{4.3.26}$$

then δ is positive and the coefficient of ω^4 at the left-hand side of (4.3.25) is also positive. So if (4.3.26) holds then (4.3.25) is satisfied. Condition (i) of Theorem 4.3.3 takes the form

$$4\frac{[\alpha_1^2(1-\beta_1)\beta_1 - \alpha_2(1-\beta_2)][\alpha_1^2(1-\beta_1) - \alpha_2(1-\beta_2)]}{[\alpha_1^2(1-\beta_1^2) - 2\alpha_2(1-\beta_2)]^2} > \nu^2, \tag{4.3.27}$$

where ν^2 is defined in (4.3.20). Thus if (4.3.27) and (4.3.26) hold then system (4.3.2), (4.3.23) is gradient-like. Consider $\varphi(\sigma) = \sin\sigma - \gamma$ with $\gamma \in (0,1)$ and $|\nu|$ calculated by (4.3.20) for this φ. It is then not difficult to establish that, for instance, for $\alpha_1 = 300$, $\beta_1 = 0.2$, $\beta_2 = 0.5$, $\alpha_2 = 5$ and $\gamma \le 0.54$ the conditions (4.3.26) and (4.3.27) are satisfied. Note that in paper [31] where the same system with α_1, α_2, β_1 and β_2 as above is considered with the help of qualitative-numerical methods the gradient-like behavoir is shown for $\gamma \le 0.6$.

Remark 4.3.1 The statement of Theorem 4.3.1 is an interconnection of a Lyapunov function assumption (i) an L^p-type assumption (ii) and a topological assumption (iii). The assumption (ii) is not always easy to verify. If we replace this assumption by an additional requirement on V of Lyapunov type we get the following convergence result in the spirit of LaSalle, which is, together with Theorem 4.3.5, due to R. W. Brockett [40].

Theorem 4.3.4 *Suppose that the following assumptions are satisfied:*

(i) *there exist a C^1-function $V : M \to \mathbf{R}$ and a closed vector field h on M such that*

$$\langle \operatorname{grad} V(u) + h(u), f(u)\rangle \le 0 \quad \text{for all } u \in M;$$

(ii) *if for a solution $\alpha(\cdot)$ of (4.1.1) there exists a $\tau > 0$ with*

$$\langle \operatorname{grad} V(\alpha(t)) + h(\alpha(t)), f(\alpha(t))\rangle = 0 \quad \text{on } [0,\tau]$$

it follows that $\alpha(\cdot)$ is a stationary solution of (4.1.1);

(iii) *there exists an $\varepsilon > 0$ such that for any closed piece-wise C^1-path γ in M, which is not contractible in M, it follows that*

$$\sup_{\bar{\gamma} \in [\gamma]} \int_{\bar{\gamma}} \frac{\langle \operatorname{grad} V + h, f\rangle}{|f|}\, ds \le -\varepsilon + \int_{\gamma} \langle h, du\rangle.$$

Then every bounded on $\mathbf{R}_+$ solution of (4.1.1) converges to an equilibrium as $t \to +\infty$.

Proof Consider the bounded solution $\alpha(\cdot, p)$ and define the compact and positively invariant set $N = \gamma^+(p) \cup \omega(p) = \overline{\gamma^+(p)}$, where $\gamma^+(p)$ is the orbit through p and $\omega(p)$ is the ω-limit set of $\alpha(\cdot, p)$. We define a function Ψ on $\gamma^+(p)$ by

$$\Psi(v) = \int_0^t \langle \operatorname{grad} V(\alpha(\tau)) + h(\alpha(\tau)), f(\alpha(\tau))\rangle\, d\tau,$$

where $\alpha(t) = v$.

Since the trajectory through p winds densely through $\omega(p)$ this defines Ψ on a dense subset of $\omega(p)$. Thus Ψ can be extended by continuity to a function on $\gamma^+(p) \cup \omega(p)$.

To see this notice that if $\alpha(t_1)$ and $\alpha(t_2)$ are two points on the trajectory of $\alpha(t) = \alpha(t,p)$ we have

$$
\begin{aligned}
V(\alpha(t_2)) - V(\alpha(t_1)) &= \int_{t_1}^{t_2} \langle \operatorname{grad} V(\alpha(\tau)), f(\alpha(\tau))\rangle\, d\tau \\
&= \int_{t_1}^{t_2} \langle \operatorname{grad} V(\alpha(\tau)) + h(\alpha(\tau)), f(\alpha(\tau))\rangle\, d\tau - \int_{t_1}^{t_2} \langle h(\alpha(\tau)), f(\alpha(\tau))\rangle\, d\tau \\
&\le \sup_{\gamma \in P[\alpha(t_1),\alpha(t_2)]} \int_{\gamma} \frac{\langle \operatorname{grad} V + h, f\rangle}{|f|}\, ds - \int_{T[\alpha(t_1),\alpha(t_2)]} \langle h, du\rangle,
\end{aligned}
$$

$$(4.3.28)$$

where $P[\alpha(t_1), \alpha(t_2)]$ denotes the set of piece-wise smooth pathes from $\alpha(t_1)$ to $\alpha(t_2)$ and $T[\alpha(t_1), \alpha(t_2)]$ the trajectory of (4.1.1) from $\alpha(t_1)$ to $\alpha(t_2)$.

There exists a positive number a such that for every $u_0 \in N$ the set $\{u \in M \; : \; d(u_0, u) \leq a\}$ is contractible.

Let $|\operatorname{grad} V(u) + h(u)| + |h(u)| \leq k$ on N and define

$$b(\varepsilon) := \min\left\{a, \frac{\varepsilon}{k}\right\}.$$

Suppose now that $\bar{t}_n$ and $\bar{\bar{t}}_n$ are two sequences with $\bar{t}_n < \bar{\bar{t}}_n$ and

$$\lim_{n \to \infty} \alpha(\bar{t}_n) = \lim_{n \to \infty} \alpha(\bar{\bar{t}}_n) = q \in \omega(p).$$

From the inequality (4.3.28) we have

$$\begin{aligned}
0 &= \lim_{n \to \infty} V(\alpha(\bar{\bar{t}}_n)) - V(\alpha(\bar{t}_n) \\
&\leq \sup_{\gamma \in P[\alpha(\bar{t}_n), \alpha(\bar{\bar{t}}_n)]} \int_\gamma \frac{\langle \operatorname{grad} V + h, f \rangle}{|f|} ds \; -- \int_{T[\alpha(\bar{t}_n), \alpha(\bar{\bar{t}}_n)]} \langle h, du \rangle.
\end{aligned}$$

Given the ε of (iii) there exists an n such that $d(\alpha(\bar{t}_n), \alpha(\bar{\bar{t}}_n)) < b(\varepsilon)$ and hence we can construct a closed curve using the trajectory from $\alpha(\bar{t}_n)$ to $\alpha(\bar{\bar{t}}_n)$ and a non-trajectory element of length at most $b(\varepsilon)$. Denote this closed curve by γ_n. Then an obvious estimates gives

$$0 \leq \sup_{\overline{\gamma} \in [\gamma_n]} \int_{\overline{\gamma}} \frac{\langle \operatorname{grad} V + h, f \rangle}{|f|} ds - \int_{\gamma_n} \langle h, du \rangle + kb(\varepsilon) \leq -\varepsilon + kb(\varepsilon) < 0,$$

so if γ_n is non-contractible a contradiction. The alternative is that for all $m \geq n$ the closed curves γ_m constructed above are contractible. We have

$$\begin{aligned}
|\Psi(\alpha(\bar{\bar{t}}_m)) - \Psi(\alpha(\bar{t}_m))| &= \left| \int_{\bar{t}_m}^{\bar{\bar{t}}_m} \langle \operatorname{grad} V(\alpha(\tau)) + h(\alpha(\tau)), f(\alpha(\tau)) \rangle \, d\tau \right| \\
&\leq -\int_{\gamma_m} \langle \operatorname{grad} V + h, du \rangle + kd(\alpha(\bar{t}_m), \alpha(\bar{\bar{t}}_m)) \\
&= kd(\alpha(\bar{t}_m), \alpha(\bar{\bar{t}}_m)).
\end{aligned}$$

This establishes the continuity of Ψ on $\omega(p)$.

Using now hypotheses (i) and (ii) of the theorem we see that we have on $N \supset \omega(p)$ a continuous function Ψ which is non-increasing along the trajectories of (4.1.1) in N and not identically zero along a non-constant trajectory. Lemma 4.2.2 allows us to conclude that the trajectory of $\alpha(\cdot, p)$ approaches an equilibrium. $\blacksquare$

Corollary 4.3.1 *Suppose that the vector field f is closed and there exists an $\varepsilon > 0$ such that for every non-contractible path γ in M we have*

$$\inf_{\overline{\gamma} \in [\gamma]} \int_{\overline{\gamma}} |f| ds > \varepsilon + \int_\gamma \langle f, du \rangle. \tag{4.3.29}$$

Then every bounded solution converges to an equilibrium.

Proof Since f is closed we can suppose that this vector field has the form $f = -\operatorname{grad} V - h$, where $V : M \to \mathbf{R}$ is C^1 and h is closed.

It follows that with the use of these V and h condition (i) of Theorem 4.3.4 takes the form

$$\langle \operatorname{grad} V(u) + h(u), f(u) \rangle = -|f(u)|^2 \leq 0$$

on M.

Suppose that for a solution $\alpha(\cdot)$ of (4.1.1) we have

$$\langle \operatorname{grad} V(\alpha(t)) + h(\alpha(t)), f(\alpha(t)) \rangle = 0$$

on $[0, \tau]$. Hence, $|f(\alpha(t))|^2 = 0$ on $[0, \tau]$. Since $\mathcal{E}$ is discrete it follows that $\alpha(t) \equiv const$ on $\mathbf{R}$.

Condition (iii) of Theorem 4.3.4 is fulfilled if there exists an $\varepsilon > 0$ such that for every non-contractible γ

$$\inf_{\overline{\gamma} \in [\gamma]} \int_{\overline{\gamma}} |f| ds > \varepsilon + \int_{\gamma} \langle -h, du \rangle$$

which is the same as in the statement of the Corollary 4.3.1. $\blacksquare$

Let us consider now a pendulum-like system in the first canonical form

$$\dot{u} = Pu + Q\varphi(\sigma), \qquad \sigma = R^* u. \tag{4.3.30}$$

We suppose that the matrices P, Q, and R are of order $n \times n$, $n \times m$ and $n \times m$, respectively, $\varphi : \mathbf{R}^m \to \mathbf{R}^m$ is of class C^1 and the Jacobi matrix $\varphi'(\cdot)$ is symmetric. Suppose also that $P = P^*$, $Q = R$ and there exist m linearly independent vectors $d_1, \ldots, d_m \in \mathbf{R}^n$ satisfying for $j = 1, \ldots, m$ the properties

$$\varphi(Q^* u + Q^* d_j) = \varphi(Q^* u) \tag{4.3.31}$$

for all $u \in \mathbf{R}^n$ and

$$P d_j = 0 \tag{4.3.32}$$

Under these conditions system (4.3.30) may be regarded as a closed vector field on $\mathbf{R}^n/\Gamma$ with $\Gamma = \left\{ \sum_{j=1}^m k_j d_j \ : \ k_j \in \mathbf{Z} \right\}$. Our aim is to deduce on the base of Corollary 4.3.1 sufficient conditions for the gradient-like behavior of (4.3.30) in the "output-space" $\{\sigma\}$.

For this reason we assume additionally that the matrix P is negative semi-definite ($P \leq 0$) and the vector $Q^* d_j$ ($j = 1, 2, \ldots, m$) form an orthogonal basis of $\mathbf{R}^m$. Define the matrix $R_0 := \lim_{s \to \infty} s Q^* (sI - P)^{-1} Q = Q^* Q$. This matrix is positive definite. Indeed, if we suppose that there exists some $d \in \mathbf{R}^m$, $d \neq 0$, with $Qd = 0$ then from the basis property it follows that $d = \sum_{j=1}^m k_j Q^* d_j$ with $k_j \in \mathbf{Z}$ and $0 = d_j^* Qd = k_j (Q^* d_j)^2$ for $j = 1, 2, \ldots, m$. Thus $k_j = 0$ for all j. It is also evident that $Q^* Q \neq 0$ ($d_j^* QQ^* d_j > 0$).

In the statement of the following theorem $\lambda_{min}(A)$ and $\lambda_{max}(A)$ denotes the minimal and maximal eigenvalue of a matrix $A = A^*$, respectively.

Theorem 4.3.5 *Assume that the equilibria of* (4.3.30) *are isolated and there exists an $\varepsilon > 0$ such that*

$$\sqrt{\lambda_{min}(Q^* Q)} \inf_{\overline{\gamma} \in [\gamma]} \int_{\overline{\gamma}} |\varphi| ds > \varepsilon \sqrt{\lambda_{max}(Q^* Q)} \int_{\gamma} \langle \varphi(u), du \rangle \tag{4.3.33}$$

for any non-contractible curve γ in $\mathbf{R}^m/\Gamma'$, where

$$\Gamma' = \left\{ \sum_{j=1}^m k_j Q^* d_j \ : \ k_j \in \mathbf{Z} \right\}.$$

Then system (4.3.30) *is gradient-like.*

Proof The desered conclusion follows from Corollary 4.3.1 if there is an $\varepsilon > 0$ such that

$$\inf_{\overline{\gamma}\in[\gamma]} \int_{\overline{\gamma}} |Pu + Q\varphi(Q^*u)|ds > \varepsilon + \int_{\gamma} \langle Q\varphi(Q^*u), du \rangle \tag{4.3.34}$$

for any non-contractible path γ in $\mathbf{R}^n/\Gamma$.

Since $P = P^*$ there exists an orthogonal change of coordinates so that P is diagonal with its last m diagonal elements zero. Clearly, in such a coordinate system

$$|Pu + Q\varphi(Q^*u)| \geq |Q_l\varphi(Q^*u)| \geq \sqrt{\lambda_{min}(Q_l^*Q_l)}|\varphi(Q^*u)|,$$

where Q_l denotes the $m \times m$ matrix consisting of the last m rows of Q. Obviously, $Q^*Q = Q_l^*Q_l$.

Thus its suffices to show that there is an $\varepsilon > 0$ such that

$$\sqrt{\lambda_{min}(Q^*Q)} \inf_{\overline{\gamma}\in[\gamma]} \int_{\overline{\gamma}} |Q\varphi(Q^*u)|ds > \varepsilon + \int_{\gamma} \langle Q\varphi(Q^*u), du \rangle \tag{4.3.35}$$

for any non-contractible path in $\mathbf{R}^n/\Gamma$.

We introduce in (4.3.35) the new variables $v := Q^*u \in \mathbf{R}^m$. Obviously, $\langle Q\varphi(Q^*u), du \rangle = \langle \varphi(v), dv \rangle$. Furthermore,

$$\sqrt{\langle dv, dv \rangle} = \sqrt{du^*QQ^*du} \leq \sqrt{\lambda_{max}(QQ^*)}ds.$$

It follows, that (4.3.35) is satisfied, if

$$\frac{\sqrt{\lambda_{min}(Q^*Q)}}{\sqrt{\lambda_{max}(QQ^*)}} \inf_{\overline{\gamma}\in[\gamma]} \int_{\overline{\gamma}} |\varphi(v)|ds > \varepsilon + \int_{\gamma} \langle \varphi(v), dv \rangle$$

for any non-contractible curve γ in $\mathbf{R}^m/\Gamma'$, where Γ' is defined above and $ds = \sqrt{\langle du, du \rangle}$. The last inequality coincides with (4.3.33). $\blacksquare$

4.4 Second Order Systems of the Josephson-Type

This section is concerned mainly with the system

$$\begin{aligned}
\ddot{\Phi}_1 + \varepsilon\dot{\Phi}_1 + \gamma(\Phi_1 - \Phi_2) + \sin\Phi_1 &= \widetilde{I} \\
\ddot{\Phi}_2 + \varepsilon\dot{\Phi}_2 + \gamma(\Phi_2 - \Phi_1) + \sin\Phi_2 &= \widetilde{I}
\end{aligned} \tag{4.4.1}$$

derscribing the dynamics of coupled current-biased Josephson junctions. In (4.4.1) Φ_j is the phase difference in the electron-wave function across the j-th junction, $\widetilde{I}$ is the constant bias current, ε denotes the dissipation and γ characterizes the strength of coupling. The parameters $\widetilde{I}$, ε and γ are supposed to be positive.

Note that system (4.4.1) is derived by a space discretization from a boundary value problem for the sine-Gordon equation see, for instance, [104]. This system also describes the motion of two coupled pendulums with forcing term and dissipation. Using the results of the previous sections we want to give sufficient conditions for the solutions to converge. In addition to this we prove a result which guarantees the existence of circular solutions in system (4.4.1).

Proceeding as in [7] we convert (4.4.1) by the new variables $u = \frac{1}{2}(\Phi_1 + \Phi_2)$ and $v = \frac{1}{2}(\Phi_1 - \Phi_2)$ into the second order system

$$\begin{aligned}
\ddot{u} + \varepsilon\dot{u} &= -\sin u \cos v + \widetilde{I}, \\
\ddot{v} + \varepsilon\dot{v} + 2\gamma v &= -\sin v \cos u
\end{aligned} \tag{4.4.2}$$

80

or the first order system

$$\begin{aligned}
\dot{u} &= w, \\
\dot{w} &= -\varepsilon w - \sin u \cos v + \widetilde{I}, \\
\dot{v} &= z, \\
\dot{z} &= -\varepsilon z - 2\gamma v - \sin v \cos u.
\end{aligned} \qquad (4.4.3)$$

Let us for various $\widetilde{I}$ consider the set of equilibria of (4.4.3). For $\widetilde{I} > 1$ there do not exist any equilibria. For $0 < \widetilde{I} \leq 1$ the set of equilibria $\mathcal{E}$ in $\mathbf{R}^4$ is infinite. More precise information about $\mathcal{E}$ gives the following proposition.

Lemma 4.4.1 *Every point of $\mathcal{E}$ is isolated.*

Proof Suppose that $(u_0, 0, v_0, 0) \in \mathcal{E}$ is an arbitrary point. Then v_0 satisfies the equation $g(v_0) = 0$, where $g(v) = \sin^2 v - \widetilde{I}^2 \tan^2 v - 4\gamma^2 v^2$. Assume for a moment that $(u_0, 0, v_0, 0)$ is not isolated. Then in every neighborhood of v_0 in $\mathbf{R}$ there exists a point $y \neq v_0$ with $g'(y) = 0$. Consequently in every neighborhood of v_0 in $\mathbf{R}$ there are also points where g'' is zero. The computation of g'' shows that $g''(v)$ can be rewritten as a third order polynomial with respect to $\tan^2 v$. It is impossible that the set of zeros of such a function has a point of accumulation. $\blacksquare$

Let us now consider the more general second order system

$$\ddot{u} + M\dot{u} + f(u) = 0, \qquad (4.4.4)$$

where $M = M^*$ is a positive definite matrix of order $n \times n$, $f : \mathbf{R}^n \to \mathbf{R}^n$ is of class C^1 and the Jacobi matrix of f is symmetric, i. e. $f'(u)^* = f'(u)$ for all $u \in \mathbf{R}^n$. Suppose also that there exist m linearly independent vectors $d_1, \ldots, d_m \in \mathbf{R}^n$ satisfying the relation

$$f(u + d_j) = f(u) \qquad \text{for all} \qquad u \in \mathbf{R}^n. \qquad (4.4.5)$$

Clearly, system (4.4.4) can also be written as a first order system in $\mathbf{R}^{2n}$

$$\begin{aligned}
\dot{u} &= v \\
\dot{v} &= -Mv - f(u).
\end{aligned} \qquad (4.4.6)$$

Because of (4.4.5), system (4.4.6) is pendulum-like with respect to the group $\Gamma = \left\{ \sum_{j=1}^{n} k_j d'_j \right\}$, where $d'_j = (d_j, 0) \in \mathbf{R}^{2n}$. It is elementary to check that system (4.4.2) is of type (4.4.6) with $m = 1$ and $d_1 = [2\pi, 0]$.

Consider at first the monostability problem for system (4.4.6).

Theorem 4.4.1 *System (4.4.6) is monostable.*

Proof Because of the symmetry and continuity of $f'(u)$ there exists a C^1-function $W : \mathbf{R}^n \to \mathbf{R}$ with $f(u) = \operatorname{grad} W(u)$. Define a function $V : \mathbf{R}^n \times \mathbf{R}^n \to \mathbf{R}$ by $V(u, v) = \frac{1}{2}v^*v + W(u)$. For the derivative of V with respect to system (4.4.6) we have

$$\dot{V}_{(4.4.6)}(u, v) = -v^*Mv \leq 0$$

for all $(u, v) \in \mathbf{R}^n \times \mathbf{R}^n$.

Suppose that $(u(\cdot), v(\cdot))$ is a solution of (4.4.6) with $V(u(0), v(0)) = V(u(\tau), v(\tau))$ for some $\tau > 0$. It follows that $v(t) \equiv 0$ on $[0, \tau]$ and consequently $u(t) \equiv c$, where $f(c) = 0$. Thus the assumptions of Theorem 1.1.3, p. 4, are satisfied and system (4.4.6) is quasi-monostable. Because the set of equilibria is discrete system (4.4.6) is in fact monostable. $\blacksquare$

Remark 4.4.1 In system (4.4.2) we have $f(u_1, u_2) = [\sin u_1 \cos u_2 - \widetilde{I}, 2\gamma u_2 + \sin u_2 \cos u_1]$. A potential for f in $\mathbf{R}^2$ is given by $W(u_1, u_2) = -\cos u_1 \cos_2 - \widetilde{I} u_1 + \gamma u_2^2$. We can look at (4.4.2) as defining a vector field on $\mathbf{R}^2/\Gamma$ with $\Gamma = \{kf_1 \; : \; k \in \mathbf{Z}, d_1 = [2\pi, 0]\}$. In this case every solution of (4.4.2) (and their derivative) is bounded in $\mathbf{R}^2/\Gamma$ on $\mathbf{R}_+$. The function W, however, cannot serve as a potential for f in $\mathbf{R}^2/\Gamma$. As a consequence we can either investigate (4.4.2) with a closed but not exact vector field on $\mathbf{R}^2/\Gamma$ or we can investigate it in $\mathbf{R}^2$ with the exact vector field f. We choose the second variant and have to derive sufficient conditions for the boundedness of solutions on $\mathbf{R}_+$.

In connection with the investigation of solution boundedness of system (4.4.2) it is useful to consider the following control system

$$
\begin{aligned}
\dot{u} &= Au + b[\varphi(\sigma) + f(\sigma, \eta)], & \sigma = c^* u, & \qquad (4.4.7) \\
\dot{v} &= Pv + q\Psi(\sigma, \eta), & \eta = r^* v, & \qquad (4.4.8)
\end{aligned}
$$

in which A is a singular $n \times n$ matrix, b and c are n-vectors.

Introduce the transfer function $\chi(s) = c^*(A - sI)^{-1}b$ and suppose that χ is non-degenerate. Assume futhermore that P is a Hurwitz matrix of order $m \times m$ and q, r are m-vectors. The function $\varphi : \mathbf{R} \to \mathbf{R}$ is continuous and Δ-periodic, the functions f, $\Psi : \mathbf{R}^2 \to \mathbf{R}$ are continuous and bounded.

Define a value ξ by

$$
\xi > \sup_{(\sigma, \eta) \in \mathbf{R}^2} |\Psi(\sigma, \eta)| \cdot \int\limits_0^{+\infty} |r^* e^{Pt} q| \, dt. \qquad (4.4.9)
$$

Theorem 4.4.2 *Suppose that there exist an integer $k \in [1, n]$ and numbers δ_1, δ_2, σ_1, σ_2, $\mu > 0$ and $\lambda > 0$ such that the following conditions are satisfied:*

(i) $\lim\limits_{s \to \infty} s^k \chi(s) > 0$ *and the functions φ and f belong to C^{k-1} on $\mathbf{R}$ and $\mathbf{R} \times \mathbf{R}$, respectively;*

(ii) $\delta_1 < f(\sigma, \eta) < \delta_2$ *for all $(\sigma, \eta) \in \mathbf{R} \times (-\xi, \xi)$;*

(iii) $(\varphi(\sigma) - \delta_i)(\sigma - \delta_i) \leq \mu(\sigma - \sigma_i)^2$ *for all $\sigma \in \mathbf{R}$ and $i = 1, 2$;*

(iv) *the matrix $A + \lambda I$ has one positive eigenvalue and $n - 1$ eigenvalues with negative real part;*

(v) $\operatorname{Re}\chi(i\omega - \lambda) + \mu|\chi(i\omega - \lambda)|^2 \leq 0$ *for all $\omega \in \mathbf{R}$.*

Then every solution of (4.4.7), (4.4.8) is bounded on $\mathbf{R}_+$.

Proof Consider an arbitrary solution $(x(\cdot), y(\cdot))$ of (4.4.7), (4.4.8). Since A is Hurwitzian and the function Ψ is bounded on $\mathbf{R}^2$, it follows tha $v(\cdot)$ is bounded on $\mathbf{R}_+$.

Employing (4.4.8) we get for η the representation

$$
\eta(t) = r^* v(t) = r^* e^{P(t-t_0)} v(t_0) + \int\limits_{t_0}^{t} r^* e^{P(t-\tau)} q\Psi(\sigma(\tau), \eta(\tau)) \, d\tau
$$

for all $t \geq t_0$.

Since A is Hurwitzian there exist constants $\delta > 0$ and $\varrho > 0$ such that

$$
|r^* e^{P(t-t_0)} v(t_0)| \leq \delta e^{-\varrho(t-t_0)} |v(t_0)|
$$

for all $t \geq t_0$.

Combining the last inequality and the estimate (4.4.9) we see that there exists a time $t_1 \geq t_0$ such that

$$-\xi < r^* v(t) < \xi \qquad \text{for all} \qquad t \geq t_1. \tag{4.4.10}$$

Because of (ii), (4.4.10) implies that $\sigma(t) = c^* u(t)$ and $\eta(t) = r^* v(t)$ satisfy the inclusion

$$\delta_1 < f(\sigma(t), \eta(t)) < \delta_2 \qquad \text{for all} \qquad t \geq t_1.$$

We consider now (4.4.7) as a non-automomous system with $g(t) := f(\sigma(t), \eta(t))$ for $t \geq t_1$. Because of (i), (iii) – (v) the assumptions of Theorem 3.2.2, p. 55, are satisfied. It follows by this theorem that $u(\cdot)$ is bounded on $[t_1, +\infty)$. $\blacksquare$

Corollary 4.4.1 *Consider system* (4.4.3) *and define*

$$\beta = \begin{cases} \dfrac{1}{2\gamma} & if \quad \dfrac{\varepsilon^2}{4} - 2\gamma > 0, \\[2ex] \dfrac{4}{\varepsilon^2 + 4\kappa^2} \coth \dfrac{\pi\varepsilon}{4\kappa} & if \quad \dfrac{\varepsilon^2}{4} - 2\gamma < 0, \end{cases} \tag{4.4.11}$$

where $\kappa = \sqrt{2\gamma - \varepsilon^2/4}$. *Suppose that* $\beta < \frac{\pi}{2}$, $\cos \beta > \tilde{I}$ *and* $\varepsilon \geq \sqrt{2}$.
Then any solution of (4.4.3) *is bounded on* $\mathbf{R}_+$ *and, hence, system* (4.4.3) *is gradient-like.*

Proof Let us write system (4.4.3) in the form (4.4.7), (4.4.8) with

$$A = \begin{bmatrix} 0 & 1 \\ 0 & -\varepsilon \end{bmatrix}, \qquad b = \begin{bmatrix} 0 \\ -1 \end{bmatrix}, \qquad c = \begin{bmatrix} 1 \\ 0 \end{bmatrix},$$

$$\varphi(\sigma) = \frac{1+a}{2} \sin \sigma, \qquad f(\sigma, \eta) = \frac{1-a}{2} \sin \sigma + (\cos \eta - 1) \sin \sigma - \tilde{I},$$

(a is to be determined)

$$P = \begin{bmatrix} 0 & 1 \\ -2\gamma & -\varepsilon \end{bmatrix}, \qquad q = \begin{bmatrix} 0 \\ -1 \end{bmatrix}, \qquad r = \begin{bmatrix} 1 \\ 0 \end{bmatrix},$$

$$\Psi(\sigma, \eta) = \sin \eta \cos \sigma.$$

It is an easy matter to show that all hypotheses about A, b, c, φ, f, P, q, r and Ψ in (4.4.7), (4.4.8) are satisfied.

Consider an arbitrary solution (u, w, v, z) of (4.4.3). The component $v(\cdot)$ can be expressed in the form

$$v(t) = e^{\lambda_1 t} c_1 + e^{\lambda_2 t} c_2 - \frac{1}{\lambda_2 - \lambda_1} \int_0^t \left[e^{\lambda_2(t-\tau)} - e^{\lambda_1(t-\tau)} \right] \sin v(\tau) \cos u(\tau) \, d\tau,$$

where $\lambda_{1,2} = -\frac{\varepsilon}{2} \pm \sqrt{\frac{\varepsilon^2}{4} - 2\gamma}$ (we suppose $\lambda_1 \neq \lambda_2$) and c_1, c_2 are certain constants. From this representation it follows that

$$\limsup_{t \to +\infty} |v(t)| \leq \beta,$$

where β is defined in (4.4.11).

We choose a number $\alpha > 0$ so small that the assumptions of Corollary 4.4.1 are satisfied with $\beta + \alpha$ instead of β. Clearly, with $\xi = \beta + \alpha$ condition (4.4.9) it is also fulfilled. Determine now the number a by $a = 1 - \cos(\alpha + \beta) + \widetilde{I}$. The transfer function χ is given by $\chi(s) = \frac{1}{s(s + \varepsilon)}$. This shows that condition (i) of Theorem 4.4.2 is valid with $k = 2$.

Under the hypotheses of Corollary 4.4.1 we get the estimate

$$\left| \frac{1 - a}{2} \sin \sigma + (\cos \eta - 1) \sin \sigma - \widetilde{I} \right| < \frac{1 + a}{2}$$

for all $\sigma \in \mathbf{R}$ and $\eta \in (-\alpha - \beta, \alpha + \beta)$.

Thus, condition (ii) of Theorem 4.4.2 is fulfilled if we take $\delta_2 = -\delta_1 = \frac{1 + a}{2}$. Finally, we choose $\lambda = \varepsilon/2$ and see that condition (iv) is satisfied. The hypothesis $\varepsilon \geq \sqrt{2}$ guarantees the frequency-domain condition (v).

Thus, all hypotheses of Theorem 4.4.2 are verified and the assertion of Corollary 4.4.1 follows from this theorem. $\blacksquare$

One can sharpen the conclusions of Corollary 4.4.1 by choosing system (4.4.7) in the form

$$\ddot{\sigma} + \beta_2 \dot{\sigma} + \beta_1 \varphi(\sigma) + f(\sigma, \eta) = 0, \tag{4.4.12}$$

where β_1 and β_2 are certain numbers. In addition to the assumptions on φ and f in system (4.4.7), (4.4.8) we suppose that φ is C^1.

Theorem 4.4.3 *Suppose, that for the numbers δ_1, δ_2 and ξ, satisfying (4.4.9), the following assumptions are fulfilled:*

(i) *$\delta_1 < f(\sigma, \eta) < \delta_2$ for all $(\sigma, \eta) \in \mathbf{R} \times (-\xi, \xi)$;*

(ii) *the functions $\sigma \mapsto \beta_1 \varphi(\sigma) + \delta_i$ $(i = 1, 2)$ possess on $[0, \Delta)$ exactly two zeros and for all $\sigma \in R$ and $i = 1, 2$ it follows that $[\beta_1 \varphi'(\sigma)]^2 + [\beta_1 \varphi(\sigma) + \delta_i]^2 > 0$;*

(iii) *every solution of*

$$\ddot{\sigma} + \beta_2 \dot{\sigma} + \beta_1 \varphi(\sigma) + \delta_i = 0$$

$(i = 1, 2)$ is bounded on $\mathbf{R}_+$.

Then every solution of (4.4.12), (4.4.8) is bounded on $\mathbf{R}_+$.

Proof We can proceed, as in the proof of Theorem 4.4.2, to conclude that for an arbitrary solution $\sigma(\cdot), \eta(\cdot)$ of (4.4.12), (4.4.8) there exists a time t_1 such that for $t \geq t_1$ (4.4.10) is satisfied. Let us define the function $g(t) = f(\sigma(t), \eta(t))$ for $t \geq t_1$. We convert with the aid of g the equation (4.4.12) into a non-autonomous second-order equation. Using the comparison principle of Chaplygin [25], we receive the conclusion of Theorem 4.4.3, $\blacksquare$

Corollary 4.4.2 *Consider system (4.4.3) and choose β according to (4.4.11). Suppose that $\beta < \frac{\pi}{2}$ and every solution of*

$$\ddot{\sigma} + \varepsilon \dot{\sigma} + \sin \sigma \pm (1 - \cos \beta + \widetilde{I}) = 0$$

is bounded on $\mathbf{R}_+$
Then every solution of (4.4.3) is bounded on $\mathbf{R}_+$ and, hence system (4.4.3) is gradient-like.

Proof As a comparison equation (4.4.12) for our system (4.4.3) we choose

$$\ddot{\sigma} + \varepsilon \dot{\sigma} + \sin \sigma + (\cos \eta - 1) \sin \sigma - \widetilde{I} = 0,$$

i. e., $f(\sigma,\eta) = (\cos\eta - 1)\sin\sigma - \widetilde{I}$. Suppose that $\alpha > 0$ is small enough that the assumptions of the corollary are also satisfied when β is replaced by $\beta + \alpha$. It is easily seen that

$$|f(\sigma,\eta)| < 1 - \cos(\alpha + \beta) + \widetilde{I}$$

for all $(\sigma,\eta) \in \mathbf{R} \times (-\xi,\xi)$. Hence, it is possible to take $\delta_2 = 1 - \cos(\alpha + \beta) + \widetilde{I} = -\delta_1$.

Since $|\delta_i| < 1$ for $i = 1,2$, condition (ii) of Theorem 4.4.3 is satisfied. Thus this theorem is applicable. ∎

As it was shown in the previous part, the dynamics of (4.4.3) may be complicated only in this case when there exist unbounded solutions. Because of this we turn over to the existence of circular solutions for (4.4.3)

Let us regard the control system

$$
\begin{aligned}
\dot{u} &= Au + b[\varphi(\sigma) - f(\sigma,\eta)], & \dot{\sigma} &= c^*u, & (4.4.13) \\
\dot{v} &= Pv + q\Psi(\sigma,\eta), & \eta &= r^*v, & (4.4.14)
\end{aligned}
$$

where (4.4.13) is a system of indirect control. We suppose that b, c, P, q, r, f and Ψ are the same as in system (4.4.7), (4.4.8) and, in addition to this, f and Ψ are Δ-periodic with respect to σ. The matrix A is arbitrary of order $n \times n$.

It follows that system (4.4.13), (4.4.14) may be regarded as a pendulum-like system with respect to the angular variable σ. Suppose that the value ξ is defined by (4.4.9) with respect to system (4.4.13), (4.4.14) and the transfer function χ is definded as above.

Theorem 4.4.4 *Assume that δ_1 and $\lambda > 0$ are numbers and the following conditions are satisfied:*

(i) $\int\limits_{0}^{\Delta} [\varphi(\sigma) - \delta_1]\, d\sigma \le 0$;

(ii) $\delta_1 \le f(\sigma,\eta)$ *for all* $(\sigma,\eta) \in \mathbf{R} \times (-\xi,\xi)$;

(iii) *the matrix $A + \lambda I$ has one positive eigenvalue and $n-1$ eigenvalues with negative real part;*

(iv) $\mathrm{Re}\,\chi(i\omega - \lambda) < 0$ *for all* $\omega \ge 0$ *and* $\lim\limits_{\omega \to \infty} \omega^2 \mathrm{Re}\,\chi(i\omega - \lambda) < 0$;

(v) $c^*b < 0$ *and the second order equation*

$$\ddot{\vartheta} + 1/\sqrt{-c^*b}\,\lambda\dot{\vartheta} + \varphi(\vartheta) - \delta_1 = 0$$

has a circular solution on $\mathbf{R}_+$.

Then system (4.4.13), (4.4.14) has a circular solution on $\mathbf{R}_+$.

Proof In a similar way as in the proof of Theorem 4.4.2 one shows that for an arbitrary solution $u(\cdot)$, $\sigma(\cdot)$, $v(\cdot)$ of (4.4.13), (4.4.14) there exists a time t_1 such that for all $t \ge t_1$ the inclusion (4.4.10) is satisfied. By condition (ii) of the theorem it follows that

$$\delta_1 \le f(\sigma(t),\eta(t)) \qquad \text{for all} \qquad t \ge t_1.$$

Setting $g(t) = \delta_1 - f(\sigma(t),\eta(t))$ for $t \ge t_1$ and $\varphi_1(\sigma) = \varphi(\sigma) - \delta_1$ for $\sigma \in \mathbf{R}$ we see that $u(\cdot)$, $\sigma(\cdot)$ is a solution of the non-autonomous system

$$\dot{u} = Au + b[\varphi_1(\sigma) + g(t)], \qquad \dot{\sigma} = c^*u. \tag{4.4.15}$$

From the results of [84], it follows that, under the conditions of Theorem 4.4.4, there exist for (4.4.15) circular solutions and, hence, for system (4.4.13), (4.4.14). ∎

Let us consider now the existence of circular solutions for system (4.4.3). It is easily verified that for $\widetilde{I} > 1$ system (4.4.3) has circular solutions. In this case, from the first equation of (4.4.2) for an arbitrary solution $u(\cdot)$, $v()$ it follows that there exist a time t_1 and constants $\varrho_1 > 0$ and $\varrho_2 >$ such that

$$
\begin{aligned}
\dot{u}(t) &= e^{-\varepsilon t}\dot{u}(0) + \int_0^t e^{-\varepsilon(t-\tau)}[\widetilde{I} - \sin u(\tau)\cos v(\tau)]\,d\tau \\
&\geq e^{-\varepsilon t}\dot{u}(0) + \frac{\varrho_1}{\varepsilon}(1 - e^{-\varepsilon t}) \geq \varrho_2 > 0
\end{aligned}
$$

for all $t \geq t_1$. Such a standard argument was used in [104].

Corollary 4.4.3 *Consider system (4.4.3) and define $\beta < \pi$ by (4.4.11). Suppose that there exist numbers $\lambda > \varepsilon$ and $a > 0$ such that*

$$
\widetilde{e} := \widetilde{I} - \max_{\eta \in [-\beta,\beta]} |a - \cos\eta| > 0
$$

and the second order equation

$$
\ddot{\vartheta} + \lambda\dot{\vartheta} + a\sin\vartheta = \widetilde{e}
$$

has circular solutions on $\mathbf{R}_+$.
Then system (4.4.3) has circular solutions.

Proof According to (4.4.3) we define a control system (4.4.13), (4.4.14) by

$$
\begin{aligned}
\dot{u} &= -\varepsilon u - [a\sin\sigma - \widetilde{I} - \sin\sigma(a - \cos v)], \\
\dot{\sigma} &= u, \\
\dot{v} &= z, \\
\dot{z} &= -\varepsilon z - 2\gamma v - \sin v\cos u.
\end{aligned}
$$

Thus, $A = -\varepsilon$, $b = -1, c = 1$, $\varphi(\sigma) = a\sin\sigma$, $f(\sigma,\eta) = \widetilde{I} + \sin\sigma(a - \cos\eta)$, $P = \begin{bmatrix} 0 & 1 \\ -2\gamma & -\varepsilon \end{bmatrix}$, $q = \begin{bmatrix} 0 \\ -1 \end{bmatrix}$, $r = \begin{bmatrix} 1 \\ 0 \end{bmatrix}$, $\Psi(\sigma,\eta) = \sin\eta\cos\sigma$.

It is easy to see that the assumptions on system (4.4.13), (4.4.14) in this form are satisfied. Besides this the value $\xi = \alpha + \beta$ satisfies (4.4.9), when we choose $\alpha > 0$ sufficiently small.

We have $f(\sigma,\eta) = \widetilde{I} + \sin\sigma(a - \cos\eta) \geq \widetilde{e} =: \delta_1$ for all $(\sigma,\eta) \in \mathbf{R} \times (-\xi,\xi)$. It is clear that $c^*b = -1$ and $\chi(s) = \frac{1}{s + \varepsilon}$, so that condition (iv) of Theorem 4.4.4 is fulfilled with $\lambda > \varepsilon$. The verification of the other conditions of Theorem 4.4.4 is obvious. ∎

In the following part we return to system (4.4.4), (4.4.5) in order to apply the results of Theorem 4.3.4 to second order systems. The material of this part is due to R. W. Brockett [40].

Define for system (4.4.4) $\mu_0 := \max_{j=1,\ldots,n} \lambda_j$, where λ_j are the eigenvalues of M, and $R(u) := f'(u)M^{-1} + M^{-1}f'(u)$ for $u \in \mathbf{R}^n$. Define also the group $\Gamma' = \{k_j d_j \;:\; k_j \in \mathbf{Z}\}$, where the d_j are defined by (4.4.6).

Theorem 4.4.5 *Suppose that there exist positive numbers β and ε such that the following conditions are satisfied:*

(i) *for any $u \in \mathbf{R}^n$ the minimal eigenvalue $\lambda(u)$ of $M - \beta R(u)$ is positive;*

(ii) *for any non-contractible path γ in $\mathbf{R}^n/\Gamma'$ it follows that*

$$\inf_{\overline{\gamma} \in [\gamma]} 2 \int_{\overline{\gamma}} \sqrt{\lambda(u)\beta\mu_0^{-1}} |f(u)| ds \geq \varepsilon + (1+\beta) \int_\gamma \langle f(u), du \rangle.$$

Then system (4.4.4) is gradient-like.

Proof We think of the second-order differential equation (4.4.4) as being defined on the tangent bundle of the n-torus $\mathbf{R}^n/\Gamma'$. Clearly $|f(u)|$ is bounded from above on $\mathbf{R}^n/\Gamma'$ and since M is positive definite all orbits of (4.4.4) enter a set of the form $|\dot{u}| \leq k$. Since $f'(u)$ is symmetric the vector field f is closed and because it is defined on $\mathbf{R}^n/\Gamma'$ it follows that there exists a constant n-vector α such that $f(u) - \alpha$ is exact. We define V, analogous to the V appearing in the proof of Theorem (4.3.4), as

$$V(\dot{u}, u) = \frac{1}{2}\langle \dot{u}, u \rangle + (1+\beta) \int_0^u \langle f(v) - \alpha, dv \rangle + \beta \langle \dot{u}, M^{-1}f(u) \rangle.$$

A computation shows that

$$\langle \operatorname{grad} V, f \rangle = \langle \dot{u}, -Mu \rangle - (1+\beta)\langle \alpha, \dot{u} \rangle - \beta \langle f(u), M^{-1}f(u) \rangle + \beta \langle \dot{u}, R(u)\dot{u} \rangle.$$

If we take $\langle h, f \rangle = (1+\beta)\langle \alpha, \dot{u} \rangle$ we see by assumption (i) that $\langle \operatorname{grad} V + h, f \rangle$ is non-positive. If for a solution $u(\cdot)$ of (4.4.4) we have

$$\langle \operatorname{grad} V + h, f \rangle = \langle \dot{u}, -[M - \beta R(u)]\dot{u} \rangle - \beta \langle f(u), M^{-1}f(u) \rangle \equiv 0$$

it immediately follows that $u(\cdot)$ is a constant u_0 and $f(u_0) = 0$.

In order to ensure that V is decreasing around "nearly closed" paths we need to show that around non-contractible paths

$$\inf_{\overline{\gamma} \in [\gamma]} \int_{\overline{\gamma}} \langle \dot{u}, [M - \beta R(u)]\dot{u} \rangle + \beta \langle f(u), M^{-1}f(u) \rangle \, dt \geq \varepsilon + (1+\beta) \int_\gamma \langle f(v), dv \rangle.$$

We have here used the fact that for the closed γ

$$\int_\gamma \langle \alpha, dv \rangle = \int_\gamma \langle f(v), dv \rangle.$$

By assumption we have

$$\langle \dot{u}, [M - \beta R(u)]\dot{u} \rangle \geq \lambda(u)\langle \dot{u}, \dot{u} \rangle$$

and

$$\langle f(u), M^{-1}f(u) \rangle \geq \frac{1}{\mu_0}|f(u)|^2.$$

Thus we get

$$\inf_{\overline{\gamma} \in [\gamma]} \int_{\overline{\gamma}} \langle \dot{u}, [M - \beta R(u)]\dot{u} \rangle + \beta \langle f(u), M^{-1}f(u) \rangle \, dt \;\geq\; \inf_{\overline{\gamma} \in [\gamma]} \int_{\overline{\gamma}} \lambda(u)\langle \dot{u}, \dot{u} \rangle + \frac{\beta}{\mu_0}|f(u)|^2 \, dt$$

$$= \inf_{\overline{\gamma} \in [\gamma]} \int_{\overline{\gamma}} \lambda(u)|\dot{u}| + \frac{\beta}{\mu_0}\frac{|f(u)|^2}{|\dot{u}|} \, ds$$

$$\geq \inf_{\overline{\gamma} \in [\gamma]} \int_{\overline{\gamma}} 2\sqrt{\lambda(u)\beta\mu_0^{-1}}|f(u)| \, ds.$$

From condition (ii) it follows that the last term is greater than $\varepsilon + (1 + \beta) \int_\gamma \langle f(u), du \rangle$.

The remainder of the proof is the same as that of Theorem 4.3.4. ∎

Example 4.4.1 Consider again the scalar second-order equation (4.3.15)

$$\ddot{\sigma} + a\dot{\sigma} + \varphi(\sigma) = 0, \tag{4.4.16}$$

where $a > 0$ and $\varphi : \mathbf{R} \to \mathbf{R}$ is of class C^1 and Δ-periodic.

Suppose also that there exists a $\beta > 0$ such that

$$a - \beta\varphi'(\sigma)\frac{1}{a} \geq 0 \tag{4.4.17}$$

for all $\sigma \in \mathbf{R}$.

Than according to Theorem 4.4.5 any solution of (4.4.16) approaches a constant provided that

$$2 \int_0^\Delta \sqrt{\left[a - \beta\varphi'(\sigma)\frac{1}{a}\right] \beta\frac{1}{a}} |\varphi(\sigma)| d\sigma > (1 + \beta) \left| \int_0^\Delta \varphi(\sigma)\, d\sigma \right|$$

or

$$2 \int_0^\Delta \sqrt{[a - \widetilde{\beta}\varphi'(\sigma)]\widetilde{\beta}} |\varphi(\sigma)|\, d\sigma > (1 + a\widetilde{\beta}) \left| \int_0^\Delta \varphi(\sigma)\, d\sigma \right| \tag{4.4.18}$$

with $\widetilde{\beta} = \beta/a$. Here we have used the fact that $\pi_1(S^1)$ has a single generator. This lets us pass from the inequality (ii) in the statement of Theorem 4.4.5 to one applying to all γ belonging to $\pi_1(S^1)$. Note that condition (4.4.17) coincides with condition (4.3.17), which was received as a frequency-domain condition for the gradient-like behavior of (4.4.16).

Chapter 5

The Method of Non-Local Reduction

In this chapter we develop a method for stability investigation of higher-dimensional systems of differential equations which employs the stability results of differential equations in the plane, playing the role of reduced equations. In contrast to the well-known reduction principles [120] our approach is non-local. By including in Lyapunov functions for higher-dimensional systems representations of special orbits of associated second order equations with a certain global behavior we are able to show the same global behavior for the original system.

5.1 A Lyapunov-Type Theorem

Consider the differential equation

$$\dot{x} = f(t, x) \tag{5.1.1}$$

where $f : \mathbf{R}_+ \times \mathbf{R}^n \to \mathbf{R}^n$ is continuous and locally Lipschitz continuous. Suppose that all solutions $x(\cdot, t_0, x_0)$ of (5.1.1) can be continued on $[t_0, +\infty)$.

Let us introduce a set $S \subset \mathbf{R}^m$ of parameters and define for any $\alpha \in S$ a domain $\Omega(\alpha) \subset \mathbf{R}^n$. Define also for $k = 1, \ldots, l$ the scalar-valued functions $(t, x; \alpha) \mapsto V_k(t, x; \alpha)$ for $(t, x, \alpha) \in \mathbf{R}_+ \times \overline{\Omega}(\alpha) \times S$ and $(x; \alpha) \mapsto W_k(x; \alpha)$ for $(x, \alpha) \in \overline{\Omega}(\alpha) \times S$. Suppose that for every fixed α these functions are continuous with respect to $t \in \mathbf{R}_+$ and $x \in \overline{\Omega}(\alpha)$.

Define for all $(t, \alpha) \in \mathbf{R}_+ \times S$ the sets

$$\Psi(t; \alpha) \; := \; \bigcap_{k=1,\ldots,l} \{x \; : \; V_k(t, x; \alpha) < 0\} \cap \Omega(\alpha),$$

$$\Phi(\alpha) \; := \; \bigcap_{k=1,\ldots,l} \{x \; : \; W_k(x; \alpha) < 0\} \cap \Omega(\alpha).$$

Now we formulate a basic result on boundedness, proved in [88], [103], namely the

Theorem 5.1.1 *Let the following conditions hold:*

(i)

$$V_k(t, x; \alpha) \geq W_k(x; \alpha) \tag{5.1.2}$$

for $(t, x, \alpha) \in \mathbf{R}_+ \times \overline{\Omega}(\alpha) \times S$, $k = 1, 2, \ldots, l$;

(ii) *for arbitrary solutions $x(\cdot)$ of (5.1.1) and an arbitrary $\alpha \in S$ the functions $t \mapsto V_k(t, x(t); \alpha)$ (for all t with $x(t) \in \overline{\Omega}(\alpha)$) are a.e. differentiable and satisfy a. e. the inequality*

$$\frac{d}{dt} V_k(t, x(t); \alpha) \leq \beta_k V_k(t, x(t); \alpha) \qquad (k = 1, 2, \ldots, l), \tag{5.1.3}$$

where β_k are certain numbers;

(iii) *for any $t \in \mathbf{R}_+$ and $x \in \mathbf{R}^n$ there exists an $\alpha \in S$ such that $x \in \Psi(t; \alpha)$;*

(iv) *for any $\alpha \in S$ the set $\Phi(\alpha)$ is bounded;*

(v) *for any $\alpha \in S$ yields $\bigcap\limits_{k=1,\ldots,l} \{x \; : \; W_k(x; \alpha) < 0\} \subset \Omega(\alpha).$*

Then equation (5.1.1) is Lagrange stable.

Proof We use the condition (iii) and choose for an arbitrary solution $x(\cdot, t_0, x_0)$ of (5.1.1) a vector $\alpha_0 \in S$ such that $x_0 \in \Psi(t_0; \alpha_0)$. In order to show that $x(t, t_0, x_0) \in \Psi(t_0; \alpha_0)$ for all $t \geq t_0$ we consider the functions

$$V_k(t) := V_k(t, x(t, t_0, x_0); \alpha_0) \quad (k = 1, 2, \ldots, l).$$

Because of assumption (ii) it follows from Lemma 3.1.1, p. 48, that

$$V_k(t) < 0 \qquad \text{for all} \qquad t \geq t_0 \text{ and } k = 1, 2, \ldots, l. \tag{5.1.4}$$

Using (5.1.4) and the assumptions (i) and (v) we see that $x(t, t_0, x_0) \in \Omega(\alpha_0)$ for all $t \geq t_0$ and, consequently,

$$x(t, t_0, x_0) \in \Psi(t; \alpha_0) \subset \Phi(\alpha_0) \qquad \text{for all} \qquad t \geq t_0.$$

Condition (iv) allows us to conclude that $x(\cdot, t_0, x_0)$ is bounded on $[t_0, +\infty)$. Thus Theorem 5.1.1 is proved. ∎

Remark 5.1.1 Let us compare Theorem 5.1.1 with Yoshizawa's theorem (Theorem 1.1.1, p. 3) in the case of differentiable in time Lyapunov functions. Suppose that the a. e. differentiable functions $V : \mathbf{R}_+ \times \mathbf{R}^n \to \mathbf{R}$ and $W : \mathbf{R}^n \to \mathbf{R}$ satisfy the conditions of Theorem 1.1.1. In order to verify the conditions of Theorem 5.1.1 of the present section we take $m = 1$, $S = \mathbf{R}$, $\Omega(\alpha) = \mathbf{R}^n$ for all $\alpha \in \mathbf{R}$, $l = 1$, $W_1(x; \alpha) = W(x) - \alpha$, $V_1(t, x; \alpha) = V(t, x) - \alpha$ for $(t, x, \alpha) \in \mathbf{R}_+ \times \mathbf{R}^n \times \mathbf{R}$ and $\beta_1 = 0$. It follows that all the conditions of Theorem 5.1.1 are satisfied. Indeed, the condition (i) of Theorem 5.1.1 coincides with that of Yoshizawa's theorem. The condition (ii) of Theorem 5.1.1 follows in the case $\beta_1 = 0$ from the condition (ii) of Yoshizawa's theorem. If we take $\alpha_0 = 2V(t_0, x_0)$ the condition (iii) of Theorem 5.1.1 is fulfilled. The condition (iv) is satisfied because $W(x) \to +\infty$ as $|x| \to +\infty$ and the condition (v) is fulfilled automatically.

5.2 The Idea of Non-Local Reduction for Autonomous Systems of Indirect Control

Theorem 5.1.1 gives the opportunity to carry out the idea of non-local reduction for a wide class of control systems

$$\dot{z} = Az + b\varphi(\sigma), \quad \dot{\sigma} = c^* z + \varrho\varphi(\sigma), \tag{5.2.1}$$

where A is a constant $n \times n$ matrix, b and c are constant n-vectors, ϱ is a number and $\varphi : \mathbf{R} \to \mathbf{R}$ is locally Lipschitz continuous and guarantees the continuation of all solutions on $[0, +\infty)$.

Let us consider the second order equation

$$\ddot{\vartheta} + a\dot{\vartheta} + \varphi(\vartheta) = 0 \quad (a > 0) \tag{5.2.2}$$

which we regard as a reduced equation for (5.2.1). It is equivalent to the system

$$\begin{aligned}
\dot{\vartheta} &= \eta, \\
\dot{\eta} &= -a\eta - \varphi(\vartheta).
\end{aligned} \tag{5.2.3}$$

We shall consider also the first order equation

$$F\frac{dF}{d\vartheta} + aF + \varphi(\vartheta) = 0 \tag{5.2.4}$$

which corresponds to (5.2.2). Certain solutions of this equation will be used in the construction of Lyapunov function for (5.2.2). So first of all it is necessary to investigate some properties of solutions of (5.2.4).

Lemma 5.2.1 *Suppose that equation* (5.2.2) *is Lagrange stable. Then for any numbers* β *and* $\gamma \geq 0$ *there exist numbers* ϑ_1 *and* ϑ_2 *such that* $\vartheta_1 < \beta < \vartheta_2$ *and on* $[\vartheta_1, \vartheta_2]$ *there exist two solutions* F_1 *and* F_2 *of* (5.2.4) *with the following properties:*

(i) $F_1(\vartheta_1) = F_2(\vartheta_1) = 0$,

(ii) $F_k(\vartheta) \neq 0$ *for* $\vartheta \neq \vartheta_k$ $(k = 1, 2)$,

(iii) $F_1^2(\beta) = F_2^2(\beta) > \gamma$.

Proof We consider the solution of (5.2.2) with initial conditions

$$\vartheta(0) = \beta, \quad \dot{\vartheta}(0) = \gamma_0 > \sqrt{\gamma}. \tag{5.2.5}$$

Consider this solution at first for $t > 0$. There are two possibilities:

1) For a certain $t = t_1 > 0$ the solution achieves an extremum (maximum) and $\vartheta(t) < \vartheta(t_1)$ for $0 < t < t_1$.

2) This solution has no extrema for $t > 0$.

In the first case we denote the extremum by ϑ_2 $(\vartheta_2 > \beta)$ and the solution of (5.2.4) with the initial condition

$$F(\beta) = \gamma_0 \tag{5.2.6}$$

by F_2.

In the second case function $\vartheta(t)$ is monoton increasing. Because ϑ is bounded there exists the limit $\lim\limits_{t\to\infty} \vartheta(t) := \vartheta_2$. Note that $\dot{\vartheta}(t) \to 0$ as $t \to +\infty$. Indeed, from $a > 0$ and the boundedness of ϑ on $\mathbf{R}_+$ it follows that $\dot{\vartheta}(t)$ is bounded and consequently $\ddot{\vartheta}(t)$ is also bounded. On the other hand we have $\dot{\vartheta}(t) \geq 0$ on $\mathbf{R}_+$ and $\dot{\vartheta} \in L^1(\mathbf{R}_+)$. From the Barbalat lemma (Theorem 2.1.3, p. 16) it follows that $\dot{\vartheta}(t) \to 0$ as $t \to +\infty$. So in this case we can also denote the solution of (5.2.4), (5.2.6) by F_2.

Let us now consider the solution of (5.2.2), (5.2.5) for $t < 0$. In the case when ϑ has an extremum for $t < 0$ at $t = t_2 < 0$ we choose $\vartheta_1 = \vartheta(t_2)$ and F_1 as the continuation of F_2 on $[\vartheta_1, \beta]$ (Figure 5.2.1).

In the case when $\vartheta(t)$ has no extrema on $(-\infty, 0]$ we consider for (5.2.2) on $(-\infty, 0]$ another Cauchy problem

$$\vartheta(0) = \beta, \quad \dot{\vartheta}(0) = -\gamma_0 \tag{5.2.7}$$

and define F_1 in the same way as we have defined F_2 for the Cauchy problem (5.2.2), (5.2.5). Thus Lemma 5.2.1 is proved. ∎

Suppose now that equation (5.2.2) is Lagrange stable. In order to use the results of the previous Section 5.1 we put $m = 2$ and $S := \{\alpha = (\vartheta_1, \vartheta_2) : \vartheta_1 < \vartheta_2\}$ where all the intervals $(\vartheta_1, \vartheta_2)$ are generated for equation (5.2.4) according to Lemma 5.2.1 by a certain pair $\gamma > 0$ and β. The corresponding solutions of the equation (5.2.4) we denote by $F_k(\cdot; \alpha)$ $(k = 1, 2)$.

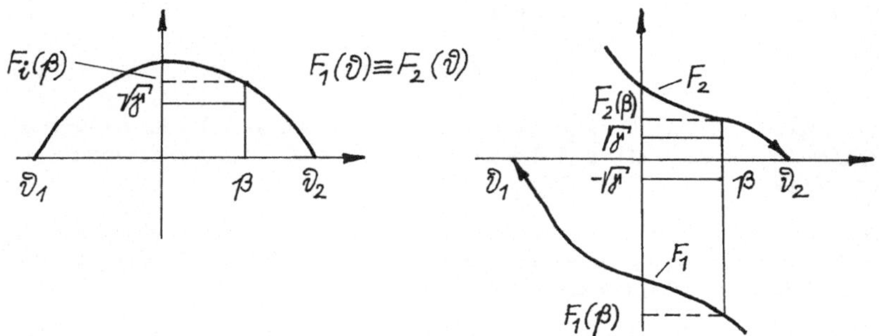

Figure 5.2.1

Let us define the set

$$\Omega(\alpha) := \{x = (z,\sigma) \; : \; z \in \mathbf{R}^n, \sigma \in (\vartheta_1,\vartheta_2)\}$$

and the functions W_k $(k = 1,2)$ by

$$W_k(z,\sigma;\alpha) := z^*Hz - \frac{1}{2}F_k^2(\sigma;\alpha), \quad (z,\sigma) \in \overline{\Omega}(\alpha), \qquad (5.2.8)$$

where $H = H^*$ is a certain $n \times n$ matrix. Let $V_k(t,z,\sigma;\alpha) \equiv W_k(z,\sigma;\alpha)$.

In the following the transfer functions

$$K(s) = c^*(A - sI)^{-1}b - \varrho \qquad \text{and} \qquad D(s) = c^*(A - sI)^{-1}b$$

are used.

Theorem 5.2.1 *Let the transfer function D be non-degenerate. Suppose there exist non-negative numbers a and λ such that the following conditions hold:*

(i) *equation (5.2.2) is Lagrange stable;*

(ii) *the matrix $A + \lambda I$ is Hurwitzian;*

(iii)

$$\operatorname{Re} K(i\omega - \lambda) - \frac{a^2}{2\lambda}|K(i\omega - \lambda)|^2 \geq 0 \qquad (5.2.9)$$

for all $\omega \in \mathbf{R}$.

Then system (5.2.1) is Lagrange stable.

Proof We have to show that under the condition of Theorem 5.2.1 the assumptions of Theorem 5.2.1 are satisfied. A direct computation shows that along the solutions of (5.2.1) we have

$$\frac{d}{dt}V_k + 2\lambda V_k = G(z,\varphi(\sigma)) - \lambda F_k^2 - F_k'F_k\dot\sigma - \frac{a^2}{4\lambda}\dot\sigma^2 - \varphi(\sigma)\dot\sigma \qquad (k = 1,2),$$

where for arbitrary $z \in \mathbf{R}^n$, $\xi \in \mathbf{R}$ the quadratic form G is defined by

$$G(z,\xi) = 2z^*H[(A + \lambda I)z + b\xi] + \xi(c^*z + \varrho\xi) + \frac{a^2}{4\lambda}(c^* + \varrho\xi)^2.$$

It is not difficult to see that along the solution z, σ of (5.2.1)

$$\dot{V}_k + 2\lambda V_k \leq G(z, \varphi(\sigma)) + \frac{\lambda}{a^2}(F_k' F_k + \varphi)^2 - \lambda F_k^2 =$$

$$= G(z, \varphi(\sigma)) + \frac{\lambda}{a^2}[F_k' F_k - aF_k + \varphi][F_k' F_k + aF_k + \varphi].$$

Using (5.2.4) we see that for $\sigma(t) \in [\vartheta_1, \vartheta_2]$

$$\dot{V}_k + 2\lambda V_k \leq G(z(t), \varphi(\sigma(t))).$$

The frequency-domain condition (5.2.9) guarantees (Theorem 1.4.1, p. 9) that there exists a matrix $H = H^*$ such that $G(z, \xi) \leq 0$ for all $z \in \mathbf{R}^n$ and $\xi \in \mathbf{R}$. From the non-degeneracy of D and (ii) it follows that $H > 0$. So the conditions (ii) (with $\beta_k = -2\lambda$, $k = 1, 2$), (iv) and (v) (because $H > 0$ and $F_k(\vartheta_k; \alpha) = 0$) of Theorem 5.1.1 are satisfied. The same is true for condition (iii) of this theorem because for any initial point $x_0 = (z_0, \sigma_0)$ we can choose $\beta = \sigma_0$ and $\gamma = 3z_0^* H z_0$. Thus Theorem 5.2.1 is proved. $\blacksquare$

Remark 5.2.1 Theorem 5.2.1 goes back to the paper [103]. It is easy to show that there are functions of the type (5.2.8) which do not satisfy the conditions of Yoshizawa's theorem (Theorem 1.1.1, p. 3).

In the following lemma essential properties of the non-local reduction method used in this book are shown. Suppose that there are given positive numbers λ and a, an interval $(\vartheta_1, \vartheta_2)$, a continuous function $\Psi : \mathbf{R} \to \mathbf{R}$, continuously differentiable functions $F_k : [\vartheta_1, \vartheta_2] \to \mathbf{R}$ ($k = 1, 2$) and continuously differentiable functions w, $\sigma : \mathbf{R} \to \mathbf{R}$.

Lemma 5.2.2 *Suppose the following conditions hold:*

(i) $F_1(\vartheta_1) = F_2(\vartheta_2) = 0$;

(ii) $F_k(\vartheta) \neq 0$ *for* $\vartheta \neq \vartheta_k$ ($k = 1, 2$);

(iii)
$$F_k'(\vartheta) F_k(\vartheta) + aF_k(\vartheta) + \Psi(\vartheta) = 0 \qquad for \qquad \vartheta \in (\vartheta_1, \vartheta_2); \tag{5.2.10}$$

(iv) $w(t) \geq 0$ *for* $t \geq 0$;

(v)
$$\dot{w}(t) + 2\lambda w(t) + \frac{a^2}{4\lambda}[\dot{\sigma}(t)]^2 + \Psi(\sigma(t))\dot{\sigma}(t) \leq 0 \tag{5.2.11}$$

for $t \geq 0$;

(vi) $\sigma(0) \in (\vartheta_1, \vartheta_2)$, $2w(0) - F_k^2(\sigma(0)) < 0$ ($k = 1, 2$).

Then function σ is bounded on $[0, +\infty)$.

Proof Consider for $k = 1, 2$ the functions

$$v_k(t) := w(t) - \frac{1}{2} F_k^2(\sigma(t)), \quad (t \geq 0). \tag{5.2.12}$$

We immediately get

$$\dot{v}_k(t) + 2\lambda v_k(t) = \dot{w}(t) + 2\lambda w(t) - F_k(\sigma(t)) F_k'(\sigma(t))\dot{\sigma}(t) - \lambda F_k^2(\sigma(t)).$$

It follows from (5.2.11) that

$$\dot{v}_k(t) + 2\lambda v_k(t) \leq -\frac{a^2}{4\lambda}\dot{\sigma}^2 - \Psi\dot{\sigma} - F_k'F_k\dot{\sigma} - \lambda F_k^2, \quad t \geq 0.$$

According to (5.2.10) for all t such that $\sigma(t) \in (\vartheta_1, \vartheta_2)$ we obtain

$$\frac{a^2}{4\lambda}\dot{\sigma}^2 + \lambda F_k^2 + F_k'F_k\dot{\sigma} + \Psi\dot{\sigma} \geq \lambda F_k^2 - \frac{\lambda}{a^2}[\Psi + F_k'F_k]^2$$

$$= \frac{\lambda}{a^2}[F_k'F_k + aF_k + \Psi][F_k'F_k - aF_k + \Psi] = 0.$$

Thus for all t such that $\sigma(t) \in (\vartheta_1, \vartheta_2)$ the estimates

$$\dot{v}_k(t) + 2\lambda v_k(t) \leq 0, \quad k = 1, 2 \tag{5.2.13}$$

hold.

According to (vi) we have $\sigma(0) \in (\vartheta_1, \vartheta_2)$. From this it follows that $\sigma(t) \in (\vartheta_1, \vartheta_2)$ for all $t > 0$. Indeed, suppose there is a value $\bar{t} > 0$ with $\sigma(\bar{t}) = \vartheta_1$ (or $\sigma(\bar{t}) = \vartheta_2$). Consequently we have

$$F_1(\sigma(\bar{t})) = 0 \qquad (\text{or} \qquad F_2(\sigma(\bar{t})) = 0). \tag{5.2.14}$$

From the condition (vi) and inequality (5.2.13) it follows from Lemma 3.1.1, p. 48, that for all $t \in [0, \bar{t}]$ we have $v_k(t) < 0$ and $w(t) < 0$. This inequality contradicts the condition (iv). Thus Lemma 5.2.2 is proved. ■

Remark 5.2.2 It is easy to prove Theorem 5.2.1 with the help of Lemma 5.2.2. Let $z(\cdot)$, $\sigma(\cdot)$ be an arbitrary solution of (5.2.1). We introduce the function $w(t) := z(t)^*Hz(t)$, by means of the matrix $H = H^*$, the numbers $\beta = \sigma(0)$ and $\gamma = 2w(0)$ and apply Lemma 5.2.1. By the latter there exist numbers ϑ_1, ϑ_2 and functions F_k ($k = 1, 2$) such that the conditions (i) – (iii) and (vi) of Lemma 5.2.2 are satisfied for $\Psi = \varphi$. The left-hand side of (5.2.11) coincides with the function $G(z, \varphi(\sigma))$ from the proof of Theorem 5.2.1. It is clear from the proof of Theorem 5.2.1 that the frequency-domain inequality (5.2.9) guarantees the conditions (v) and (iv) of Lemma 5.2.2. So $\sigma(\cdot)$ is bounded on $\mathbf{R}_+$. Since A is Hurwitzian the solution $z(\cdot)$ is also bounded on $\mathbf{R}_+$.

Remark 5.2.3 We could substitute the conditions (i) – (iii) and (vi) of the lemma by the requirement that the equation $\ddot{\vartheta} + a\dot{\vartheta} + \Psi(\vartheta) = 0$ should be Lagrange stable.

5.3 Gradient-Like Behavior of Pendulum-Like Systems

In the first part of this section we consider the autonomous pendulum-like system

$$\dot{z} = Az + b\varphi(\sigma), \quad \dot{\sigma} = c^*z + \varrho\varphi(\sigma) \tag{5.3.1}$$

where A is a constant $n \times n$ matrix, b and c are constant n-vectors, ϱ is a number and $\varphi : \mathbf{R} \to \mathbf{R}$ is continuously differentiable and satisfies $\varphi(\sigma + \Delta) = \varphi(\sigma)$ for all $\sigma \in \mathbf{R}$. We suppose also that

$$\varphi(\sigma)^2 + [\varphi'(\sigma)]^2 \neq 0 \quad \text{for} \quad \sigma \in [0, \Delta) \tag{5.3.2}$$

and that φ has on the interval $[0, \Delta)$ exactly two zeros σ_1 and σ_2 with $\sigma_1 < \sigma_2$.

In order to give a basis free formulation of the results we use the transfer functions χ and K defined by

$$\chi(s) = \frac{1}{s}K(s) = \frac{1}{s}[c^*(A - sI)^{-1}b - \varrho].$$

We suppose that $\chi(\cdot)$ is non-degenerate.

It is clear from the previous section that Theorem 5.2.1 is applicable to the pendulum-like system (5.3.1). But we can obtain more precise information for this system by using supplementary properties of φ. Thus, let μ_1 and μ_2 with $\mu_1 < 0 < \mu_2$ be such that

$$\mu_1 \leq \varphi'(\sigma) \leq \mu_2 \qquad \text{for} \qquad \sigma \in \mathbf{R}. \tag{5.3.3}$$

As an associated second order equation we want to use the equation

$$\ddot{\vartheta} + a\dot{\vartheta} + \varphi(\vartheta) = 0 \tag{5.3.4}$$

where φ is the nonlinearity from (5.3.1) and $a > 0$ is a parameter.

We begin with a supplementary reduction result, which is crucial for the further theory. Let us introduce positive numbers λ and ε, a Δ-periodic C^1-function $\Psi : \mathbf{R} \to \mathbf{R}$ and the C^1-functions $w, \sigma : \mathbf{R}_+ \to \mathbf{R}$. Suppose that Ψ has exactly two zeros on $[0, \Delta)$ and

$$\Psi(\vartheta)^2 + [\Psi'(\vartheta)]^2 \neq 0 \qquad \text{for} \qquad \vartheta \in \mathbf{R}. \tag{5.3.5}$$

The following lemma and the Theorems 5.3.1, 5.3.2 are taken from [49].

Lemma 5.3.1 *Suppose the following conditions are fulfilled:*

(i) *any solution of*

$$\ddot{\vartheta} + 2\sqrt{\lambda\varepsilon}\dot{\vartheta} + \Psi(\vartheta) = 0 \tag{5.3.6}$$

is bounded on $[0, +\infty)$;

(ii) *if for a* $t \in \mathbf{R}_+$ *we have* $\Psi(\sigma(t)) = 0$ *and* $\Psi'(\sigma(t)) < 0$ *then* $w(t) \geq 0$;

(iii) *for all* $t \geq 0$ *we have*

$$\dot{w}(t) + 2\lambda w(t) + \varepsilon\dot{\sigma}(t)^2 + \Psi(\sigma(t))\dot{\sigma}(t) \leq 0. \tag{5.3.7}$$

Then function $\sigma(\cdot)$ *is bounded on* $\mathbf{R}_+$.

Proof Let us consider the equation

$$F\frac{dF}{d\vartheta} + 2\sqrt{\lambda\varepsilon}F + \Psi(\vartheta) = 0, \tag{5.3.8}$$

which corresponds to (5.3.6). The condition (i) of Lemma 5.3.1 allows us to make use of Lemma 5.2.1. In addition to this we can employ the periodicity of Ψ. From Section 2.2 it follows that equation (5.3.8) has a solution F_0 such that for the sequence of functions $\{F_k\}_{k\in\mathbf{Z}}$ defined by

$$F_k(\vartheta) = F_0(\vartheta + k\Delta), \quad \vartheta \in \mathbf{R}, \tag{5.3.9}$$

we have the following properties:

1) $F_k(\vartheta) = 0$ if and only if $\vartheta = \vartheta_2 - k\Delta$, where $\Psi(\vartheta_2) = 0$ and $(\vartheta_2, 0)$ is an unstable equilibrium of (5.3.8);

2) $\lim\limits_{\vartheta \to \pm\infty} F_k(\vartheta)^2 = +\infty$.

Consider now for $k \in \mathbf{Z}$ the functions

$$v_k(t) := w(t) - \frac{1}{2}F_k(\sigma(t))^2 \quad (t \geq 0).$$

As in the proof of Lemma 5.2.2 we can establish that for all $t \geq 0$ and $k \in \mathbf{Z}$:

$$
\begin{aligned}
\dot{v}_k(t) + 2\lambda v_k(t) &\leq -\varepsilon\dot{\sigma}^2 - \Psi\dot{\sigma} - F_k'F_k\dot{\sigma} - \lambda F_k^2 \leq -\lambda F_k^2 + \frac{1}{4\varepsilon}[\Psi + F_k'F_k]^2 \\
&= \frac{1}{4\varepsilon}[F_k'F_k + 2\sqrt{\lambda\varepsilon}F_k + \Psi][F_k'F_k - 2\sqrt{\lambda\varepsilon}F_k + \Psi] = 0.
\end{aligned}
\tag{5.3.10}
$$

From (5.3.9) and the property 2) for F_k it follows that for any $w(0)$ it is possible to find an integer k_0 such that

$$v_{k_0}(0) < 0, \quad v_{-k_0}(0) < 0, \quad |\sigma(0) - \vartheta_2| < k_0\Delta. \tag{5.3.11}$$

From (5.3.10) by Lemma 3.1.1, p. 48, we get that for all $t \geq 0$

$$v_{k_0}(t) < 0 \quad \text{and} \quad v_{-k_0}(t) < 0. \tag{5.3.12}$$

Let us demonstrate that the last property of (5.3.11) implies that $|\sigma(t) - \vartheta_2| < k_0\Delta$ for $t \geq 0$. Suppose to the contrary, that there exists a $t^* > 0$ such that $|\sigma(t^*) - \vartheta_2| = k_0\Delta$. As $(\vartheta_2 + k\Delta, 0)$ for $k \in \mathbf{Z}$ are unstable equilibria of (5.3.6) we can assume that $\Psi(\sigma(t^*)) = 0$, $\Psi'(\sigma(t^*)) < 0$ and $F_{k_0}(\sigma(t^*))F_{-k_0}(\sigma(t^*)) = 0$.

From (5.3.12) we have $w(t^*) < 0$, a contradiction to (ii). Lemma 5.3.1 is proved. ∎

Lemma 5.3.1 may be applied for stability investigation of various classes of differential equations. In the case of system (5.3.1) we get the following result.

Theorem 5.3.1 *Suppose there exist numbers $\lambda > 0$, $\varepsilon > 0$, $\tau \geq 0$ and $\kappa \neq 0$ such that the following conditions are satisfied:*

(i) *any solution of the equation*

$$\ddot{\vartheta} + 2\sqrt{\lambda\varepsilon}\dot{\vartheta} + \kappa\varphi(\vartheta) = 0 \tag{5.3.13}$$

is bounded on $[0, +\infty)$;

(ii) *$A + \lambda I$ is a Hurwitz matrix;*

(iii)

$$
\begin{aligned}
\mathrm{Re}\,\{\kappa K(i\omega - \lambda) - \varepsilon|K(i\omega - \lambda)|^2 \\
-\tau[K(i\omega - \lambda) + \mu_1^{-1}(i\omega - \lambda)]^*[K(i\omega - \lambda) + \mu_2^{-1}(i\omega - \lambda)]\} &\geq 0
\end{aligned}
\tag{5.3.14}
$$

for all $\omega \in \mathbf{R}$.

Then (5.3.1) is Lagrange stable.

Proof Let us define

$$Q = \begin{bmatrix} A & b \\ 0 & 0 \end{bmatrix}, \quad l = \begin{bmatrix} 0 \\ 1 \end{bmatrix}, \quad h = \begin{bmatrix} 0 \\ \varrho \end{bmatrix},$$

where Q is an $(n+1) \times (n+1)$ matrix and l, h are $n+1$-vectors.

Consider the nonlinear system for $y \in \mathbf{R}^{n+1}$ and $\sigma \in \mathbf{R}$

$$\dot{y}(t) = Qy + l\xi, \quad \dot{\sigma}(t) = h^*y(t), \quad \xi = \frac{d}{dt}\varphi(\sigma(t)). \tag{5.3.15}$$

Clearly, if $z(\cdot)$, $\sigma(\cdot)$ is a solution of (5.3.1) then $y(t) = [z(t), \varphi(\sigma(t))]^\top$ is a solution of (5.3.15).

Note that for any function $\sigma \in C^1(\mathbf{R}_+, \mathbf{R})$ we have by assumption (5.3.3) for $\xi(t) := \frac{d}{dt}\varphi(\sigma(t))$

$$\mu_1 \dot{\sigma}(t)^2 \leq \xi(t)\dot{\sigma}(t) \leq \mu_2 \dot{\sigma}(t)^2, \quad t \geq 0. \tag{5.3.16}$$

In the sequel we use the properties of Q, l, h and K obtained in the proof of Theorem 4.3.3, p. 72.

We want to apply Lemma 5.3.1 and put, for a fixed arbitrary solution $y(\cdot)$ of (5.3.15), $w(t) = y(t)^* H y(t)$, where $H = H^*$ is an $(n+1) \times (n+1)$ matrix to be defined. Furthermore we define $\Psi(\sigma) = \kappa\varphi(\sigma)$ for $\sigma \in \mathbf{R}$. We show that under the frequency-domain condition (5.3 14) there exists a matrix H such that w satisfies (5.3.7). Note that for all $t \geq 0$

$$\dot{w}(t) + 2\lambda w(t) = 2y(t)^* H[(Q + \lambda I)y(t) + l\xi(t)].$$

Consider the quadratic form for $y \in \mathbf{R}^{n+1}$ and $\xi \in \mathbf{R}$

$$G(y,\xi) = 2y^* H[(Q + \lambda I)y + l\xi] + \varepsilon(h^* y)^2 + \kappa y^* l h^* y - \tau[h^* y - \mu_1^{-1}\xi][\mu_2^{-1}\xi - h^* y]. \tag{5.3.17}$$

Since $\chi(\cdot)$ is non-degenerate the pair (A, b) in (5.3.1) is controllable and consequently (see the proof of Theorem 4.3.3, p. 72) the pair (Q, l) is controllable. It is easy to see that the pair $(Q + \lambda I, l)$ for $\lambda > 0$ is also controllable. Consider now the Hermitian form for $y \in \mathbf{C}^{n+1}$ and $\xi \in \mathbf{C}$

$$F(y,\xi) = \mathrm{Re}\left\{\varepsilon y^* h h^* y + \kappa y^* l h^* y - \tau(h^* y - \mu_1^{-1}\xi)^*(\mu_2^{-1}\xi - h^* y)\right\}.$$

By Theorem 1.4.1, p. 9, the inequality

$$F\left[(i\omega I - \lambda I - Q)^{-1}l\xi, \xi\right] \leq 0, \quad \omega \in \mathbf{R}, \ \xi \in \mathbf{C} \tag{5.3.18}$$

guarantees the existence of a certain matrix $H = H^*$ such that

$$G(y,\xi) \leq 0 \quad \text{for all} \quad y \in \mathbf{R}^{n+1} \text{ and } \xi \in \mathbf{R}. \tag{5.3.19}$$

By the identity (4.3.7), p. 73,

$$h^*(Q - sI)^{-1}l = \frac{1}{s}K(s)$$

we find that

$$h^*[(Q + \lambda I) - i\omega I]^{-1}l = \frac{1}{i\omega - \lambda}K(i\omega - \lambda). \tag{5.3.20}$$

It is an easy matter to show that (5.3.18) is equivalent to (5.3.14). If $z(\cdot)$, $\sigma(\cdot)$ satisfies (5.3.1) then

$$G(y(t), \xi(t)) + \tau[h^* y(t) - \mu_1^{-1}\xi(t)][\mu_2^{-1}\xi(t) - h^* y(t)]$$

coincides with the left-hand side of (5.3.7). Using (5.3.19) and (5.3.16) we see that (5.3.7) is satisfied. Thus (5.3.14) guarantees the assumption (iii) of Lemma 5.3.1.

The assumption (ii) of Lemma 5.3.1 is also fulfilled. Indeed, let us represent H in the form

$$H = \begin{bmatrix} H_{11} & H_{12} \\ H_{21} & H_{22} \end{bmatrix},$$

where H_{11} is an $n \times n$ matrix, $H_{12} = H_{21}^*$ is an n-vector and H_{22} is a number. Setting in (5.3.19) $\xi = 0$ and $y = [z, 0]^\mathsf{T}$ we get the inequality

$$2z^* H_{11}(A + \lambda I)z \leq -(\varepsilon + \tau)(c^* z)^2 \quad \text{for all} \quad z \in \mathbf{R}^n.$$

Since χ is non-degenerate the pair (A, c) is observable. Then by (ii) it follows from Lemma 1.2.1, p. 6, that H_{11} is positive definite. Suppose now that for a certain $\bar{t} \geq 0$ we have $\varphi(\sigma(\bar{t})) = 0$.

It follows that $w(\bar{t}) = z(\bar{t})^*H_{11}z(\bar{t})$. So as $\kappa \neq 0$ and $H_{11} > 0$ the condition (ii) of Lemma 5.3.1 is satisfied. Condition (i) of this lemma coincides with condition (i) of the theorem. Thus we can apply Theorem 5.3.1 to conclude that $\sigma(\cdot)$ is bounded on $\mathbf{R}_+$. Since A is Hurwitzian and φ is bounded any solution component $z(\cdot)$ of (5.3.1) is also bounded on $\mathbf{R}_+$. Theorem 5.3.1 is proved. ∎

Theorem 5.3.1 in common with various monostability criteria can be used to approximate parameter regions of system (5.3.1) where this system shows gradient-like behavior. Sometimes Theorem 5.3.1 supplies not only with boundedness conditions but also with gradient-like behavior conditions.

Theorem 5.3.2 *Let all the conditions of Theorem 5.3.1 be fulfilled for certain numbers $\lambda > 0$, $\varepsilon > 0$, $\tau \geq 0$ and $\kappa \neq 0$. If in addition to this we have*

$$\tau \mu_1^{-1}\mu_2^{-1} = 0 \qquad and \qquad \tau(\mu_1^{-1} + \mu_2^{-1})\varrho \leq 0 \tag{5.3.21}$$

then system (5.3.1) is gradient-like.

Proof We can proceed exactly as in the proof of Theorem 5.3.1 up to the inequality (5.3.19). Because of assumptions (5.3.21), we can represent the quadratic form G in the following way

$$\begin{aligned} G(y,\xi) \;=\; & 2y^*H(Q + \lambda I)y + \tau(h^*y)^2 + \varepsilon(h^*y)^2 + \kappa y^* lh^*y + \\ & y^*[2Hl - \tau(\mu_1^{-1} + \mu_2^{-1})h]\xi, \qquad y \in \mathbf{R}^{n+1},\ \xi \in \mathbf{R}. \end{aligned} \tag{5.3.22}$$

Note that the quadratic form (5.3.22) is linear with respect to ξ. From this and (5.3.19) we deduce that

$$2Hl = \tau(\mu_1^{-1} + \mu_2^{-1})h. \tag{5.3.23}$$

Using the representation of H as in the proof of Theorem 5.3.1 we get from (5.3.23)

$$2H_{12} = \tau(\mu_1^{-1} + \mu_2^{-1})c, \quad 2H_{22} = \tau(\mu_1^{-1} + \mu_2^{-1})\varrho. \tag{5.3.24}$$

Hence

$$w(t) \;=\; z(t)^*H_{11}z(t) + \tau(\mu_1^{-1} + \mu_2^{-1})c^*z(t)\varphi(\sigma(t)) + \frac{1}{2}\tau(\mu_1^{-1} + \mu_2^{-1})\varrho\varphi(\sigma(t))^2. \tag{5.3.25}$$

The function w satisfies (5.3.7). Let us substitute (5.3.25) into (5.3.7). It is elementary to check that for $t \geq 0$

$$\frac{d}{dt}\left\{ w(t) + [2\lambda\tau(\mu_1^{-1} + \mu_2^{-1}) + \kappa]\int_0^{\sigma(t)} \varphi(\sigma)\,d\sigma \right\} \leq \tag{5.3.26}$$
$$\leq \lambda\tau(\mu_1^{-1} + \mu_2^{-1})\varrho\varphi(\sigma(t))^2 - 2\lambda z(t)^*H_{11}z(t) - \varepsilon\dot{\sigma}(t)^2.$$

Let $U : \mathbf{R}^{n+1} \times \mathbf{R} \to \mathbf{R}$ be defined by

$$U(z,\sigma) = z^*H_{11}z + \tau(\mu_1^{-1} + \mu_2^{-1})c^*z\varphi(\sigma) + \frac{1}{2}\tau(\mu_1^{-1} + \mu_2^{-1})\varrho\varphi(\sigma)^2 +$$

$$+ [2\lambda\tau(\mu_1^{-1} + \mu_2^{-1}) + \kappa]\int_0^{\sigma} \varphi(\vartheta)\,d\vartheta.$$

Because of (5.3.26), the derivative of U along the solution $z(\cdot)$, $\sigma(\cdot)$ of (5.3.1) satisfies the inequality

$$\frac{d}{dt}U(z(t),\sigma(t)) \leq \lambda\tau(\mu_1^{-1} + \mu_2^{-1})\varrho\varphi(\sigma(t))^2 - 2\lambda z(t)^*H_{11}z(t) - \varepsilon\dot{\sigma}(t)^2$$

for $t \geq 0$. We conclude from this and (5.3.21) that

$$\frac{d}{dt}U(z(t),\sigma(t)) \leq -2\lambda z(t)^* H_{11} z(t) \qquad \text{for} \qquad t \geq 0. \tag{5.3.27}$$

We want to apply Theorem 1.1.3, p. 4, to our system (5.3.1). The first assumption of this theorem is fulfilled since $H_{11} > 0$. In order to verify the second assumption we suppose that $z(\cdot)$, $\sigma(\cdot)$ is a solution of (5.3.1) such that $U(z(0),\sigma(0)) = U(z(T),\sigma(T))$ for some $T > 0$. Then from (5.3.27) it follows that $z(t) = 0$ on $[0,T]$. Because of (5.3.1) this implies that $b\varphi(\sigma(t)) = 0$ on $[0,T]$. Since by assumption (A,b) is controllable it is clear that $b \neq 0$. This proves $\varphi(\sigma(t)) = 0$ and, consequently, $\dot{\sigma}(t) = 0$ on $[0,T]$. Finally we show that the solution $z(\cdot)$, $\sigma(\cdot)$ is an equilibrium. Thus, all hypotheses of Theorem 1.1.3, p. 4 are verified. This theorem says then, that system (5.3.1) is quasi-monostable. Since by Theorem 5.3.1 every solution of (5.3.1) is bounded on $\mathbf{R}_+$ and all the equilibria of (5.3.1) are isolated the proof is complete. $\blacksquare$

Theorems 5.3.1 and 5.3.2 give the opportunity to extend to higher-dimensional systems any result on gradient-like behavior known for associated second-order equations (5.3.4). The next two theorems give an example of such extension.

Consider equation (5.3.4) with the nonlinearity

$$\varphi(\vartheta) = \sin(\vartheta + \vartheta_0) - \sin\vartheta_0 \quad (\vartheta \in \mathbf{R},\ \vartheta_0 \in (0,\frac{\pi}{2}) - \text{fixed}). \tag{5.3.28}$$

A well-known result on the convergence of solutions for this pendulum equation is due to C. Böhm [37] and W. D. Hayes [58].

Theorem 5.3.3 (Böhm-Hayes) *Consider the equation* (5.3.4), (5.3.28) *and assume* $a \geq 2\sin(\frac{1}{2}\vartheta_0)$. *Then equation* (5.3.4), (5.3.28) *is gradient-like.*

In applying this theorem together with the Theorems 5.3.1, 5.3.2 we get for system (5.3.1) with the nonlinearity (5.3.28) the following higher-dimensional analogue of the Böhm-Hayes theorem.

Theorem 5.3.4 (Generalized Böhm-Hayes theorem) *Suppose that* $\mu_1 \leq -1$, $\mu_2 \geq 1$ *and there exist numbers* $\lambda > 0$, $\varepsilon > 0$, $\tau \geq 0$ *and* $\kappa > 0$ *such that the conditions (i) and (ii) of Theorem 5.3.1 and the relations*

$$\tau\mu_1^{-1}\mu_2^{-1} = 0, \quad \tau(\mu_1^{-1} + \mu_2^{-1})\varrho \leq 0, \quad \sqrt{\frac{\lambda\varepsilon}{\kappa}} \geq \sin(\frac{1}{2}\vartheta_0)$$

are satisfied. Then system (5.3.1) *with* φ *from* (5.3.28) *is gradient-like.*

Remark 5.3.1 Note that in fact Böhm and Hayes have proved that $a_{cr} \leq 2\sin(\frac{1}{2}\vartheta_0)$, where a_{cr} is the global bifurcation value of a for (5.3.4) (See Sect. 2.2). Thus Theorem 5.3.3 is a corollary of this result.

Remark 5.3.2 In order to verify the assumptions of Theorem 5.3.3 for (5.3.1), (5.3.28), we take in (5.3.13) the change of variables $\bar{t} = t/\sqrt{\kappa}$ to get the equation (5.3.13) in the form

$$\ddot{\vartheta} + 2\sqrt{\frac{\lambda\varepsilon}{\kappa}}\dot{\vartheta} + \varphi(\vartheta) = 0. \tag{5.3.29}$$

5.4 Gradient-Like Behavior of Equations studied in the Theory of Phase-locked loops

In this section we apply the results of Sect. 5.3 to the investigation of concrete equations studied in the theory of PLL's. The chief phenemenon in such systems is to guarantee the locking in of an arbitrary solution to an equilibrium, that is the gradient-like behavior.

Example 5.4.1 Consider the system (5.3.1) with $n = 1$, $\varphi \in C^1$ and the transfer function

$$K(s) := T\frac{1 + mTs}{1 + Ts}, \tag{5.4.1}$$

where constants sastisfy $T > 0$ and $m \in (0, 1)$.

This system describes an autonomous second-order phase-locked loop with proportionally integrating filter.

In order to apply Theorem 5.3.2 we take $\tau = 0$ and $\kappa = 1$. So the relations (5.3.21) are fulfilled automatically. Now we have to choose the parameters ε and λ. Note that condition (i) of Theorem 5.3.1 is equivalent to the inequality $2\sqrt{\lambda\varepsilon} > a_{cr}$, where a_{cr} is the bifurcation value of the parameter a in (5.3.4). Note also that by the condition (ii) of Theorem 5.3.1 we have to choose $\lambda < 1/T$. Condition (iii) of Theorem 5.3.1 takes the form

$$\omega^2(mT^2 - \varepsilon m^2 T^3) + (1 - mT\lambda)[1 - T\lambda - \varepsilon T(1 - mT\lambda)] \geq 0 \qquad \text{for all} \qquad \omega \in \mathbf{R}. \tag{5.4.2}$$

The greatest possible ε that guarantees the non-negativeness of the second member of the left-hand side of (5.4.2) is

$$\varepsilon = \frac{1 - T\lambda}{T(1 - mT\lambda)}. \tag{5.4.3}$$

If we take $\lambda = \alpha/T$ with $\alpha < 1$ it is easy to show that the greatest possible λ is

$$\lambda = \frac{1}{T(1 + \sqrt{1 - m})}. \tag{5.4.4}$$

So choosing ε and λ by (5.4.3) and (5.4.4) the condition (ii) and (iii) of Theorem 5.3.1 are satisfied. It follows that for the gradient-like behavior of (5.3.1), (5.4.1) it is sufficient that the second order equation

$$\ddot{\vartheta} + \frac{2}{T(1 + \sqrt{1 - m})}\dot{\vartheta} + \varphi(\vartheta) = 0 \tag{5.4.5}$$

is gradient-like.

Equation (5.4.5) with $\varphi(\vartheta) = \sin\vartheta - \gamma$, where $\gamma \in (0, 1)$, and $m = 0$ was investigated in [33]. The domain Γ in parameters space (γ, T) where the system is gradient-like is shown in Figure 5.4.1. The region Γ is bounded from above by the solid curve, denoted by 0. With the help of the parameter region for (5.4.5) and Theorem 5.3.2 it is easy to determine for (5.3.1), (5.4.1) parameter regions of gradient-like behaviors for various $m \in (0, 1)$. For $m = 0.05$, $m = 0.1$, $m = 0.2$, $m = 0.5$ and $m = 0.7$ such regions are shown in Figure 5.4.2.

In Figure 5.4.1 are also shown for system (5.3.1), (5.4.1) the global convergence parameter regions obtained by Theorem 4.3.3, p. 72, (γ positive and bounded above by the solid curve, denoted by m) and the global convergence parameter regions obtained in [33] by qualitative-numerical methods (γ positive and bounded above by the dotted curve).

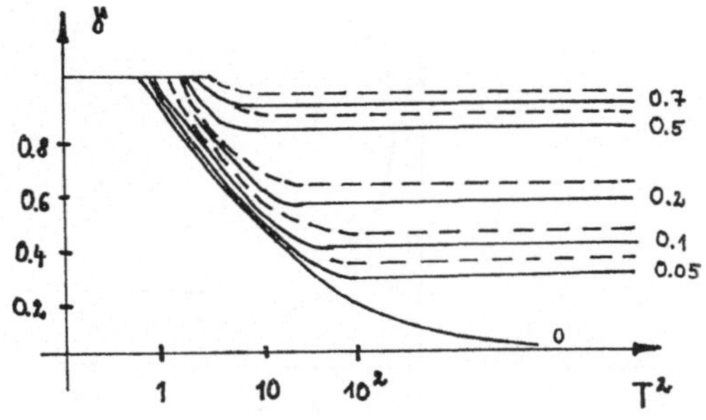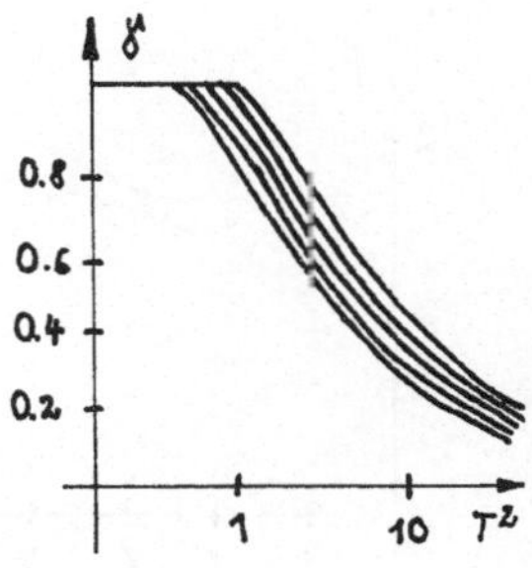

Figure 5.4.1   **Figure 5.4.2**

Example 5.4.2 Consider now system (5.3.1) with $n = 2$ and

$$K(s) = \frac{1}{s^2 + \alpha s + \beta}, \tag{5.4.6}$$

where $\alpha > 0$, $\beta > 0$ and $\alpha\beta > 1$.

Suppose that φ is C^1 and satisfies (5.3.3) with $\mu_1 = -\infty$ and $\mu_2 < \alpha\beta$. This equation describes the dynamics of autonomous phase-locked loops with RCRC – and RLC filters. We want to apply Theorem 5.3.2 with $\tau = \mu_2$. The relations (5.3.21) are fulfilled automatically (note that $\varrho = 0$). Condition (iii) of Theorem 5.3.1 may be written as follows

$$(\alpha - 3\lambda - \kappa)\omega^2 + (\kappa + \lambda)(\lambda^2 - \alpha\lambda + \beta) - \mu_2 - \varepsilon \geq 0 \qquad \text{for} \qquad \omega \in \mathbf{R}. \tag{5.4.7}$$

Taking into account that $\varepsilon\lambda$ must have the greatest permissible value we choose at first

$$\varepsilon = (\kappa + \lambda)(\lambda^2 - \alpha\lambda + \beta) - \mu_2 \qquad \text{and} \qquad \kappa = \alpha - 3\lambda. \tag{5.4.8}$$

It follows that

$$\varepsilon = (\alpha - 2\lambda)(\lambda^2 - \alpha\lambda + \beta) - \mu_2. \tag{5.4.9}$$

This guarantees that the condition (iii) of Theorem 5.3.1 is satisfied. Condition (ii) of this theorem takes the form

$$\lambda < \frac{1}{2}\mathrm{Re}\left(\alpha - \sqrt{\alpha^2 - 4\beta}\right). \tag{5.4.10}$$

If we suppose additionally that $3\lambda < \alpha$ we have that $\kappa > 0$. Taking $t = \sqrt{\kappa}t'$ we transform (5.3.13) to

$$\ddot{\vartheta} + 2\sqrt{\frac{\lambda\varepsilon}{\kappa}}\dot{\vartheta} + \varphi(\vartheta) = 0.$$

Thus the system under consideration is gradient-like if the equation

$$\ddot{\vartheta} + g(\alpha,\beta)\dot{\vartheta} + \varphi(\vartheta) = 0 \tag{5.4.11}$$

is gradient-like. In (5.4.11) we use the notation

$$g(\alpha,\beta) = \sup_{\lambda \in \Gamma} 2\sqrt{\lambda}[(\alpha - 2\lambda)(\lambda^2 + \alpha\lambda - \beta) - \mu_2](\alpha - 3\lambda)^{-1},$$

where

$$\Gamma := \left\{\lambda \in (0, \frac{\alpha}{3}) \ : \ \lambda < \frac{1}{2}\mathrm{Re}\left(\alpha - \sqrt{\alpha^2 - 4\beta}\right), \ (\alpha - 2\lambda)(\lambda^2 - \alpha\lambda + \beta) - \mu_2 > 0\right\}.$$

101

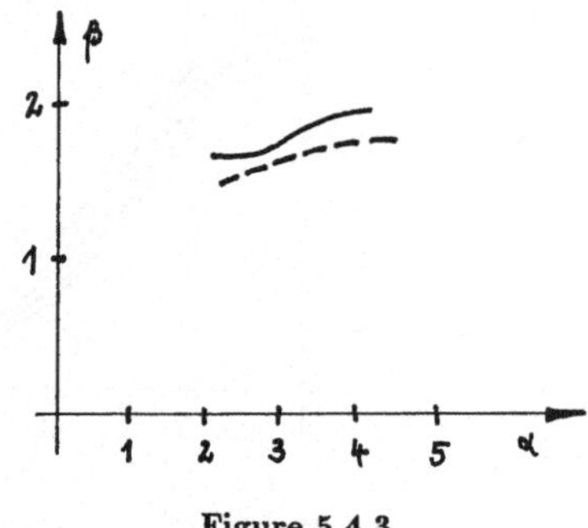
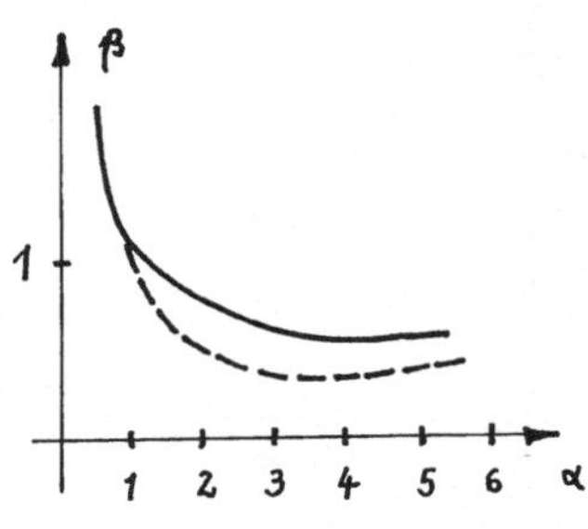

Figure 5.4.3　　　　　　　　　Figure 5.4.4

For the nonlinearity $\varphi(\vartheta) = \sin\vartheta - \gamma$ with $\gamma \in (0,1)$ we can choose $\mu_2 = 1$. In this case we can use for (5.4.11) the parameter domain for gradient-like behavior of (5.4.11) with $m = 0$ (Example 5.4.1, Figure 5.4.1).

They give domains in the α, β-space for (5.3.1), (5.4.6) where this system is gradient-like. If $m = 0.9$ such a domain is shown in Figure 5.4.3 (β positive, bounded above by the solid curve). The dotted line in Figure 5.4.3 shows the boundary of the domain of gradient-like behavior obtained in [137] numerically via computers. Combining the domains of gradient-like behavior received in this section for $\gamma = 0.2$ with the domains obtained for the same system in Section 4 we get the domain in Figure 5.4.4. The dotted line shows the boundary of the domain with gradient-like behavior which has been received for system (5.3.11), (5.4.10) with $\varphi(\vartheta) = \sin\vartheta - \gamma$ in [12] by approximate methods.

Example 5.4.3 Let in (5.3.1) be $n = 2$ and

$$K(s) = \frac{\Gamma s + 1}{s^2 + \alpha s + \beta} \tag{5.4.12}$$

where α, β and Γ positive parameters with $1 - \alpha\Gamma + \beta\Gamma^2 \neq 0$.

Suppose that φ is C^1 and satisfies (5.3.3) with $\mu_1 = -\infty$ and $\mu_2 > 0$. This system describes a phase-locked loop with a filter of type "1/2".

Let us apply Theorem 5.3.2 again. Note that the relations (5.3.21) are fulfilled because $\varrho = \mu_1^{-1} = 0$. We choose $\tau = \mu_2$ and the condition (iii) of Theorem 5.3.1 may be rewritten as follows:

$$\begin{aligned}
&\Gamma\omega^4 + [(\kappa + \lambda)(\alpha\Gamma - \lambda\Gamma - 1) - (1 - \lambda\Gamma)(2\lambda - \alpha) - \Gamma(\lambda^2 - \alpha\lambda + \beta) \\
&-(\mu_2 + \varepsilon)\Gamma^2]\omega^2 + (\kappa + \lambda)(1 - \lambda\Gamma)(\lambda^2 - \alpha\lambda + \beta) - (\mu_2 + \varepsilon)(1 - \Gamma\lambda)^2 \geq 0
\end{aligned} \tag{5.4.13}$$

for all $\omega \in \mathbf{R}$.

Now we take $\lambda < 1/\Gamma$ and the parameters κ and ε in such a way that the coefficent of ω^2 and the ω free term are equal to zero:

$$\kappa = \kappa(\alpha, \beta, \Gamma, \lambda) = [-(1 - \lambda\Gamma)(3\lambda + \beta\Gamma - \alpha - \alpha\Gamma\lambda) - \lambda\Gamma^2(\lambda^2 - \lambda\alpha + \beta)](\Gamma^2\beta - \alpha\Gamma + 1)^{-1}, \tag{5.4.14}$$

$$\varepsilon = \varepsilon(\alpha, \beta, \Gamma, \lambda) = (\lambda^2 - \lambda\alpha + \beta)(\lambda^2\Gamma - \Gamma\beta - 2\lambda + \alpha)(\Gamma^2\beta - \alpha\Gamma + 1)^{-1} - \mu_2. \tag{5.4.15}$$

Thus the system(5.3.1), (5.4.12) is gradient-like if the second-order equation

$$\ddot{\vartheta} + g(\alpha, \beta, \Gamma)\dot{\vartheta} + \varphi(\vartheta) = 0 \tag{5.4.16}$$

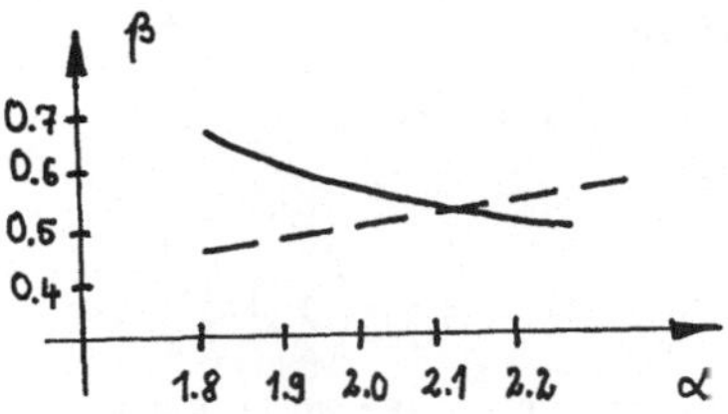

Figure 5.4.5

is gradient-like. In this equation we use the notation

$$g(\alpha, \beta, \Gamma) := \sup_{\lambda \in G} 2\sqrt{\lambda \varepsilon(\alpha, \beta, \Gamma, \lambda)}/\kappa(\alpha, \beta, \Gamma, \lambda)$$

where $\kappa(\alpha, \beta, \Gamma, \lambda)$ and $\varepsilon(\alpha, \beta, \Gamma, \lambda)$ are defined by (5.4.14), (5.4.15) and

$$G = \left\{ \lambda \in (0, \Gamma^{-1}) \ : \ \lambda < \frac{1}{2}\mathrm{Re}\left(\alpha - \sqrt{\alpha^2 - 4\beta}\right), \ \kappa(\alpha, \beta, \Gamma, \lambda) \geq 0, \ \varepsilon(\alpha, \beta, \Gamma, \lambda) \geq 0 \right\}.$$

Using well-known results for (5.4.16) we can construct as in Example 5.4.2 parameter domains for system (5.3.1), (5.4.12) with $\varphi(\vartheta) = \sin \vartheta - \gamma$, $\gamma \in (0, 1)$, where this system is gradient-like. In Figure 5.4.5 is shown the boundary of this domain for system (5.3.1), (5.4.12) with $\Gamma = 0.5$ and $\gamma = 0.5$. The boundary in the positive β-direction of the domain of gradient-like behavior received for this system by Yu. N. Bakaev [13] by means of approximate methods is drawn by the dotted line.

Example 5.4.4 Consider system (5.3.1) with arbitrary n and the transfer function

$$K(s) = \sum_{i=1}^{n} \frac{\Gamma_i}{s + \alpha_i}, \tag{5.4.17}$$

where all Γ_i and α_i are positive and $\alpha_i \neq \alpha_j$ for $i \neq j$.

This system describes the dynamics of a phase-locked loop with a low-pass filter which consists of n parallel connected RC-chains. In order to use Theorem 5.3.2 we take $\kappa = 1$ and $\tau = 0$. Condition (iii) of Theorem 5.3.1 means that

$$\sum_{i=1}^{n} \left[\Gamma_i(\alpha_i - \lambda) - \varepsilon \Gamma_i^2 \right] \left[(\alpha_i - \lambda)^2 + \omega^2 \right]^{-1} \geq 0. \tag{5.4.18}$$

for all $\omega \in \mathbf{R}$.

A sufficient condition for (5.4.18) is

$$\alpha_j - \lambda - \varepsilon \sum_{i=1}^{n} \Gamma_i \geq 0 \qquad \text{for} \qquad j = 1, 2, \ldots, n. \tag{5.4.19}$$

103

The condition (ii) of Theorem 5.3.1 takes the form

$$\lambda < \min_{i=1,\dots,n} \alpha_i =: \alpha_{min}. \tag{5.4.20}$$

If we choose now

$$\lambda = \frac{1}{2}\alpha_{min} \qquad \text{and} \qquad \varepsilon = \lambda \left(\sum_{i=1}^{n} \Gamma_i\right)^{-1}$$

we can guarantee that the inequalities (5.4.19) and (5.4.20) are fulfilled. Thus system (5.3.1), (5.4.17) is gradient-like if the second order equation

$$\ddot{\vartheta} + \alpha_{min} \left(\sum_{i=1}^{n} \Gamma_i\right)^{-\frac{1}{2}} \dot{\vartheta} + \varphi(\vartheta) = 0$$

is gradient-like.

5.5 Non-Local Reduction of Higher dimensional Systems to Autonomous Systems in the Plane

In the previous sections of this chapter we have received a number of convergence and bound-edness results for system (5.3.1) using the non-local reduction principle and the properties of associated second-order equations (5.3.4). In this section, whose results go back to the paper [89], we give the extension of this method on the use of associated two-dimensional systems

$$\begin{aligned}
\dot{\vartheta} &= \quad\;\; \eta \;+\; \beta\varphi(\vartheta), \\
\dot{\eta} &= -\alpha\eta \;-\; \varphi(\vartheta)
\end{aligned} \tag{5.5.1}$$

where α and β are parameters and $\varphi : \mathbf{R} \to \mathbf{R}$ is a locally Lipschitz continuous and Δ-periodic function. Note, that equation (5.3.4), written as a system of first order equations, is a particular case of (5.5.1).

At the beginning we state and prove a technical lemma concerning the non-local reduction procedure. Suppose that λ and $\nu > 0$ are numbers, Ψ and f are scalar-valued C^1-functions on $\mathbf{R}$ and $w, \sigma : \mathbf{R}_+ \to \mathbf{R}$ are C^1. Consider an interval $(\vartheta_1, \vartheta_2)$, where ϑ_1 may be $-\infty$ or ϑ_2 may be $+\infty$. Assume, moreover, that there exists a continuous real-valued function F, defined on the closure of $(\vartheta_1, \vartheta_2)$, which is C^1 on $(\vartheta_1, \vartheta_2)$.

Lemma 5.5.1 *Let the following conditions hold:*

(i) $F(\vartheta) \geq 0$ *for all* $\vartheta \in (\vartheta_1, \vartheta_2)$;

(ii) *if* ϑ_k $(k = 1, 2)$ *is finite then* $F(\vartheta_k) = 0$;

(iii) $F'(\vartheta)F(\vartheta) + \Psi(\vartheta) \leq 0$ *for all* $\vartheta \in (\vartheta_1, \vartheta_2)$;

(iv) $F'(\vartheta)[F(\vartheta) - \sqrt{2\nu}f(\vartheta)] + \lambda\sqrt{2\nu}F(\vartheta) + \Psi(\vartheta) = 0$ *for all* $\vartheta \in (\vartheta_1, \vartheta_2)$;

(v) $w(t) \geq \nu[\dot{\sigma}(t) + f(\sigma(t))]^2$ *for all* $t \in \mathbf{R}$;

(vi) $\dot{w}(t) + 2\lambda w(t) + \Psi(\sigma(t))[\dot{\sigma}(t) + f(\sigma(t))] \leq 0$ *for all* $t \in \mathbf{R}_+$;

(vii) $\sigma(0) \in (\vartheta_1, \vartheta_2)$, $2w(0) - F(\sigma(0))^2 < 0$.

Then $\sigma(t) \in (\vartheta_1, \vartheta_2)$ for all $t \geq 0$.

Proof Suppose that the conclusion of Lemma 5.5.1 is false. As σ is continuous and (vii) is satisfied there exists a time $T > 0$ such that $\sigma(t) \in (\vartheta_1, \vartheta_2)$ for $t \in [0,T)$ and $\sigma(T) = \vartheta_k$, where ϑ_k is finite. Consider the function $v(t) := w(t) - \frac{1}{2}F(\sigma(t))^2$ on $[0,T]$. We want to show that $v(t) < 0$ for all $t \in [0,T]$. Suppose that this is not the case. Then by (vii) there exists a $T_1 \in (0,T]$ such that $v(t) < 0$ at $[0,T_1)$ and $v(T_1) = 0$. So for all $t \in [0,T_1]$ we get

$$w(t) - \nu[\dot{\sigma}(t) + f(\sigma(t))]^2 \leq \frac{1}{2}F(\sigma(t))^2 - \nu[\dot{\sigma}(t) + f(\sigma(t))]^2. \tag{5.5.2}$$

Using this and condition (v) we get

$$F(\sigma(t))^2 \geq 2\nu[\dot{\sigma}(t) + f(\sigma(t))]^2 \quad (t \in [0,T_1]). \tag{5.5.3}$$

The inequality (5.5.3) together with (i) imply that

$$F(\sigma(t))(2\nu)^{-1/2} \geq \dot{\sigma}(t) + f(\sigma(t)) \qquad \text{on} \qquad [0,T_1]. \tag{5.5.4}$$

Use the representation

$$\dot{v} + 2\lambda v = \dot{w} + 2\lambda w - F'F\dot{\sigma} - \lambda F^2$$

to obtain by (vi) the inequality

$$\dot{v} + 2\lambda v \leq -\lambda F^2 - F'F\dot{\sigma} - \Psi[\dot{\sigma} + f] = -\lambda F^2 - [F'F + \Psi](\dot{\sigma} + f) + F'Ff$$

on $[0,T_1]$, where F, Ψ and f are functions of $\sigma(t)$.

Combining (5.5.4) and the conditions (iii) and (iv) we see that on $[0,T_1]$

$$\dot{v} + 2\lambda v \leq -\frac{1}{\sqrt{2\nu}}F\left[F'F - \sqrt{2\nu}F'f + \Psi + \lambda\sqrt{2\nu}F\right] = 0. \tag{5.5.5}$$

This immediately implies

$$v(T_1) \leq v(0)e^{-\lambda T_1} < 0,$$

a contradiction to $v(T_1) = 0$.

This proves $v(t) < 0$ for all $t \in [0,T]$. Finally combining (ii) and the assumption $\sigma(T) = \vartheta_k$ we find that $w(T) < 0$. This contradicts (v) because from (v) we have $w(t) \geq 0$ for all $t \in \mathbf{R}_+$. Thus Lemma 5.5.1 is proved. ∎

As in Sect. 5.2 we will find a function F satisfying Lemma 5.5.1 as a solution of a certain reduced equation for system (5.3.1).

But at first we shall analyse the more general system in the plane

$$\begin{aligned} \dot{\vartheta} &= \xi, \\ \dot{\xi} &= G(\vartheta, \xi). \end{aligned} \tag{5.5.6}$$

Suppose that $G : \mathbf{R} \times \mathbf{R} \to \mathbf{R}$ is continuous and the uniqueness of solutions for (5.5.6) is given. Assume also that G satisfies the following conditions:

(a) there exists a $\Delta > 0$ such that $G(\vartheta + \Delta, \xi) = G(\vartheta, \xi)$ for all $(\vartheta, \xi) \in \mathbf{R} \times \mathbf{R}$;

(b) $G(\cdot, 0)$ has exactly the two zeros ϑ_0 and Ξ_0 on $[0, \Delta)$. For definiteness we suppose that $\Xi_0 < \vartheta_0$ and $G(\vartheta, 0) > 0$ on $(\vartheta_0 - \Delta, \Xi_0)$, $G(\vartheta, 0) < 0$ on (Ξ_0, ϑ_0);

(c) there exist numbers l_1 and l_2 such that $|G(\vartheta, \xi)| \leq l_1|\xi| + l_2$ for all $(\vartheta, \xi) \in \mathbf{R} \times \mathbf{R}$.

Lemma 5.5.2 *Suppose that system (5.5.6) is Lagrange stable. Suppose further that there exist exactly two integral curves $(\widetilde{\vartheta}_0(\cdot), \widetilde{\xi}_0(\cdot))$ and $(\widetilde{\widetilde{\vartheta}}_0(\cdot), \widetilde{\widetilde{\xi}}_0(\cdot))$ of (5.5.6) such that the following relations are true:*

1) $\displaystyle\lim_{t \to +\infty} \widetilde{\vartheta}_0(t) = \lim_{t \to +\infty} \widetilde{\widetilde{\vartheta}}_0(t) = \vartheta_0,$

2) $\displaystyle\lim_{t \to +\infty} \widetilde{\xi}_0(t) = \lim_{t \to +\infty} \widetilde{\widetilde{\xi}}_0(t) = 0,$

3) there exists a sufficiently large $T > 0$ such that

$$\widetilde{\xi}_0(t) > 0 \ \text{ and } \ \widetilde{\widetilde{\xi}}_0(t) < 0 \tag{5.5.7}$$

for all $t \geq T$.

Then the inequality (5.5.7) is valid for all $t \in \mathbf{R}$ and

$$\lim_{t \to -\infty} \widetilde{\xi}_0(t) = +\infty \ \text{ and } \ \lim_{t \to -\infty} \widetilde{\widetilde{\xi}}_0(t) = -\infty. \tag{5.5.8}$$

Proof The proof is identical to the one given for Propositions 2.2.10, p. 25, 2.2.15, p. 29, and Corollary 2.2.3, p. 26. ∎

Let us pass now to the system

$$\begin{aligned}
\dot{\vartheta} &= \eta - u(\vartheta) \\
\dot{\eta} &= -a\eta - \Psi(\vartheta),
\end{aligned} \tag{5.5.9}$$

where $a > 0$ is a parameter and u, $\Psi : \mathbf{R} \to \mathbf{R}$ are Δ-periodic C^1-functions such that the function $r(\vartheta) := \Psi(\vartheta) + au(\vartheta)$, $\vartheta \in \mathbf{R}$ has exactly two zeros on $[0, \Delta)$ and

$$r(\vartheta)^2 + r'(\vartheta)^2 \neq 0$$

for all $\vartheta \in \mathbf{R}$.

Let us show that system (5.5.9) can be transformed to system (5.5.6). Denote by ϑ_0 that zero of $r(\cdot)$ on $[0, \Delta)$ for which the inequality

$$r'(\vartheta_0) < 0 \tag{5.5.10}$$

is valid. Denote furthermore by Σ_0 the nearest to ϑ_0 zero of $r(\cdot)$ such that $\Sigma_0 < \vartheta_0$. It is not difficult to see that $(\vartheta_0, u(\vartheta_0))$ is an equilibrium of (5.5.9) of the saddle-type. The two ω-separatrices $(\widetilde{\vartheta}_0(\cdot), \widetilde{\eta}_0(\cdot))$ and $(\widetilde{\widetilde{\vartheta}}_0(\cdot), \widetilde{\widetilde{\eta}}_0(\cdot))$ approach the point $(\vartheta_0, u(\vartheta_0))$ for $t \to +\infty$ and for $T > 0$ sufficiently large we have

$$\widetilde{\eta}_0(t) - u(\widetilde{\vartheta}_0(t)) > 0$$

and

$$\widetilde{\widetilde{\eta}}_0(t) - u(\widetilde{\widetilde{\vartheta}}_0(t)) > 0$$

for all $t \geq T$.

Let us introduce new coordinates by $\vartheta := \vartheta$ and $\xi := \eta - u(\vartheta)$. It follows that for a solution $(\vartheta(\cdot), \eta(\cdot))$ of (5.5.9) we have

$$\dot{\xi} = \dot{\eta} - u'(\vartheta)\dot{\vartheta} = -a\eta - \Psi(\vartheta) - u'(\vartheta)\xi = -a(\xi + u(\vartheta)) - \Psi(\vartheta) - u'(\vartheta)\xi.$$

106

If we put

$$G(\vartheta, \xi) := -a\xi - au(\vartheta) - \Psi(\vartheta) - u'(\vartheta)\xi \qquad (5.5.11)$$

we get for (ϑ, ξ) a system in the form (5.5.6). Define the functions

$$\widetilde{\xi}_0(t) = \widetilde{\eta}_0(t) - u(\widetilde{\vartheta}_0(t))$$

and

$$\widetilde{\widetilde{\xi}}_0(t) = \widetilde{\widetilde{\eta}}_0(t) - u(\widetilde{\widetilde{\vartheta}}_0(t))$$

for all $t \in \mathbf{R}$. Then $(\widetilde{\vartheta}_0(\cdot), \widetilde{\xi}_0(\cdot))$ and $(\widetilde{\widetilde{\vartheta}}_0(\cdot), \widetilde{\widetilde{\xi}}_0(\cdot))$ are integral curves of (5.5.6), (5.5.11) satisfying the hypotheses of Lemma 5.5.2. Hence if the system (5.5.9) is Lagrange stable the following relations are true:

$$\frac{d}{dt}\widetilde{\vartheta}_0(t) > 0 \qquad \text{and} \qquad \frac{d}{dt}\widetilde{\widetilde{\vartheta}}_0(t) < 0 \qquad \text{on} \qquad [0, +\infty), \qquad (5.5.12)$$

$$\lim_{t \to -\infty} \frac{d}{dt}\widetilde{\vartheta}_0(t) = +\infty \quad \text{and} \quad \lim_{t \to -\infty} \frac{d}{dt}\widetilde{\widetilde{\vartheta}}_0(t) = -\infty. \qquad (5.5.13)$$

Consider now the first order equation

$$\frac{dF}{d\vartheta} = \frac{-aF - \Psi(\vartheta)}{F - u(\vartheta)} \qquad (5.5.14)$$

associated with system (5.5.9).

Lemma 5.5.3 *Suppose there exist numbers ϑ_1 and ϑ_2 with*

$$\Psi(\vartheta)u(\vartheta) > 0 \qquad (5.5.15)$$

on $(\vartheta_1, \vartheta_2)$. Suppose also that F is a solution of (5.5.14) defined on $(\vartheta_1, \vartheta_2)$ and satisfying the inequality

$$F(\vartheta) > u(\vartheta) \qquad (5.5.16)$$

on $(\vartheta_1, \vartheta_2)$.
Then

$$F'(\vartheta)F(\vartheta) + \Psi(\vartheta) \leq 0 \qquad \text{for all} \qquad \vartheta \in (\vartheta_1, \vartheta_2). \qquad (5.5.17)$$

Proof It follows directly from (5.5.14) that

$$F'(\vartheta)F(\vartheta) + \Psi(\vartheta) = \frac{-aF(\vartheta)^2 - \Psi(\vartheta)u(\vartheta)}{F(\vartheta) - u(\vartheta)}$$

on $(\vartheta_1, \vartheta_2)$. Hence by (5.5.15) and (5.5.16) we receive the assertion of Lemma 5.5.3. ∎

Lemma 5.5.4 *Let the following conditions be true:*

(i)

$$\{\vartheta \ : \ \Psi(\vartheta) < 0\} \supset \{\vartheta \ : \ u(\vartheta) < 0\}. \qquad (5.5.18)$$

(ii) $\{\vartheta \ : \ u(\vartheta) = 0\} \neq \emptyset$.

(iii) $F(\cdot)$ *is a solution of (5.5.14) on $(\vartheta_1, \vartheta_2)$ satisfying (5.5.16) and one of the following conditions holds:*

$$F(\vartheta_2) \geq 0 \qquad \text{for} \qquad \vartheta_1 = -\infty \text{ and } \vartheta_2 \neq +\infty \qquad (5.5.19)$$

or

$$F(\vartheta_1) \geq 0 \qquad \text{for} \qquad \vartheta_1 \neq -\infty \text{ and } \vartheta_2 = +\infty. \qquad (5.5.20)$$

Then $F(\vartheta) > 0$ for all $\vartheta \in (\vartheta_1, \vartheta_2)$.

Proof Let us consider the case when (5.5.19) is true (the case (5.5.20) goes through in an analogous manner). Suppose to the contrary that there exists a $\vartheta_3 \in (-\infty, \vartheta_2)$ such that $F(\vartheta_3) \leq 0$. We show now that in this case there exists a $\vartheta_4 \in (-\infty, \vartheta_3)$ such that $F(\vartheta_4) \geq 0$. Indeed, let $F(\vartheta) < 0$ for all $\vartheta \in (-\infty, \vartheta_3)$. Then we have $u(\vartheta) < 0$ for $\vartheta \in (-\infty, \vartheta_3)$ and, according to (5.5.18), $\Psi(\vartheta) < 0$ on $(-\infty, \vartheta_3)$. It follows that the function r has no zeros on $(-\infty, \vartheta_3)$ which contradicts the assumption about $r(\cdot)$.

We conclude from this that we have $F(\vartheta_2) \geq 0$, $F(\vartheta_3) \leq 0$ and $F(\vartheta_4) \geq 0$. Because F is C^1 there exists a $\vartheta_5 \in (\vartheta_4, \vartheta_2)$ with $F'(\vartheta_5) = 0$ and $F(\vartheta_5) \leq 0$.

This immediately implies $aF(\vartheta_5) + \Psi(\vartheta_5) = 0$. Employing this we get

$$\Psi(\vartheta_5) = -aF(\vartheta_5) \geq 0 \tag{5.5.21}$$

and on the other hand

$$u(\vartheta_5) < F(\vartheta_5) \leq 0. \tag{5.5.22}$$

The inequalities (5.5.21) and (5.5.22) contradict assumption (i) and Lemma 5.5.4 is proved. ∎

Lemma 5.5.5 *Suppose that the functions $u(\cdot)$ and $\Psi(\cdot)$ satisfy the conditions (i) and (ii) of Lemma 5.5.4 and the inequality (5.5.15) for all $\sigma \in \mathbf{R}$. Suppose also that $\{\vartheta : \Psi(\vartheta) = 0\} \subset \{\vartheta : u(\vartheta) = 0\}$ and system (5.5.3) is Lagrange stable.*

Under these conditions, (5.5.14) has a solution F on $(-\infty, \infty)$ so that the following properties hold:

$$F(\vartheta_0) = 0, \tag{5.5.23}$$
$$\lim_{\vartheta \to \pm\infty} F(\vartheta)^2 = \infty, \tag{5.5.24}$$
$$F(\vartheta) > 0 \qquad for \qquad \vartheta < \vartheta_0, \tag{5.5.25}$$
$$F'(\vartheta)F(\vartheta) + \Psi(\vartheta) \leq 0 \qquad for \qquad \vartheta < \vartheta_0. \tag{5.5.26}$$

Proof Because of (5.5.15), we have $\vartheta_0 \in \{\vartheta : u(\vartheta) = 0\}$. Hence $(\vartheta_0, 0)$ is a saddle point of (5.5.9). Suppose that $(\widetilde{\vartheta}_0(\cdot), \widetilde{\eta}_0(\cdot))$ and $(\widetilde{\widetilde{\vartheta}}_0(\cdot), \widetilde{\widetilde{\eta}}_0(\cdot))$ are two ω-separatrices through this saddle point. From (5.5.13) and the first equation of (5.5.9) it follows that F, defined by the two separatrices, satisfies $\lim_{\vartheta \to \infty} |F(\vartheta)| = +\infty$. Let us now use the Lemmas 5.5.3 and 5.5.4 with $\vartheta_1 = -\infty$ and $\vartheta_2 = \vartheta_0$. The first inequality of (5.5.12) guarantees that (5.5.16) is satisfied for $\vartheta \in (-\infty, \vartheta_2)$. The remaining assumptions of Lemmas 5.5.3 and 5.5.4 are satisfied under the conditions of Lemma 5.5.5. From the inequalities (5.5.25) and (5.5.26) the assertation of Lemma 5.5.5 follows. ∎

Remark 5.5.1 It is clear now that in the case when Ψ and u have certain properties and the system (5.5.9) is Lagrange stable the conditions (i), (ii) and (iii) of Lemma 5.5.1 may be satisfied for a certain solution of (5.5.14). More that that, number a and function u may be defined in such a way that condition (iv) may also be satisfied.

Suppose that in the following lemma λ and ν are numbers, $\sigma(\cdot)$, $w(\cdot)$ and $f, \Psi : \mathbf{R} \to \mathbf{R}$ are C^1.

Lemma 5.5.6 *Let in system (5.5.9) $u(\vartheta) = \sqrt{2\nu} f(\vartheta)$ and $f(\vartheta)\Psi(u(\vartheta)) \geq 0$ for all $\vartheta \in \mathbf{R}$. Suppose that $\{\vartheta : f(\vartheta) = 0\} \neq \emptyset$ and the following conditions are satisfied:*

(i) *for $a = \lambda\sqrt{2\nu}$ and $u(\vartheta) = \sqrt{2\nu} f(\vartheta)$ the system (5.5.9) is Lagrange stable;*

(ii) $w(t) \geq \nu[\dot{\sigma}(t) + f(\sigma(t))]^2$ for all $t \geq 0$.

Then the function $\sigma(\cdot)$ is bounded from above on $[0, +\infty)$.

Proof The phase portrait of system (5.5.9) in the (ϑ, η)-plane is periodic with respect to ϑ. Using this fact we get from condition (i) and Lemma 5.5.5 the existence of a number $\vartheta^* > \sigma(0)$ and a solution $F(\cdot)$ of (5.5.14) such that $F(\vartheta^*) = 0$, $2w(0) - F(\sigma(0))^2 < 0$, $F(\vartheta) > 0$ for $\vartheta < \vartheta^*$ and $F'(\vartheta)F(\vartheta) + \Psi(\vartheta) \leq 0$ for all $\vartheta < \vartheta^*$. So for these ϑ^* and $F(\cdot)$ the conditions (i) – (iv) and (vii) of Lemma 5.5.1 are satisfied for $\vartheta_1 = -\infty$ and $\vartheta_2 = \vartheta^*$. The remaining conditions of Lemma 5.5.1 are given by the assumption (ii) and (iii) of Lemma 5.5.6. It follows that $\sigma(t) \leq \vartheta^*$ for all $t \in [0, +\infty)$ and Lemma 5.5.6 is proved. $\blacksquare$

Let us now consider the control system (5.3.1) and deduce for it a criterion for global convergence, combining the frequency-domain method and the non-local properties of a reduced system (5.5.9) associated with (5.3.1).

Let us assume in addition to (5.3.2) that

$$\int_0^\Delta \varphi(\sigma)d\sigma \leq 0. \tag{5.5.27}$$

Introduce for (5.3.1) the transfer function $D(s) = c^*(A - sI)^{-1}b$ and suppose that $D(\cdot)$ is non-degenerate. Assume that

$$\Gamma := \lim_{s \to \infty} sD(s) > 0 \qquad \text{and} \qquad \varrho \leq 0. \tag{5.5.28}$$

Theorem 5.5.1 *Suppose there exists a number $\lambda > 0$ such that the following conditions hold:*

(i) *the system*

$$\begin{aligned}
\dot{\vartheta} &= \eta + \frac{\varrho}{\sqrt{\Gamma}}\varphi(\vartheta), \\
\dot{\eta} &= -\frac{\lambda}{\sqrt{\Gamma}}\eta - \varphi(\vartheta)
\end{aligned} \tag{5.5.29}$$

is Lagrange stable.

(ii) $\operatorname{Re} D(i\omega - \lambda) > 0$ *for all* $\omega \in \mathbf{R}$ *and* $\lim_{\omega \to \infty} \omega^2 \operatorname{Re} D(i\omega - \lambda) > 0$.

(iii) $A + \lambda I$ *is a Hurwitz matrix.*

Then system (5.3.1) is gradient-like.

Proof Because of (ii) and (iii) the Yakubovich-Kalman theorem (Theorem 1.4.2, p. 9) guarantees the existence of a matrix $H = H^* > 0$ and a positive number ε such that

$$(A + \lambda I)^* H + H(A + \lambda I) \leq -\varepsilon I, \tag{5.5.30}$$

$$2Hb = -c. \tag{5.5.31}$$

Using the property $\det(I + pq^*) = 1 + p^*q$, where p and q are arbitrary n vectors, we obtain

$$\det[H + (2c^*b)^{-1}cc^*] = \det H[1 + (2c^*b)^{-1}c^*H^{-1}c].$$

Applying now (5.5.31) we receive $\det[H + (2c^*b)^{-1}cc^*] = 0$. Using this together with $H > 0$ and $cc^* \geq 0$ we get

$$H + (2c^*b)^{-1}cc^* \geq 0. \tag{5.5.32}$$

Let $z(\cdot)$, $\sigma(\cdot)$ be an arbitrary solution of (5.3.1). In order to apply Lemma 5.5.6 we introduce the functions

$$w(t) = z(t)^* H z(t), \quad \Psi(\sigma) = \varphi(\sigma), \quad f(\sigma) = -\varrho\varphi(\sigma)$$

and the number

$$\nu = -(2c^* b)^{-1} = (2\Gamma)^{-1}.$$

From (5.5.32) it follows that condition (ii) of Lemma (5.5.6) is satisfied for ν, $w(t)$ and $f(\sigma(t))$. Because of (5.5.30) and (5.5.31) the condition (iii) of Lemma 5.5.6 is fulfilled. Condition (i) of this lemma coincides with the condition (i) of Theorem 5.5.1. We immediately get the boundedness from above of $\sigma(\cdot)$.

Let us now consider the function

$$u(t) = z(t)^* H z(t) + \int\limits_0^{\sigma(t)} \varphi(\vartheta)\,d\vartheta.$$

From the inequalities $H > 0$ and (5.5.27) and from the boundedness of $\sigma(\cdot)$ from above it follows that the function $u(\cdot)$ is bounded from below on $[0, +\infty)$. Since $H > 0$ and $\varrho \leq 0$ it follows from (5.5.30) and (5.5.31) that

$$\dot{u}(t) \leq -\varepsilon|z(t)|^2 \tag{5.5.33}$$

for all $t \geq 0$. Using the boundedness of $u(\cdot)$ we get by integrating (5.5.33)

$$\int\limits_0^{+\infty} |z(t)|^2\,dt < +\infty. \tag{5.5.34}$$

Since by assumption A is Hurwitzian and $\varphi(\cdot)$ is periodic the solution component $z(\cdot)$ and its derivative $\dot{z}(\cdot)$ are bounded on $[0, +\infty)$, Thus we have the boundedness of $\frac{d}{dt}|z(t)|^2$ on $[0, +\infty)$ and by the Corollary 2.1.1, p. 17, of Barbalat's lemma (Theorem 2.1.3, 16) we receive

$$\lim_{t\to\infty} z(t) = 0. \tag{5.5.35}$$

From (5.5.33) it follows that $u(\cdot)$ is decreasing. Using this fact, the boundedness from below of $u(\cdot)$ and (5.5.35), we get the existence of the finite limit

$$\lim_{t\to+\infty} \int\limits_0^{\sigma(t)} \varphi(\vartheta)\,d\vartheta.$$

Recalling that (5.3.2) is satisfied we see that $\sigma(t)$ has a finite limit for $t \to +\infty$ (compare with the proof of Theorem 5.3.1 and the proof of Theorem 4.4.3, p. 84). Theorem 5.5.1 is proved. $\blacksquare$

Example 5.5.1 Let us consider system (5.3.1) with $n = 2$ and

$$K(s) = \frac{s+1}{0.8.s^2 + s + \beta}, \tag{5.5.36}$$

where $\beta > 0$ is a parameter. Note that here $\varrho = 0$ and $K(s) \equiv D(s)$. The direct computation shows that

$$\operatorname{Re} D(i\omega - \lambda) = \frac{\omega^2[0.2 - 0.8\lambda] + (1 - \lambda)[0.8\lambda^2 - \lambda + \beta]}{|0.8(i\omega - \lambda)^2 + (i\omega - \lambda) + \beta|^2}.$$

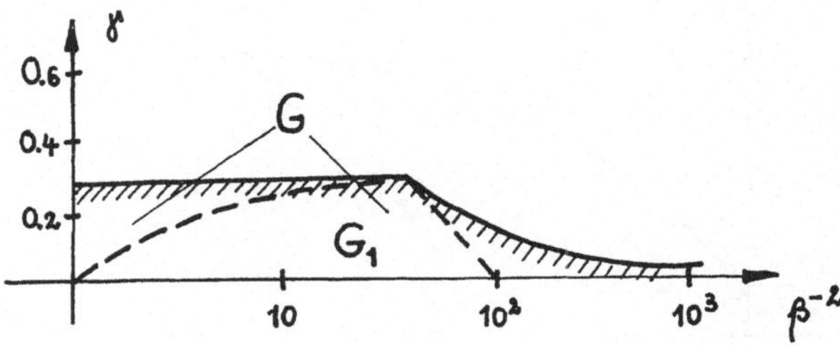

Figure 5.5.1

So the condition (ii) of Theorem 5.5.1 is fulfilled if

$$\lambda < \frac{1}{4} \quad \text{and} \quad 0.8\lambda^2 - \lambda + \beta > 0. \tag{5.5.37}$$

It follows that we can take

$$\begin{aligned}
\lambda &< \tfrac{1}{4} && \text{for} && \beta \geq \tfrac{1}{5} && \text{and} \\
\lambda &< \frac{1 - \sqrt{1 - 3.2\beta}}{1.6} && \text{for} && \beta < \tfrac{1}{5}.
\end{aligned} \tag{5.5.38}$$

It is not difficult to see that under (5.5.38) the condition (iii) of Theorem 5.5.1 is also satisfied. System (5.5.29) is equivalent here to the second-order equation

$$\ddot{\vartheta} + \sqrt{0.8}\lambda\dot{\vartheta} + \varphi(\vartheta) = 0, \tag{5.5.39}$$

which is gradient-like for $\sqrt{0.8}\lambda > a_{cr}$, where the critical value a_{cr} depends on φ.

Thus for $\beta \geq \tfrac{1}{5}$ the system (5.3.1), (5.5.36) is gradient-like if the equation

$$\ddot{\vartheta} + \sqrt{0.8}/4\dot{\vartheta} + \varphi(\vartheta) = 0$$

is gradient-like.

For $\beta < \tfrac{1}{5}$ the system (5.3.1), (5.5.36) is gradient-like if the equation

$$\ddot{\vartheta} + \left(1 - \sqrt{1 - 3.2\beta}\right)\left(2\sqrt{0.8}\right)^{-1}\dot{\vartheta} + \varphi(\vartheta) = 0$$

has this property.

Consider the particular case $\varphi(\sigma) = \sin\sigma - \gamma$ with $\gamma \in (0,1)$. The region G in the (β^{-2}, γ)-space where system (5.3.1) with this nonlinearity is gradient-like lies below the solid curve in Figure 5.5.1. By G_1 we have denoted the region of gradient-like behavior of (5.3.1) obtained in [139] by using the method of two-dimensional comparison systems based on vector field comparison principles.

Example 5.5.2 Let us consider the system (5.3.1) with $n = 2$ and

$$D(s) = \frac{\beta_1}{s + \alpha_1} + \frac{\beta_2}{s + \alpha_2}. \tag{5.5.40}$$

111

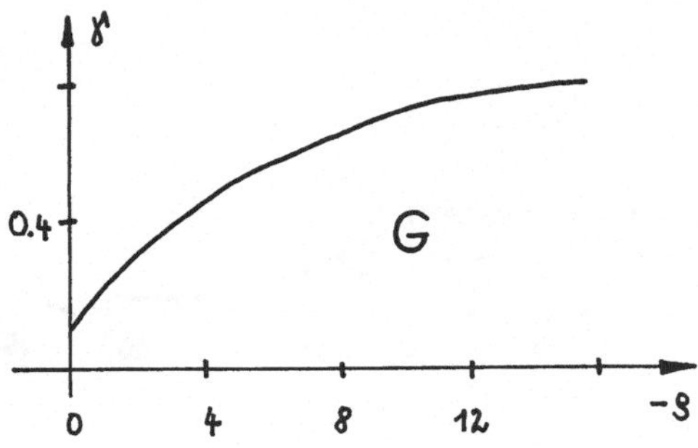

Figure 5.5.2

Here α_1, α_2, β_1, β_2 are positive parameters with $\alpha_1 \neq \alpha_2$. System (5.3.1) with a transfer function (5.5.40) describes the dynamics of PLL's with two sequentiel or parallel connected proportionally-integrating filters [140].

Let us apply Theorem 5.5.1. It is clear that if $\lambda \in (0, \min\{\alpha_1, \alpha_2\})$ the conditions (ii) and (iii) of Theorem 5.5.1 are satisfied. Consider again $\varphi(\vartheta) = \sin\vartheta - \gamma$, $\gamma \in (0,1)$. Then for any permissible α_i, β_i $(i = 1, 2)$ the results of the paper [33] concerning second-order equations are available. In Figure 5.5.2 it is shown the case $\alpha_1 = 0.1$, $\alpha_2 = 0.2$, $\beta_1 = 2$, $\beta_1 = 1$ in the parameter domain with gradient-like behavior systems (5.3.1) obtained by Theorem 5.5.1. The boundary of G in the positive γ-direction gives an approximation from below of the boundary of the full convergence domain in $(\gamma, -\varrho)$ space. In the next chapter the boundary of the full convergence domain of this example will be approximated from above.

5.6 Non-Local Reduction in Non-Autonomous Pendulum-Like Systems

In this section we shall use the non-local reduction principle to study certain non-autonomous systems. We start with a linear non-autonomous second order equation, deal with a wide class of control systems with non-stationary nonlinearities and study forced pendulum-like systems.

Consider the scalar nonlinear equation

$$\ddot{\sigma}(t) + p(t)\dot{\sigma}(t) + q(t)\varphi(\sigma(t)) = 0 \tag{5.6.1}$$

where $p : \mathbf{R}_+ \to \mathbf{R}$ is continuous, $q : \mathbf{R}_+ \to \mathbf{R}$ belongs to C^1 and $\varphi : \mathbf{R} \to \mathbf{R}$ is continuous.

Theorem 5.6.1 *Assume that there exist non-negative numbers λ and M such that:*

(i) $0 < q(t) \leq M$ *for all* $t \in \mathbf{R}_+$;

(ii) $p(t) + \dfrac{\dot{q}(t)}{2q(t)} \leq \lambda \left(1 + \dfrac{q(t)}{4}\right)$ *for all* $t \in \mathbf{R}_+$;

(iii) *the equation*

$$\ddot{\vartheta}(t) + \lambda\dot{\vartheta}(t) + \varphi(\vartheta) = 0 \tag{5.6.2}$$

is Lagrange stable.

112

Then equation (5.6.1) *is Lagrange stable.*

Proof Let us consider the system

$$\begin{aligned}
\dot{\sigma} &= \eta, \\
\dot{\eta} &= -p(t)\eta - q(t)\varphi(\sigma)
\end{aligned}$$
(5.6.3)

which is equivalent to (5.6.1). In order to reduce the Lagrange stability problem of (5.6.3) to the investigation of (5.6.2) we consider an associated to (5.6.2) first order equation

$$F\frac{dF}{d\vartheta} + \lambda F + \varphi(\vartheta) = 0.$$
(5.6.4)

Define for arbitrary $(\sigma, \eta) \in \mathbf{R}^2$, $t \geq 0$, $k = 1,2$ and parameter values $\alpha = (\vartheta_1, \vartheta_2)$ with $\vartheta_1 < \vartheta_2$ the functions

$$V_k(t, \sigma, \eta; \alpha) = \frac{\eta^2}{2q(t)} - \frac{1}{2}F_k(\sigma; \alpha)^2$$

and

$$W_k(\sigma, \eta; \alpha) = \frac{\eta^2}{2M} - \frac{1}{2}F_k(\sigma; \alpha)^2.$$

The continuous functions $F_k(\cdot\,; \alpha)$ are defined on $[\vartheta_1, \vartheta_2]$, satisfying on $(\vartheta_1, \vartheta_2)$ the Cauchy problem for (5.5.4) with $F_k(\vartheta_k) = 0$. Denote by S the set of all $\alpha = (\vartheta_1, \vartheta_2)$ with the required properties. Define $\Omega(\alpha) = \{(\sigma, \eta) \; : \; \sigma \in (\vartheta_1, \vartheta_2)\}$. Clearly, functions V_k and W_k are defined on $\mathbf{R}_+ \times \Omega(\alpha) \times S$. Let us verify the assumptions of Theorem 5.1.1. Assumption (i) of Theorem 5.1.1 follows from condition (i) of Theorem 5.5.1. Let us establish the validity of inequality (5.1.3) with $\beta_k = -2\lambda$. Suppose there is a solution $\sigma(\cdot), \eta(\cdot)$ of (5.6.3). For t with $\sigma(t) \in [\vartheta_1, \vartheta_2]$ we have

$$\begin{aligned}
\frac{d}{dt}V_k(t, \sigma(t), \eta(t); \alpha) + 2\lambda V_k(t, \sigma(t), \eta(t); \alpha) &= \frac{\eta\dot{\eta}}{q} - \frac{\dot{q}}{2q^2}\eta^2 - F_k'F_k\dot{\sigma} + \frac{\lambda\eta^2}{q} - \lambda F_k^2 \\
&= -\frac{p}{q}\eta^2 - \eta\varphi - F_k'F_k\eta - \lambda F_k^2 + \frac{\lambda}{q}\eta^2 - \frac{\dot{q}}{2q^2}\eta^2 \\
&= -\lambda F_k^2 - \frac{1}{q}\left[p - \lambda + \frac{\dot{q}}{2q}\right]\eta^2 - [\varphi + F_k'F_k]\,\eta.
\end{aligned}$$

Note by (5.6.4) we have

$$F_k'(\sigma(t))F_k(\sigma(t)) + \varphi(\sigma(t)) = -\lambda F_k(\sigma(t))$$

and thus

$$\frac{d}{dt}V_k - \beta_k V_k = -\lambda F_k^2 + \lambda F_k\eta - \frac{1}{q}\left[p - \lambda + \frac{\dot{q}}{2q}\right]\eta^2.$$

Using this and assumption (ii) we get

$$\frac{d}{dt}V_k - \beta_k V_k \leq -\lambda F_k^2 + \lambda F_k\eta - \frac{\lambda}{4}\eta^2 = -\lambda\left(F_k - \frac{1}{2}\eta\right)^2 \leq 0.$$

This implies, that assumption (ii) of Theorem 5.1.1 is satisfied. It is obvious that the conditions (iv) and (v) of this theorem are fulfilled. We use Lemma 5.2.1 to show that assumption (iii) is also given. Indeed, for any t_0, $\sigma_0 = \sigma(t_0)$ and $\eta_0 = \eta(t_0)$ we define $\beta_0 = \sigma(t_0)$ and $\gamma_0 = \eta(t_0)/q(t_0)$. According to Lemma 5.2.1 we can find such a vector $\alpha_0 = (\vartheta_1^0, \vartheta_2^0)$ and solution F_k $(k = 1, 2)$ of equation (5.6.4), such that $F_1(\vartheta_1^0) = F_2(\vartheta_2^0) = 0$ and $F_1(\beta_0)^2 = F_2(\beta_0)^2 > \gamma_0$. From this it follows that $V_k(t_0, \sigma_0, \eta_0; \alpha_0) < 0$. Thus Theorem 5.6.1 is proved. ∎

From Theorem 5.6.1 we get in a direct way a result of Liang [106], namely,

Corollary 5.6.1 *Suppose that the conditions (i) and (ii) of Theorem 5.6.1 are satisfied with* $\lambda = 0$. *Suppose also that*

$$\lim_{\sigma \to +\infty} \int_0^\sigma \varphi(\vartheta)\, d\vartheta = +\infty.$$ (5.6.5)

Then equation (5.6.1) is Lagrange stable.

Proof It is well known that (5.6.5) implies the property (iii). ∎

Example 5.6.1 On the basis of Theorem 5.6.1 we can deduce important global properties of equations describing the dynamics of synchronization problems. So equation (5.6.1) describes the dynamics both PLL's with perturbed coefficients and an integrating filter with variable parameters [140] and two-vibrator synchronization systems [119]. For the latter system we have $p(t) \equiv K/I$, $q(t) = \frac{me}{I}(g - A\omega^2 \sin \omega t)$, $\varphi(\sigma) = \cos \sigma$, where I is the moment of inertia of the rotor, m is its mass, e is the eccentricity, A is the amplitude of vertical oscillations of the suspension, ω is the frequency of oscillations of the vibrator engaged in the wiring.

It is well known that the equation $\ddot{\vartheta} + \lambda\dot{\vartheta} + \cos\vartheta = 0$ is Lagrange stable for any $\lambda > 0$. The conditions (i) – (iii) of Theorem 5.6.1 are satisfied if $g > A\omega^2$ and $2k > IA\omega^2/(g - A\omega^2)$. Under these conditions a self-synchronization of two vibrators attached at a common suspension is impossible. In paper [119] a sufficient condition for self-synchronization obtained by approximate methods is given by $2k < meA\omega$. It is easy to see that this condition for certain parameters may give false results.

We consider now the higher-dimensional systems

$$\begin{aligned} \dot{z} &= Az + bu(t)\varphi(\sigma), \\ \dot{\sigma} &= c^*z. \end{aligned}$$ (5.6.6)

Here A is a constant $n \times n$ matrix, b and c are n-vectors, $u : \mathbf{R}_+ \to \mathbf{R}$ is differentiable and satisfies the inequalities

$$0 < u(t) \leq M, \qquad t \geq 0.$$ (5.6.7)

The transfer function of system (5.6.6) from input $u\varphi$ to the output $(-\dot{\sigma})$ is

$$D(s) = c^*(A - sI)^{-1}b.$$

We shall use the notation

$$\Gamma := \lim_{s \to \infty} sD(s).$$

Theorem 5.6.2 *Suppose that* $\Gamma > 0$ *and* $D(\cdot)$ *is non-degenerate. Suppose also that there exist non-negative numbers* λ *and* ε *such that the following conditions are true:*

   (i) $A + \lambda I$ *is a Hurwitz matrix;*

   (ii) $\mathrm{Re}\, D(i\omega - \lambda) \geq 0$ *for all* $\omega \in \mathbf{R}$;

   (iii) *the equation*

$$\ddot{\vartheta} + \sqrt{\lambda\varepsilon}\,\dot{\vartheta} + \varphi(\vartheta) = 0$$ (5.6.8)

   is Lagrange stable;

   (iv) $\lambda + \dfrac{\dot{u}(t)}{u(t)} - \varepsilon\Gamma u(t) \geq 0$ *for all* $t \in \mathbf{R}_+$.

Then system (5.6.6) is Lagrange stable.

Proof From conditions (i) and (ii) it follows by Corollary 1.4.1, p. 9, that there exists a positive $n \times n$ matrix $H = H^* > 0$ such that

$$(A + \lambda I)^* H + H(A + \lambda I) \leq 0, \tag{5.6.9}$$

$$2Hb = -c. \tag{5.6.10}$$

Note that (5.6.10) coincides with (5.5.31). So by the same argument as in the proof of Theorem 5.5.1 we see that

$$H + \frac{1}{2c^*b}cc^* \geq 0, \qquad \text{i. e.}$$

$$2\Gamma z^* H z \geq (c^* z)^2 \quad \text{for all} \quad z \in \mathbf{R}^n. \tag{5.6.11}$$

Let us apply now Theorem 5.1.1 to system (5.6.6), considering (5.6.8) as an associated reduced equation. Define the parameters $\alpha = (\vartheta_1, \vartheta_2)$ with $\vartheta_1 < \vartheta_2$ and consider the first order equation

$$\frac{dF}{d\vartheta}F + \sqrt{\lambda\varepsilon}F + \varphi(\vartheta) = 0 \tag{5.6.12}$$

which corresponds to (5.6.8). We assume that $F_k : [\vartheta_1, \vartheta_2] \to \mathbf{R}$ are functions satisfying (5.6.12) with the initial condition $F_k(\vartheta_k) = 0$ $(k = 1, 2)$. Denote by S the set of all such α for which functions F_k exist and by $\Omega(\alpha)$ with $\alpha = (\vartheta_1, \vartheta_2)$ the set

$$\Omega(\alpha) = \{(z, \sigma) \ : \ z \in \mathbf{R}^n, \ \sigma \in (\vartheta_1, \vartheta_2)\}.$$

Let us consider the functions $V_k : \mathbf{R}_+ \times \Omega(\alpha) \times S \to \mathbf{R}$ and $W_k : \Omega(\alpha) \times S \to \mathbf{R}$ defined by

$$V_k(t, z, \sigma; \alpha) = \frac{1}{u(t)}z^* H z - \frac{1}{2}F_k(\sigma; \alpha)^2,$$

$$W_k(z, \sigma; \alpha) = \frac{1}{M}z^* H z - \frac{1}{2}F_k(\sigma; \alpha)^2$$

$(k = 1, 2)$.

It is not difficult to check the conditions of Theorem 5.1.1. Condition (i) of this theorem follows from (5.5.7). In order to show condition (ii) we put $\beta_k = -\lambda$ $(k = 1, 2)$ and consider $\sigma(\cdot)$ with $\sigma(t) \in [\vartheta_1, \vartheta_2]$. We have (omitting some arguments)

$$\frac{d}{dt}V_k(t, z(t), \sigma(t); \alpha) - \beta_k V_k(t, z(t), \sigma(t); \alpha) = \frac{2}{u}z^* H(A + \frac{\lambda}{2}I)z + 2z^* Hb\varphi - F_k' F_k \dot\sigma - \frac{\lambda}{2}F_k^2 - \frac{\dot u}{u^2}z^* H z$$

for $t \geq 0$.

With the help of (5.6.10) we obtain

$$\frac{d}{dt}V_k - \beta_k V_k = \frac{2}{u}z^* H(A + \frac{\lambda}{2}I)z - (F_k' F_k + \varphi)c^* z - \frac{\lambda}{2}F_k^2 - \frac{\dot u}{u^2}z^* H z$$

for $t \geq 0$.

Now using this, the inequality (5.6.9) and the relation (5.6.12) we get

$$\frac{d}{dt}V_k - \beta_k V_k \leq -\frac{\lambda}{u}z^* H z + \sqrt{\lambda\varepsilon}F_k c^* z - \frac{\lambda}{2}F_k^2 - \frac{\dot u}{u^2}z^* H z.$$

By using now the condition (iv) of Theorem 5.6.2 and the inequality (5.6.11) we receive

$$\frac{d}{dt}V_k - \beta_k V_k \leq -\frac{\lambda}{2}F_k^2 + \sqrt{\lambda\varepsilon}F_k c^* z - \frac{\varepsilon}{2}(c^* z)^2 \leq 0$$

for ≥ 0. Thus inequality (5.6.3) holds. We put now for an arbitrary $t_0 \geq 0$ $\beta = \sigma(t_0)$, $\gamma = \frac{2}{u(t_0)} z(t_0)^* H z(t_0)$ and apply Lemma 5.2.1. Then there exists an $\alpha_0 = (\vartheta_1^0, \vartheta_2^0)$ such that $V_k(t_0, z(t_0), \sigma(t_0); \alpha_0) \leq 0$ and condition (iii) is satisfied. The remaining conditions (iv) and (v) of Theorem 5.1.1 are true because $H > 0$ and $F_k(\vartheta_k; \alpha) = 0$ $(k = 1, 2)$. Now Theorem 5.1.1 allows us to conclude that system (5.6.6) is Lagrange stable. ∎

In the next section we state and prove a criterion for Lagrange stability of the pendulum-like systems

$$\begin{aligned} \dot{z} &= Az + b[\varphi(\sigma) + g(t)], \\ \dot{\sigma} &= c^* z + \varrho[\varphi(\sigma) + g(t)], \end{aligned} \qquad (5.6.13)$$

where as above A is a matrix of order n, b and c are n-vectors, ϱ is a number, $g : \mathbf{R}_+ \to \mathbf{R}$ and $\varphi : \mathbf{R} \to \mathbf{R}$ belong to C^1 and φ is Δ-periodic. We follow here the presentation in [84].

Theorem 5.6.3 *Suppose that the transfer function $D(p) = c^*(A - pI)^{-1}b$ is non-degenerate. Suppose also that there exist numbers $\lambda > 0$, $\varepsilon > 0$, γ_1 and γ_2 such that the following requirements are satisfied:*

(i) *$[\varphi'(\vartheta)]^2 + [\varphi(\vartheta) + \gamma_i]^2 > 0$ for all $\vartheta \in \mathbf{R}$ and $i = 1, 2$;*

(ii) *$\operatorname{Re} D(i\omega - \lambda) - \varrho - \varepsilon |D(i\omega - \lambda) - \varrho|^2 \geq 0$ for all $\omega \in \mathbf{R}$;*

(iii) *$A + \lambda I$ is a Hurwitz matrix;*

(iv) *the equations*

$$\ddot{\vartheta} + 2\sqrt{\lambda\varepsilon}\dot{\vartheta} + \varphi(\vartheta) + \gamma_i = 0 \qquad (i = 1, 2)$$

are Lagrange stable;

(v) *$\gamma_2 \leq g(t) + \frac{1}{2\lambda}\dot{g}(t) \leq \gamma_1$ for all $t \in \mathbf{R}_+$;*

(vi) *$\gamma_2 \leq g(0) \leq \gamma_1$.*

Then system (5.6.13) is Lagrange stable.

In the proof of this theorem the following reduction principle is useful. Let us introduce the C^1-functions $\Psi : \mathbf{R} \to \mathbf{R}$ (which is Δ-periodic) and $f, w, \sigma : \mathbf{R}_+ \to \mathbf{R}$ and the real parameters $\lambda > 0$, $\varepsilon > 0$, γ_1 and γ_2. Let us assume that $\Psi(\cdot) + \gamma_i$ has two zeros on $[0, \Delta)$ and $[\Psi'(\vartheta)]^2 + [\Psi(\vartheta) + \gamma_i]^2 > 0$ for $\vartheta \in \mathbf{R}$ and $i = 1, 2$.

Lemma 5.6.1 *Assume that the following conditions are satisfied:*

(i) *the equations*

$$\ddot{\vartheta} + 2\sqrt{\lambda\varepsilon}\dot{\vartheta} + \Psi(\vartheta) + \gamma_i = 0 \qquad (i = 1, 2) \qquad (5.6.14)$$

are Lagrange stable;

(ii) *if $\Psi(\sigma(t)) + \gamma_i = 0$ $(i = 1, 2)$ and $\Psi'(\sigma(t)) < 0$ for some $t \geq 0$, then $w(t) \geq 0$;*

(iii) *$\dot{w}(t) + 2\lambda w(t) + \varepsilon\dot{\sigma}(t)^2 + [f(t) + \Psi(\sigma(t))]\dot{\sigma}(t) \leq 0$ for all $t \geq 0$;*

(iv) *$\gamma_2 \leq f(t) + \frac{1}{2\lambda}\dot{f}(t) \leq \gamma_1$ for all $t \in \mathbf{R}_+$;*

(v) *$\gamma_2 \leq f(0) \leq \gamma_1$.*

Then $\sigma(\cdot)$ is bounded on $\mathbf{R}_+$.

116

Proof Let us consider the two first order equations corresponding to (5.6.14)

$$F(\vartheta)\frac{dF}{d\vartheta} + 2\sqrt{\lambda\varepsilon}F(\vartheta) + \Psi(\vartheta) + \gamma_1 = 0 \qquad (5.6.15)$$

and

$$G(\vartheta)\frac{dG}{d\vartheta} + 2\sqrt{\lambda\varepsilon}G(\vartheta) + \Psi(\vartheta) + \gamma_2 = 0. \qquad (5.6.16)$$

Denote by $\vartheta_i \in [0, \Delta)$ a zero of $\Psi(\cdot) + \gamma_i$ such that $\Psi'(\vartheta_i) < 0$. It follows that $\Psi'(\vartheta_i) \neq 0$ ($i = 1, 2$). Using the results of Section 5.2 we can assume that there exists a family of solutions $\{F_k\}_{k\in\mathbf{Z}}$ and $\{G_k\}_{k\in\mathbf{Z}}$ of (5.6.15) resp. (5.6.16) defined by

$$F_k(\vartheta) = F_0(\vartheta + k\Delta) \quad \text{and} \quad G_k(\vartheta) = G_0(\vartheta + k\Delta) \qquad (\vartheta \in \mathbf{R}, \ k \in \mathbf{Z}),$$

where F_0 and G_0 are solutions of (5.6.15) resp. (5.6.16). They have the following properties:

1) $F_k(\vartheta_1 - k\Delta) = 0$, $G_k(\vartheta_2 - k\Delta) = 0$ for $k \in \mathbf{Z}$;

2)

$$\lim_{\vartheta\to\pm\infty} F_k(\vartheta)^2 = +\infty, \quad \lim_{\vartheta\to\pm\infty} G_k(\vartheta)^2 = +\infty. \qquad (5.6.17)$$

Let us consider for any $k \in \mathbf{Z}$ the functions $v_k, u_k : \mathbf{R}_+ \to \mathbf{R}$

$$v_k(t) = w(t) - \frac{1}{2}F_k(\sigma(t))^2 - [\gamma_1 - f(t)][\sigma(t) - \vartheta_1 - k\Delta],$$

$$u_k(t) = w(t) - \frac{1}{2}G_k(\sigma(t))^2 - [\gamma_2 - f(t)][\sigma(t) - \vartheta_2 - k\Delta].$$

From condition (iii) of the present lemma it follows that

$$\dot{v}_k(t) + 2\lambda v_k(t) \leq -\varepsilon\dot{\sigma}(t)^2 - [\Psi(\sigma(t)) + f(t)]\dot{\sigma}(t) - F_k'(\sigma(t))F_k(\sigma(t))\dot{\sigma}(t) - \lambda F_k(\sigma(t))^2 - \\ -[\gamma_1 - f(t)]\dot{\sigma}(t) + [\dot{f}(t) - 2\lambda(\gamma_1 - f(t))][\sigma(t) - \vartheta_1 - k\Delta]$$

for $t \geq 0$. Note that

$$\varepsilon\dot{\sigma}(t)^2 + [f(t) + \Psi(\sigma(t))]\dot{\sigma}(t) + [\gamma_1 - f(t)]\dot{\sigma}(t) + F_k'(\sigma(t))F_k(\sigma(t))\dot{\sigma}(t)$$

$$\geq -\frac{1}{4\varepsilon}[\Psi(\sigma(t)) + \gamma_1 + F_k'(\sigma(t))F_k(\sigma(t))]^2$$

for $t \geq 0$.

We can write for $t \geq 0$

$$\begin{aligned}
\dot{v}_k(t) + 2\lambda v_k(t) \leq{}& \frac{1}{4\varepsilon}[\Psi(\sigma(t)) + \gamma_1 + F_k'(\sigma(t))F_k(\sigma(t))]^2 - \lambda F_k(\sigma(t))^2 + \\
&+ [\dot{f}(t) - 2\lambda(\gamma_1 - f(t))][\sigma(t) - \vartheta_1 - k\Delta] \\
={}& \frac{1}{4\varepsilon}\left[F_k'(\sigma(t))F_k(\sigma(t)) + 2\sqrt{\lambda\varepsilon}F_k(\sigma(t)) + \Psi(\sigma(t)) + \gamma_1\right] \cdot \\
&\cdot \left[F_k'(\sigma(t))F_k(\sigma(t)) - 2\sqrt{\lambda\varepsilon}F_k(\sigma(t)) + \Psi(\sigma(t)) + \gamma_1\right] + \\
&+ \left[\dot{f}(t) - 2\lambda(\gamma_1 - f(t))\right][\sigma(t) - \vartheta_1 - k\Delta].
\end{aligned}$$

Since F_k satisfies (5.6.15) we conclude that

$$\dot{v}_k(t) + 2\lambda v_k(t) \leq \left[\dot{f}(t) - 2\lambda(\gamma_1 - f(t))\right][\sigma(t) - \vartheta_1 - k\Delta].$$

From this and from condition (iv) we get that

$$\dot{v}_k(t) + 2\lambda v_k(t) \le 0 \tag{5.6.18}$$

for all $t \in T_k^1 := \{t \,:\, \sigma(t) > \vartheta_1 + k\Delta\}$.

Proceeding exactly as above and using the conditions (iii), (iv) and equation (5.6.16) we can show that for all $k \in \mathbf{Z}$ we have

$$\dot{u}_k(t) + 2\lambda u_k(t) \le 0 \tag{5.6.19}$$

for all $t \in T_k^2 := \{t \,:\, \sigma(t) \le \vartheta_2 + k\Delta\}$. From Lemma 3.1.1, p. 48, it follows that $v_k(0) < 0$ and $[0, t] \subset T_k^1$ imply that $v_k(t) < 0$.

In a similar way one shows that $u_k(0) < 0$ and $[0, t] \subset T_k^2$ imply the inequality $u_k(t) < 0$.

Using (5.6.17) together with condition (v) we get the existence of such an integer j that

$$\vartheta_1 - j\Delta \le \sigma(0) \le \vartheta_2 + j\Delta, \quad v_{-j}(0) < 0, \quad u_j(0) < 0. \tag{5.6.20}$$

Because of (5.6.20) the relations

$$\vartheta_1 - j\Delta \le \sigma(t) \le \vartheta_2 + j\Delta, \quad v_{-j}(t) < 0, \quad u_j(t) < 0. \tag{5.6.21}$$

hold for all $t \ge 0$.

In order to prove this we suppose to the contrary that there exists some $\tau > 0$ such that (5.6.21) is not satisfied. We may assume that

$$[\sigma(\tau) - \vartheta_2 - j\Delta][\sigma(\tau) - \vartheta_1 + j\Delta] = 0$$

and

$$[\sigma(t) - \vartheta_2 - j\Delta][\sigma(t) - \vartheta_1 + j\Delta] \le 0$$

for $t \in [0, \tau]$. This shows that either

$$F_{-j}(\sigma(\tau)) = 0, \quad \Psi(\sigma(\tau)) + \gamma_1 = 0, \quad \Psi'(\sigma(\tau)) < 0$$

or

$$G_j(\sigma(\tau)) = 0, \quad \Psi(\sigma(\tau)) + \gamma_2 = 0, \quad \Psi'(\sigma(\tau)) < 0$$

Consequently, we have for $i = 1$ or $i = 2$

$$w(\tau) < 0, \quad \Psi(\sigma(\tau)) + \gamma_i = 0, \quad \Psi'(\sigma(\tau)) = 0,$$

a contradiction to (ii).

It follows that for all $t \ge 0$ the inclusion $\vartheta_1 - j\Delta \le \sigma(t) \le \vartheta_2 + j\Delta$ takes place. ∎

Proof of Theorem 5.6.3 Since the frequency-domain condition (ii) is satisfied we may apply Theorem 1.4.2, p. 9, to conclude that there exists a real matrix $H = H^*$ of order n such that for all $z \in \mathbf{R}^n$ and $\xi \in \mathbf{R}$

$$2z^* H[(A + \lambda I)z + b\xi] + \xi(c^* z + \varrho\xi) + \varepsilon(c^* z + \varrho\xi)^2 \le 0. \tag{5.6.22}$$

Setting $\xi = 0$ we get from (5.6.22) the inequality

$$2z^* H(A + \lambda I)z \le -\varepsilon(c^* z)^2$$

for all $z \in \mathbf{R}^n$. Using the fact that $D(\cdot)$ is non-degenerate and $A + \lambda I$ is Hurwitzian we get from Lemma 1.2.1, p. 6, that $H > 0$. In order to use Lemma 5.6.1 we consider an arbitrary

118

solution $\sigma(\cdot)$, $z(\cdot)$ of (5.6.13) and introduce the functions w, $f : \mathbf{R}_+ \to \mathbf{R}$ and $\Psi : \mathbf{R} \to \mathbf{R}$ by $w(t) = z(t)^* H z(t)$, $f(t) = g(t)$ and $\Psi(\vartheta) = \varphi(\vartheta)$.

We can quickly verify that for these functions and the numbers λ, ε, γ_1 and γ_2 of the present theorem the requirements (i), (iv) and (v) of Lemma 5.6.1 are satisfied. Because of $H > 0$ condition (ii) is also valid. From (5.6.22) it follows that

$$\dot{w}(t) + 2\lambda w(t) + \varepsilon \dot{\sigma}(t)^2 + [\Psi(\sigma(t)) + f(t)]\dot{\sigma}(t) \leq 0$$

for all $t \geq 0$. This means that condition (iii) of Lemma 5.6.1 is verified. Lemma 5.6.1 guarantees the boundedness of $\sigma(\cdot)$ on $\mathbf{R}_+$. The boundedness of $z(\cdot)$ on $\mathbf{R}_+$ follows from the fact that A is Hurwitzian and $\varphi(\cdot)$ is bounded. $\blacksquare$

Remark 5.6.1 The non-local reduction principle is applicable to various other problems also. In the paper [85] on the basis of non-local reduction a new freqency-domain criterion for absolute stability of control systems is obtained. In certain cases this criterion gives a wider absolute stability sector than the circle criterion (Chapter 1).

Furthermore, by the reduction method a theorem of A. Yu. Levin [105] on the stability of the second order equation $\ddot{\vartheta} + a\dot{\vartheta} + u(t)\vartheta = 0$ is extended in [87] to systems (5.6.6) with $\varphi(\vartheta) \equiv \vartheta$.

A. A. Voronov [151] has formulated the following conjecture: If the nonlinearity of a single-loop control system lies near the origin in the instability sector and lies for large arguments in the stability sector then self-oscillations occur. Using the non-local reduction principle it is shown in [88] that for a certain class of control systems the conjecture of A. A. Voronov is true.

5.7 A Constructive Approach for Determining Lagrange Stability Criteria in the Non-Autonomous Case

In this section we want to demonstrate a more constructive technique than in the previous sections which allows us to derive various types of reduced equations and Lagrange stability criteria for autonomous and non-autonomous systems like (5.3.1) and (5.5.6), respectively. It will be shown that this technique leads in a natural manner to the reduced system (5.5.1) as the simplest comparison system.

The results of this section are due to N. Koksch (see also [70]).

Let us consider the differential equation

$$
\begin{aligned}
\dot{z} &= Az &+ \quad bu(t)\varphi(\sigma), \\
\dot{\sigma} &= c^* z &+ \quad \varrho v(t)\varphi(\sigma),
\end{aligned}
\tag{5.7.1}
$$

where A is an $n \times n$ Hurwitz matrix, b and c are n-vectors, ϱ is a number, $u : \mathbf{R}_+ \to (0, +\infty)$ is a continuously differentiable and bounded function, $v : \mathbf{R}_+ \to \mathbf{R}$ is continuous, $\varphi : \mathbf{R} \to \mathbf{R}$ is of class $\mathbf{C}^1$ and Δ-periodic. Suppose also that φ has at least one zero and $(\varphi(\sigma))^2 + (\varphi'(\sigma))^2 \neq 0$ for all $\sigma \in \mathbf{R}$. We assume furthermore that $\Gamma := -c^* b > 0$.

The basic idea of the method, which will be pointed out in this section, is as follows. There will be constructed in the space (t, z, σ) a family of sets M_α, which are bounded with respect to (z, σ), cover the whole set $\mathbf{R}_+ \times \mathbf{R}^n \times \mathbf{R}$ and which are positively invariant for (5.7.1).

If such a family of sets exists, it follows that (5.7.1) is Lagrange stable.

Positively invariant sets for (5.7.1) are characterized by the property, that solutions of (5.7.1) can intersect the boundary of such a set only in one direction. In the sequel the sets M_α, having the mentioned properties, will be constructed in the form

$$\{(t, z, \sigma) \ : \ V(t, z, \sigma) \leq \Phi_i(z), \ i = 1, 2\},$$

where V and Φ_i, $i = 1, 2$, are functions to be determined.

The following theorem is true.

Theorem 5.7.1 *Suppose that the sets T, X, Y, $\Sigma \subset \mathbf{R}$ and $Z \subset \mathbf{R}^n$ are given, $V : T \times Z \times \Sigma \to \mathbf{R}_+$ is C^1 and $P : \Sigma \times X \times Y \to \mathbf{R}$ is continuous and Lipschitz continuous in the second argument. Suppose also that*

$$P(\sigma, x, y) \geq \underline{P}(\sigma, x, y) := \sup_{\substack{(t, z) \, \in \, T \times Z \\ z^* H z \, = \, x}} \left[\dot{V}_{(5.7.1)}(t, z, \sigma) - y\left(c^* z + \varrho v(t) \varphi(\sigma)\right) \right]$$

for all $(\sigma, x, y) \in \Sigma \times X \times Y$. Suppose that $\Phi : [\vartheta_1, \vartheta_2] \subset \Sigma \to X$ with $\Phi'(\sigma) \in Y$ for $\sigma \in [\vartheta_1, \vartheta_2]$ is a solution of

$$P(\sigma, \Phi(\sigma), \Phi'(\sigma)) = 0 \tag{5.7.2}$$

on $(\vartheta_1, \vartheta_2)$.

If $(z(\cdot), \sigma(\cdot))$ is an arbitrary solution of (5.7.1) on $[t_0, +\infty)$ with $V(t_0, z(t_0), \sigma(t_0)) \leq \Phi(\sigma(t_0))$ and $\sigma(t) \in (\vartheta_1, \vartheta_2)$, $z(t) \in Z$ for all $t \in [t_0, t_1] \subset T$ then it follows that

$$V(t, z(t), \sigma(t)) \leq \Phi(\sigma(t))$$

for all $t \in [t_0, t_1]$.

Proof Suppose $(z(\cdot), \sigma(\cdot))$ is a solution described in the statement of the theorem. If we set $w(t) := V(t, z(t), \sigma(t)) - \Phi(\sigma(t))$ we have

$$
\begin{aligned}
\dot{w}(t) &= \frac{d}{dt} V(z, z(t), \sigma(t)) - \Phi'(\sigma(t))\dot{\sigma}(t) \\
&\leq \underline{P}(\sigma(t), \Phi(\sigma(t)) + w(t), \Phi'(\sigma(t))) \\
&\leq P(\sigma(t), \Phi(\sigma(t)) + w(t), \Phi'(\sigma(t))) := F(t, w(t)).
\end{aligned}
$$

Since $w_0(t) \leq 0$ and $w_0(t) \equiv 0$ is a solution of $\dot{w} = F(t, w)$, where F is Lipschitz continuous in the second argument, it follows that $w(t) \leq 0$ for all $t \in [t_0, t_1]$. $\blacksquare$

In order to use Theorem 5.7.1 for the boundedness investigation of system (5.7.1) we set $T := [t_0, +\infty)$, $Z := \mathbf{R}^n$, $\Sigma := \mathbf{R}$, $X := \mathbf{R}_+$, $Y := \mathbf{R}$ and define the function $V : T \times Z \times \Sigma \to \mathbf{R}_+$ by $V(t, z, \sigma) = \kappa(t) z^* H z$, where $H = H^*$ is a positive definite $n \times n$ matrix and $\kappa : T \to \mathbf{R}_+$ is a continuous function.

Let us start our investigation of (5.7.1) with the case $u(t) \equiv v(t) \equiv 1$ and choose $\kappa(t) \equiv 1$, $t_0 = 0$.

Using the ansatz from above for V we can write for arbitrary $(\sigma, x, y) \in \Sigma \times X \times Y$

$$
\begin{aligned}
\underline{P}(\sigma, x, y) &= \sup_{z \, : \, z^* H z = x} \left[\dot{V}_{(5.7.1)}(t, z, \sigma) - y(c^* z + \varrho \varphi(\sigma)) \right] \\
&= \sup_{z \, : \, z^* H z = x} \left[2 z^* H (A z + b\varphi(\sigma)) - y(c^* z + \varrho \varphi(\sigma)) \right].
\end{aligned}
$$

It is easy to see that for all $(\sigma, x, y) \in \Sigma \times X \times Y$ we have

$$\underline{P}(\sigma, x, y) \leq \underline{P}_1(\sigma, x, y) \leq \underline{P}_2(\sigma, x, y), \tag{5.7.3}$$

where

$$\underline{P}_1(\sigma, x, y) := \sup_{z \, : \, z^* H z = x} 2 z^* H A z + \sup_{z \, : \, z^* H z = x} \left[2 z^* H b\varphi(\sigma) - y c^* z \right] - y \varrho \varphi(\sigma),$$

$$\underline{P}_2(\sigma, x, y) := \sup_{z \, : \, z^* H z = x} 2 z^* H A z + \sup_{z \, : \, z^* H z = x} 2 z^* H b\varphi(\sigma) - \inf_{z \, : \, z^* H z = x} y c^* z - y \varrho \varphi(\sigma).$$

Clearly, that

$$\sup_{z\,:\,z^*Hz=x} 2z^*HAz \le 2\gamma x$$

with $\gamma := \frac{1}{2}\lambda_{min}(A+H^{-1}AH)$, where $\lambda_{min}(\cdot)$ is the smallest eigenvalue of the matrix in brackets. If we know additionally that for a real λ

$$H(A+\lambda I)+(A+\lambda I)^*H \le 0 \tag{5.7.4}$$

we get immediately that for the γ from above yields $\gamma \le \lambda$.

According to the linear parts we can write

$$\sup_{z\,:\,z^*Hz=x}\left[z^*(2Hb\varphi(\sigma)-yc)\right]^2 = \left[2Hb\varphi(\sigma)-yc\right]^*H^{-1}\left[2Hb\varphi(\sigma)-yc\right]x =: \underline{\beta}(\sigma,y)\cdot x.$$

Suppose now that λ satisfies (5.7.4) and $\beta : \mathbf{R}\times\mathbf{R} \to \mathbf{R}$ is a continuous function with $\beta(\sigma,y) \ge \underline{\beta}(\sigma,y)$ for all $(\sigma,y)\in\mathbf{R}\times\mathbf{R}$. Therefore, the function

$$P_1(\sigma,x,y) := -2\lambda x + \sqrt{\beta(\sigma,y)}\sqrt{x} - y\varrho\varphi(\sigma) \ge \underline{P}_1(\sigma,x,y)$$

yields the assumption of Theorem 5.7.1.

The comparision equation $P_1(\sigma,\Phi(\sigma),\Phi'(\sigma)) = 0$ can be transformed by setting $\Phi(\sigma) := \frac{1}{2}F(\sigma)^2$ to an implicit differential equation for F.

We get a simpler comparison equation if we calculate the expression $\underline{P}_2(\sigma,x,y)$. But this system is far away from the optimal one and not defined on the whole set $\mathbf{R}\times\mathbf{R}_+\times\mathbf{R}$.

As an alternative to this situation we want to assume an additional property for H:

$$2Hb = -c. \tag{5.7.5}$$

This equation immediately implies that

$$c^*H^{-1}c = -2c^*H^{-1}Hb = -2c^*b = \Gamma > 0$$

and, that we can choose β in the form

$$\begin{aligned}
\beta(\sigma,y) &:= \underline{\beta}(\sigma,y)\\
&= [\varphi(\sigma)+y]^2c^*H^{-1}c = -2c^*b[\varphi(\sigma)+y]^2.
\end{aligned}$$

Thus, the function P_1 defined by

$$P_1(\sigma,x,y) = -2\lambda x + \sqrt{-2c^*b}|\varphi(\sigma)+y|\sqrt{x} - y\varrho\varphi(\sigma)$$

satisfies the assumptions of Theorem 5.7.1. Taking the ansatz $\Phi(\sigma) = F(\sigma)^2/2$ we get the comparison equation

$$\left[\frac{\varrho}{\sqrt{\gamma}}\varphi(\sigma) - s(\sigma,F(\sigma),F'(\sigma))F(\sigma)\right]F'(\sigma) = -\frac{\lambda}{\sqrt{\Gamma}}F(\sigma) + s(\sigma,F(\sigma),F'(\sigma))\varphi(\sigma), \tag{5.7.6}$$

where $s(\sigma,F(\sigma),F'(\sigma)) = \operatorname{sgn}\left[F(\sigma)(\varphi(\sigma)+F(\sigma)F'(\sigma))\right]$. This equation is a piece-wise explicit differential equation for F and corresponds in the cases

$$s(\sigma,F(\sigma),F'(\sigma)) \equiv 1 \qquad \text{or} \qquad s(\sigma,F(\sigma),F'(\sigma)) \equiv -1$$

to the two-dimensional system

$$\begin{aligned}
\dot\sigma &= -s\eta &&+ \frac{\varrho}{\sqrt{\Gamma}}\varphi(\sigma)),\\
\dot\eta &= -\frac{\lambda}{\sqrt{\Gamma}}\eta &&+ s\varphi(\sigma))
\end{aligned} \tag{5.7.7}$$

with $s = 1$ or $s = -1$, respectively. We remark, that (5.7.7) is identically to (5.5.29) for $s = -1$.

Suppose for a time that there exist two scalar functions F_1 and F_2 satisfying the following properties:

(i) F_1 and F_2 are defined on $(-\infty, \vartheta_1]$ and $[\vartheta_2, \infty) \to \mathbf{R}$, respectively;

(ii) $F_1(\vartheta_1) = F_2(\vartheta_2) = 0$;

(iii) $\lim\limits_{\vartheta \to -\infty} |F_1(\vartheta)| = +\infty$ and $\lim\limits_{\vartheta \to +\infty} |F_2(\vartheta)| = +\infty$;

(iv) F_1 and F_2 are solutions of (5.7.6).

$$(5.7.8)$$

If there exist such functions, then the functions $F_i(\vartheta_i + k\Delta)$, $k \in \mathbf{Z}$, are also solutions of (5.7.6) and the sets of the type

$$M(k_1, k_2) := \left\{ (t, z, \sigma) \,:\, t \in \mathbf{R}_+, \ \sigma \in [\vartheta_2 + k_2\Delta, \vartheta_1 + k_1\Delta], \ 2z^* H z \leq F_i(\sigma - k\Delta)^2, \ i = 1, 2 \right\}$$
$$(5.7.9)$$

are non-empty in the case $\vartheta_2 + k_2\Delta \leq \vartheta_1 + k_1\Delta$, are positively invariant for (5.7.1) and bounded with respect to (z, σ). For any solution of (5.7.1) k_1 and k_2 may be chosen, in such a way that the initial point lies in the set $M(k_1, k_2)$. It follows, that in this case system (5.7.1) is Lagrange stable.

Our aim is to find such solutions F_i of (5.7.6) which have the property (5.7.8) and for which $s(\vartheta, F_i(\vartheta), F_i'(\vartheta)) \equiv s$ with $s = 1$ or $s = -1$. For these solutions we can write (5.7.6) and (5.7.7) in the form

$$- sF(\sigma)F'(\sigma) + \frac{\varrho}{\sqrt{\Gamma}}F'(\sigma) + \frac{\lambda}{\sqrt{\Gamma}}F(\sigma) - s\varphi(\sigma) = 0 \tag{5.7.10}$$

and

$$\begin{aligned} \dot{\sigma} &= -s\eta &&+ \ \frac{\varrho}{\sqrt{\Gamma}}\varphi(\sigma)), \\ \dot{\eta} &= -\frac{\lambda}{\sqrt{\Gamma}}\eta &&+ \ s\varphi(\sigma)). \end{aligned} \tag{5.7.11}$$

As (5.7.8) and (5.7.10) are invariant with respect to the transformation $F \to -F$, $s \to -s$, we consider, w.l.o.g, the case $s = -1$, e.g. we have to find solutions F_i of (5.7.10) with $s = 1$ and

$$\operatorname{sgn}\left[F_i(\sigma)\big(\varphi(\sigma) + F_i(\sigma)F_i'(\sigma)\big)\right] \equiv -1. \tag{5.7.12}$$

for $i = 1, 2$.

A necessary condition for this is $\lambda > 0$. If $\lambda\varrho \neq \Gamma$ every stationary point of (5.7.11) has the form $(\sigma_0, 0)$ with $\varphi(\sigma_0) = 0$.

Note that in cases when φ has no zeros at least one of the solutions described in (5.7.8) does not exist.

We suppose that φ has a zero on $\mathbf{R}$. From $\varphi(\sigma)^2 + \varphi'(\sigma)^2 \neq 0$ on $\mathbf{R}$ and the periodicity of φ it follows then that φ has on $[0, \Delta)$ at least one zero where φ' is positive and one zero where φ' is negative. A direct computation shows that the equilibrium $(\vartheta_0, 0)$ is a saddle if $(\lambda\varrho - \Gamma)\varphi'(\vartheta_0) > 0$. Let us assume that $(\vartheta_0, 0)$ denotes this saddle-point.

We shall determine the functions F_1 and F_2 as representations of a part of the stable manifold of certain saddle-points of (5.7.11).

The investigation of the phase portrait in a neighborhood of a saddle-point shows, that the stable manifold crosses the σ-axis with increasing σ- and dercreasing η-coordinates.

Hence let us introduce the following assumption:

There exist numbers

$$\sigma_0 \in \mathbf{R}, \quad \eta_0 > \max\left\{\max_{\vartheta} -\sqrt{\Gamma}\varphi(\vartheta)/\lambda, \ \max_{\vartheta} -\varrho\varphi(\vartheta)/\sqrt{\Gamma}\right\} \tag{5.7.13}$$

and $t_1 \in \mathbf{R}_+ \cup \{+\infty\}$ such that the solution $\sigma(\cdot; \sigma_0, \eta_0)$, $\eta(\cdot, \sigma_0, \eta_0)$ of (5.7.11) satisfies for $t \in (0, t_1)$ the inequality $\eta(t; \sigma_0, \eta_0) > 0$ and $\lim_{t \to t_1} \eta(t; \sigma_0, \eta_0) = 0$.

Note that because of the periodicity of φ this assumption is necessary for the existence of a function F_1 satisfying (5.7.8). We remark, that $\eta = -\varrho\varphi(\sigma)/\sqrt{\Gamma}$ and $\eta = -\sqrt{\Gamma}\varphi(\sigma)/\lambda$ describe the curves on which $\dot{\sigma} = 0$ or $\dot{\eta} = 0$, respectively.

The next lemma shows that (5.7.13) is also sufficient for the existence of a function F_1 which satisfies the first three requirements of (5.7.8) and, additionally, $F_1(\sigma) > 0$ for all $\sigma < \vartheta_1$.

Lemma 5.7.1 *Suppose that for* (5.7.11) *assumption* (5.7.13) *is satisfied. Then there exists a saddle-point* $(\vartheta_1, 0)$ *of* (5.7.11) *and a solution* $(\widetilde{\sigma}_1(\cdot), \widetilde{\eta}(\cdot))$ *of* (5.7.11) *satisfying the following properties:*

(i) $\lim_{t \to \infty} \widetilde{\eta}_1(t) = \vartheta_1$ *and* $\lim_{t \to \infty} \widetilde{\eta}_1(t) = 0$;

(ii) $\lim_{t \to -\infty} \widetilde{\sigma}_1(t) = -\infty$ *and* $\lim_{t \to -\infty} \widetilde{\sigma}_1(t) = +\infty$;

(iii) $\widetilde{\eta}_1(t) > 0$ *for all* $t \in \mathbf{R}$;

(iv) $\dot{\widetilde{\sigma}}(t) > 0$ *for all* $t \in \mathbf{R}$.

Proof The proof of this lemma uses a similar argument as the corresponding proofs in Section 2.2. ∎

With the use of Lemma 5.7.1 we can construct in a correct way a function $F_1 : (-\infty, \vartheta_1) \to \mathbf{R}$ by setting $F_1(\widetilde{\sigma}(t)) = \widetilde{\eta}_1(t)$ on $\mathbf{R}$, which satisfies the first three requirements of (5.7.8) and the inequalities $F_1(\vartheta) + \varrho\varphi(\vartheta)/\sqrt{\Gamma} > 0$ and $F_1(\vartheta) > 0$ on $(-\infty, \vartheta_1)$. This function is a solution of (5.7.10) with $s = -1$.

If we now have, additionally, the property

$$\lambda F_1^2(\vartheta) > \varrho\varphi^2(\vartheta) \tag{5.7.14}$$

for all $\vartheta < \vartheta_1$, it follows from (5.7.10) that

$$F_1(\vartheta)F_1'(\vartheta) + \varphi(\vartheta) = \frac{-\lambda F_1^2(\vartheta) + \varrho\varphi^2(\vartheta)}{\sqrt{\Gamma}(F_1(\vartheta) + \varrho\varphi(\vartheta)/\sqrt{\Gamma})} < 0$$

for all $\vartheta < \vartheta_1$. Thus function F_1 also satisfies (5.7.12) with $i = 1$. It follows that F_1 is really a solution of (5.7.6) and satisfies also the last requirement of (5.7.8).

In the case $\varrho \leq 0$ property (5.7.14) is immediately satisfied.

The case $\varrho > 0$ is more complicate. But in the case $\lambda\varrho > \Gamma$, it is easy to see that the assumption

$$|\varphi(\sigma)| \leq \sqrt{\frac{\lambda\varrho}{\Gamma}} \max_{\vartheta} \varphi(\vartheta) \tag{5.7.15}$$

for all $\sigma \in \mathbf{R}$ is sufficient for (5.7.14).

In an analogous manner one can formulate a lemma which guarantees the existence of a function F_2 satisfying (5.7.8). For this purpose we state for (5.7.11) the assumption:

There exist numbers

$$\sigma_0 \in \mathbf{R}, \qquad \eta_0 < \min\left\{ \min_{\vartheta} -\sqrt{\Gamma}\varphi(\vartheta)/\lambda,\ \min_{\vartheta} -\varrho\varphi(\vartheta)/\sqrt{\Gamma} \right\} \tag{5.7.16}$$

and $t_1 \in \mathbf{R}_+ \cup \{+\infty\}$ such that the solution $\sigma(\cdot; \sigma_0, \eta_0)$, $\eta(\cdot, \sigma_0, \eta_0)$ of (5.7.11) satisfies for $t \in (0, t_1)$ the inequality $\eta(t; \sigma_0, \eta_0) < 0$ and $\lim_{t \to t_1} \eta(t; \sigma_0, \eta_0) = 0$.

Under the assumption (5.7.16) an result, analogous to Lemma 5.7.1, guarantees the existence of a solution F_2 of (5.7.10) satisfying the first three properties of (5.7.8). In case $\varrho > 0$ the property (5.7.12) with $i = 2$ is fulfilled obviously.

Assuming that in the case $\lambda\varrho > \Gamma$

$$|\varphi(\sigma)| \le -\sqrt{\frac{\lambda\varrho}{\Gamma}}\,\min_{\vartheta}\varphi(\vartheta) \tag{5.7.17}$$

for all $\sigma \in \mathbf{R}$, what is sufficient for (5.7.14), we get that F_2 also satisfies (5.7.12) with $i = 2$. It follows that F_2 is a solution of (5.7.6) and all requirements of (5.7.8) are fulfilled.

It is easy to see that the assumptions (5.7.13) and (5.7.16) together are equivalent to the Lagrange stability of (5.7.11). Conditions (5.7.15) and (5.7.17) are satisfied if

$$|\varphi(\sigma)| \le \sqrt{\frac{\lambda\varrho}{\Gamma}}\,\min\left\{ \max_{\vartheta}\varphi(\vartheta),\ -\min_{\vartheta}\varphi(\vartheta) \right\}. \tag{5.7.18}$$

Thus, our previous results may be summerized in the following

Theorem 5.7.2 *Consider system (5.7.11) with $s = -1$, $\lambda > 0$. If this system is Lagrange stable and either $\varrho \le 0$ or, in the case $\lambda\varrho > \Gamma$, condition (5.7.18) is satisfied, then there exist two functions F_1 and F_2 satisfying (5.7.8).*

In the following theorem the existence of a matrix H yielding (5.7.4) and (5.7.5) follows from frequency-domain conditions. Clearly, the comparison equation (5.7.2) takes the form (5.7.11).

Recall that the transfer function $D(s) = c^*(A - sI)^{-1}b$ of (5.7.1) is supposed to be non-degenerate and that $\Gamma = -c^*b = \lim_{s\to\infty} sD(s) > 0$.

By using the comparison technique of this section one obtains the following refinement of Theorem 5.5.1.

Theorem 5.7.3 *Suppose that the assumptions (i) – (iii) of Theorem 5.5.1 are satisfied and either $\varrho \le 0$ or $\lambda\varrho > \Gamma$ and inequality (5.7.18) is true. Then system (5.7.1) is Lagrange stable.*

Proof According to Theorem 1.4.2, p. 9, there exists a matrix $H = H^*$ such that (5.7.4) and (5.7.5) are satisfied. Define $V(t, z, \sigma) = z^*Hz$. From the remarks taken above it follows that (5.7.11) is a comparison system in the spirit of this section. Theorem 5.7.2 guarantees the existence of functions F_1 and F_2 satisfying (5.7.8). Using these functions we get a family of sets $M(k_1, k_2)$, which are positively invariant for (5.7.1), bounded with respect to (z, σ) and which cover the set $\mathbf{R}_+ \times \mathbf{R}^n \times \mathbf{R}$. ∎

Let us now consider the case when $u(\cdot)$ and $v(\cdot)$ in (5.7.1) and $\kappa(\cdot)$ are not necessarily constant. Suppose further that for (5.7.1) there exists a matrix $H = H^*$ satisfying (5.7.4) and (5.7.5). Define a function $V(t, z, \sigma) := \kappa(t)z^*Hz$ where $\kappa : \mathbf{R}_+ \to \mathbf{R}$ is still unknown.

Let us determine comparison systems in the form (5.7.11). We have

$$\underline{P}(\sigma, x, y) = \sup_t Q_1(t, \sigma, x, y)$$

with

$$Q_1(t, \sigma, x, y) = \sup_{z\,:\,\kappa(t)z^*Hz=x} \Big[\kappa(t)z^*(HA + A^*H)z + \dot{\kappa}(t)z^*Hz -$$
$$z^*(2\kappa(t)Hbu(t)\varphi(\sigma) - yc) - y\varrho v(t)\varphi(\sigma)\Big].$$

Because of (5.7.4) it is true that

$$\sup_{z\,:\,\kappa(t)z^*Hz=x} [\kappa(t)z^*(HA + A^*H)z + \dot{\kappa}(t)z^*Hz] \leq -2\left(\lambda - \frac{\dot{\kappa}(t)}{2\kappa(t)}\right) x,$$

and, because of (5.7.5),

$$\sup_{z\,:\,\kappa(t)z^*Hz=x} z^* [2\kappa(t)Hbu(t)\varphi(\sigma) - yc] = |\kappa(t)u(t)\varphi(\sigma) + y|\sqrt{\Gamma/\kappa(t)}\sqrt{2x}.$$

It follows that

$$Q_1(t, \sigma, x, y) \leq Q_2(t, \sigma, x, y)$$

for all $t \geq 0$, $\sigma \in \mathbf{R}$, $x \in \mathbf{R}_+$, $y \in \mathbf{R}$, where

$$Q_2(t, \sigma, x, y) = -2\left(\lambda - \frac{\dot{\kappa}(t)}{2\kappa(t)}\right) x + |\kappa(t)u(t)\varphi(\sigma) + y|\sqrt{\Gamma/\kappa(t)}\sqrt{2x} - y\varrho v(t)\varphi(\sigma).$$

In order to estimate the supremum over t we give several possible realizations of this estimation.

1. Assume that $v(t) = \sqrt{u(t)}$ for all $t \geq 0$. Let us choose $\kappa(t) \equiv 1/u(t)$ and write

$$Q_2(t, \sigma, x, y) = \sqrt{\Gamma u(t)}\left[-2x\frac{\lambda + \dot{u}(t)/(2u(t))}{\sqrt{\Gamma u(t)}} + |\varphi(\sigma) + y|\sqrt{2x} - \frac{\varrho}{\Gamma}\varphi(\sigma)y\right].$$

Now take a $\delta > 0$ such that

$$\delta < \delta_0(t) := \frac{\lambda + \dot{u}(t)/(2u(t))}{\sqrt{u(t)}}$$

for all $t \geq t_0$, t_0 sufficiently large.

It follows for $T = [t_0, +\infty)$ and $0 < \underline{u} \leq u(t) \leq \overline{u}$ that

$$\underline{P}(\sigma, x, y) \leq \sqrt{\Gamma \overline{u}}\left[-2\frac{\delta}{\sqrt{\Gamma}}x + |\varphi(\sigma) + y|\sqrt{2x} - \frac{\varrho}{\sqrt{\Gamma}}\varphi(\sigma)y\right], \tag{5.7.19}$$

proposed that $-2\frac{\delta}{\sqrt{\Gamma}}x + |\varphi(\sigma) + y|\sqrt{2x} - \frac{\varrho}{\sqrt{\Gamma}}\varphi(\sigma)y \geq 0$, and $\underline{P}(\sigma, x, y) \leq 0$ in the remaining cases. Thus, taking $\Phi(\sigma) = F(\sigma)^2/2$ we get the comparison equation

$$F'(\sigma)\left[-F(\sigma) - \frac{\varrho}{\sqrt{\Gamma}}\varphi(\sigma)\right] - \frac{\delta}{\sqrt{\Gamma}}F(\sigma) - \varphi(\sigma) = 0,$$

where we have to require that a solution $F(\cdot)$, which is used for the construction of a positively invariant set, has to satisfy the inequality $\varphi(\sigma) + F'(\sigma)F(\sigma) < 0$ for all σ.

In connection with equation (5.7.19) we consider the system

$$\begin{aligned}
\dot{\sigma} &= \eta && + \frac{\varrho}{\sqrt{\Gamma}}\varphi(\sigma)), \\
\dot{\eta} &= -\frac{\delta}{\sqrt{\Gamma}}\eta && - \varphi(\sigma)).
\end{aligned} \tag{5.7.20}$$

It follows that we have proved the

Theorem 5.7.4 *Suppose that the function $u(\cdot)$ in (5.7.1) is bounded from below by a positive constant and v is given by $v(t) = \sqrt{u(t)}$ for $t \geq t_0$. Suppose also that there exists a $\lambda > 0$ and a $\delta > 0$ such that the following conditions hold:*

(i) *$\varrho \leq 0$ or $\delta\varrho > \Gamma$ and $|\varphi(\sigma)| \leq \sqrt{\delta\varrho/\Gamma} \cdot \min\left\{\max_{\vartheta}\varphi(\vartheta), -\min_{\vartheta}\varphi(\vartheta)\right\}$ for all $\sigma \in \mathbf{R}$;*

(ii) *system (5.7.20) is Lagrange stable;*

(iii) *$\operatorname{Re} D(i\omega - \lambda) \geq 0$ for all $\omega \in \mathbf{R}$;*

(iv) *D is non-degenerate and $A + \lambda I$ is Hurwitz;*

(v) *$\delta\sqrt{u(t)} \leq \lambda + \dot{u}(t)/(2u(t))$ for all $t \geq t_0$ for a certain t_0.*

Then system (5.7.1) is Lagrange stable.

Remark 5.7.1 For $\varrho = 0$ Theorem 5.7.4 corresponds with Theorem 5.6.2. But in the present theorem, in contrast to Theorem 5.6.2, some conditions on the parameters of (5.7.1) are relaxed.

2. Let us suppose now $v(t) = u(t)$ and choose $\kappa(t) = 1/u(t)$. For Q_2 we get the representation

$$Q_2(t,\sigma,x,y) = u(t)\sqrt{\Gamma}\left[-2x\frac{\lambda + \dot{u}(t)/(2u(t))}{\sqrt{\Gamma}u(t)} + |\varphi(\sigma) + y|\sqrt{1/u(t)}\sqrt{2x} - \frac{\varrho}{\Gamma}\varphi(\sigma)y\right].$$

Using the inequalities $0 < \underline{u} \leq u(t) \leq \overline{u}$ and

$$\delta < [\lambda + \dot{u}(t)/(2u(t))]/u(t)$$

for sufficiently large t, we get

$$\underline{P}(\sigma,x,y) \leq \overline{u}\sqrt{\Gamma}\left[-2\frac{\delta}{\sqrt{\Gamma}}x + |\varphi(\sigma) + y|\sqrt{1/\underline{u}}\sqrt{2x} - \frac{\varrho}{\sqrt{\Gamma}}\varphi(\sigma)y\right],$$

proposed that $-2\frac{\delta}{\sqrt{\Gamma}}x + |\varphi(\sigma) + y|\sqrt{1/\underline{u}}\sqrt{2x} - \frac{\varrho}{\sqrt{\Gamma}}\varphi(\sigma)y \geq 0$, and

$$\underline{P}(\sigma,x,y) \leq \underline{u}\sqrt{\Gamma}\left[-2\frac{\delta}{\sqrt{\Gamma}}x + |\varphi(\sigma) + y|\sqrt{1/\underline{u}}\sqrt{2x} - \frac{\varrho}{\sqrt{\Gamma}}\varphi(\sigma)y\right]$$

in the remaining cases.

As a result we get the comparison system

$$\begin{aligned}
\dot{\sigma} &= \eta && + \varrho\frac{u}{\sqrt{\Gamma}}\varphi(\sigma), \\
\dot{\eta} &= -\delta\frac{u}{\sqrt{\Gamma}}\eta && - \varphi(\sigma).
\end{aligned} \tag{5.7.21}$$

We can conclude that the following theorem is true.

126

Theorem 5.7.5 *Suppose that $u(t) = v(t) \geq \underline{u} > 0$ for all $t \geq t_0$. Suppose also that there exist $\lambda > 0$ and $\delta > 0$ such that the following conditions hold:*

(i) $\varrho \leq 0$ or $\delta\varrho\underline{u} > \Gamma$ and $|\varphi(\sigma)| \leq \sqrt{\delta\varrho\underline{u}/\Gamma} \cdot \min\left\{\max_{\vartheta}\varphi(\vartheta), -\min_{\vartheta}\varphi(\vartheta)\right\}$ for all $\sigma \in \mathbf{R}$;

(ii) *system (5.7.21) is Lagrange stable;*

(iii) $\operatorname{Re} D(i\omega - \lambda) \geq 0$ *for all* $\omega \in \mathbf{R}$;

(iv) *D is non-degenerate and $A + \lambda I$ is Hurwitzian;*

(v) $\delta u(t) \leq \lambda + \dot{u}(t)/(2u(t))$ *for all $t \geq t_1$, where t_1 is a certain number.*

Then system (5.7.1) is Lagrange stable.

Chapter 6

Circular Solutions and Cycles

In this chapter we continue the investigation of asymptotic properties of pendulum-like systems. In order to get necessary conditions for global convergence we provide theorems which ensure the existence of circular solutions and of cycles of various types. The proofs are based upon the construction of Poincaré maps which map convex and compact sets into itself and upon the use of Brouwer's fixed-point theorem.

6.1 Pendulum-Like Systems with a Single Sign-Constant Nonlinearity

Consider the pendulum-like system

$$
\begin{aligned}
\dot{z} &= Az + b\varphi(\sigma), \\
\dot{\sigma} &= c^*z + \varrho\varphi(\sigma),
\end{aligned}
\tag{6.1.1}
$$

in which A is a regular $n \times n$ matrix, b and c are n-vectors and ϱ is a number. The function $\varphi : \mathbf{R} \to \mathbf{R}$ is supposed to be Δ-periodic and continuous. It is also assumed that the transfer function of the linear part of (6.1.1) given by $\chi(s) = \frac{1}{s}[c^*(A - sI)^{-1}b - \varrho]$ is non-degenerate.

In the next theorem which is taken from [109], [92], we suppose additionally that φ has no zeros and $c^*A^{-1}b - \varrho \neq 0$. For definiteness we assume $\varphi(\sigma) < 0$ on $\mathbf{R}$ and $c^*A^{-1}b - \varrho > 0$.

Theorem 6.1.1 *Suppose that the matrix A has no eigenvalues with real part zero. Then system (6.1.1) has a cycle of the second kind.*

Proof Let us introduce the function $V : \mathbf{R}^n \times \mathbf{R} \to \mathbf{R}$ by $V(z,\sigma) = \sigma - c^*A^{-1}z$ and the two sets $\Omega_1 = \{(z,\sigma) : V(z,\sigma) = 0\}$ and $\Omega_2 = \{(z,\sigma) : V(z,\sigma) = \Delta\}$. The derivative of V along the solutions of (6.1.1) satisfies

$$
\dot{V}(z,\sigma) = -(c^*A^{-1}b - \varrho)\varphi(\sigma) > \delta > 0
\tag{6.1.2}
$$

for all $\sigma \in \mathbf{R}$, where $\delta > 0$ is a certain number.

It follows from (6.1.2) that for any $x := (\eta, \xi) \in \Omega_1$ there exists a time $t = t(x)$ such that $(z(t(x), x), \sigma(t(x), x)) \in \Omega_2$. It is easy to see that $t(x) < \Delta/\delta$. Thus there is defined a mapping $T : \Omega_1 \to \Omega_2$ which is continuous by the theorem of the continuous dependence of initial data and the fact that the set Ω_2 is by (6.1.2) without contact with the vector field. Using the continuous mapping $Q : \Omega_2 \to \Omega_1$ defined by $Q(z,\sigma) = (z, \sigma - \Delta)$ we see that $(Q \circ T)\Omega_1 \subset \Omega_1$.

Let us write now the initial data for (6.1.1) in the form $x = (\eta, c^* A^{-1}\eta) \in \Omega_1$. By the first equation of (6.1.1) we get for the first component of $Q(x)$

$$z(t(x), x) = e^{At(x)}\eta + \int_0^{t(x)} e^{A(t(x)-s)} b\varphi(\sigma(s,x))ds =: U_1(\eta)\eta + U_2(\eta). \qquad (6.1.3)$$

To see that the continuous mapping $U : \mathbf{R}^n \to \mathbf{R}^n$ defined by (6.1.3) has a fixed-point we consider the auxiliary mapping $F : \mathbf{R}^n \to \mathbf{R}^n$ defined by

$$F(\eta) = [I - U_1(\eta)]^{-1}[U(\eta) - U_2(\eta)]. \qquad (6.1.4)$$

This definition is possible since A has no eigenvalues on the imaginary axis. Note that every fixed point of (6.1.4) is also a fixed point of U.

Because $t(\cdot)$ and $\varphi(\cdot)$ are bounded the function $U_2(\cdot)$ is also bounded. Furthermore there exists a constant c_1 such that

$$|[I - U_1(\eta)]^{-1}| \le c_1$$

for all $\eta \in \mathbf{R}^n$. Here $|\cdot|$ denotes the operator norm of a matrix. Thus we have

$$\lim_{|\eta| \to \infty} \frac{|F(\eta)|}{|\eta|} \le c_1, \qquad \lim_{|\eta| \to \infty} \frac{|U_1(\eta)|}{|\eta|} = 0.$$

Hence we can find a constant $c_2 > 0$ such that

$$|F(\eta)| \le c_2 + \frac{1}{2}|\eta| \qquad (6.1.5)$$

for all $\eta \in \mathbf{R}^n$.

Let us introduce now the ball

$$K := \{\eta \; : \; |\eta| \le 2c_2\}.$$

It follows from (6.1.5) that F maps K into itself. By Brouwer's theorem there exists a fixed point η_0 of F in K which is also a fixed point of U. We get that the mapping $Q \circ T$ defined above also has a fixed point and the solution of (6.1.1) originating in $x_0 := (\eta_0, c^* A^{-1}\eta_0)$ is a cycle of the second kind since $z(t(x_0), x_0) = \eta_0$ and $\sigma(t(x_0), x_0) = c^* A^{-1}\eta_0 + \Delta$. $\blacksquare$

6.2 Frequency-domain Conditions for Existence of Circular Solutions and Cycles of the Second Kind

In this section the non-local reduction principle will be used to prove the existence of cycles. Again we consider the planar system

$$\begin{aligned}
\dot{\vartheta} &= \quad\;\; \eta \;\; - \;\; u(\vartheta), \\
\dot{\eta} &= -a\eta \;\; - \;\; \Psi(\vartheta),
\end{aligned} \qquad (6.2.1)$$

which will be employed in its capacity as reduced system for some higher dimensional original system. As usual we start with an auxiliary result, which comes from [89].

Let be $\Psi, f : \mathbf{R} \to \mathbf{R}$ continuous functions and $w, \sigma : \mathbf{R}_+ \to \mathbf{R}$ C^1-functions. Suppose also that on the closure of $(\vartheta_1, \vartheta_2)$ there is given a continuous scalar-valued function F which is differentiable in $(\vartheta_1, \vartheta_2)$.

Let λ and $\nu > 0$ be certain parameters and $(t_1, t_2) \subset [0, \infty)$ a certain time interval.

Lemma 6.2.1 *Suppose that the following conditions are fulfilled:*

(i) $F(\vartheta) > 0$ and $F(\vartheta) > \sqrt{2\nu}\, f(\vartheta)$ for all $\vartheta \in (\vartheta_1, \vartheta_2)$;

(ii) $\vartheta_1 \neq \infty$ and $F(\vartheta_1) \geq 0$;

(iii) $F'(\vartheta)F(\vartheta) + \Psi(\vartheta) \leq 0$ for all $\vartheta \in (\vartheta_1, \vartheta_2)$;

(iv) $F'(\vartheta)[F(\vartheta) - \sqrt{2\nu}\, f(\vartheta)] + \lambda\sqrt{2\nu}\, F(\vartheta) + \Psi(\vartheta) = 0$ for all $\vartheta \in (\vartheta_1, \vartheta_2)$;

(v) $w(t) \geq -\nu[\dot{\sigma}(t) + f(\sigma(t))]^2$ for all $t \in [t_1, t_2)$;

(vi) $\dot{w}(t) + 2\lambda w(t) - \Psi(\sigma(t))[\dot{\sigma}(t) + f(\sigma(t))] \leq 0$ for all $t \in [t_1, t_2)$;

(vii) $\sigma(t_1) \geq \vartheta_1$, $\dot{\sigma}(t_1) > 0$, $\dot{\sigma}(t_1) + f(\sigma(t_1)) > 0$, $w(t_1) + \frac{1}{2}F(\sigma(t_1))^2 < 0$ and $\sigma(t) < \vartheta_2$ for all $t \in (t_1, t_2)$.

Then

$$\dot{\sigma} \geq \left(\sqrt{2\nu}\right)^{-1} F(\sigma(t)) - f(\sigma(t)) > 0 \tag{6.2.2}$$

for all $t \in (t_1, t_2)$ and

$$w(t) + \frac{1}{2}F(\sigma(t))^2 < 0 \tag{6.2.3}$$

for all $t \in [t_1, t_2]$.

Proof Let us consider the function $v : \mathbf{R}_+ \to \mathbf{R}$ defined by

$$v(t) = w(t) + \frac{1}{2}F(\sigma(t))^2.$$

It follows from condition (vii) that $\dot{\sigma}(t_1) > 0$ and $v(t_1) < 0$. Thus we can assume that $v(t) \leq 0$ for all $t \in [t_1, T] \subset [t_1, t_2]$. We get

$$w(t) + [\dot{\sigma}(t) + f(\sigma(t))]^2 \leq -\frac{1}{2}F(\sigma(t))^2 + \nu[\dot{\sigma}(t) + f(\sigma(t))]^2$$

for all $t \in [t_1, T)$. Using this and (v) it follows that

$$F(\sigma(t))^2 \leq 2\nu[\dot{\sigma}(t) + f(\sigma(t))] \tag{6.2.4}$$

for all $t \in [t_1, T)$. From (i) and (vii) we find that

$$\dot{\sigma}(t) + f(\sigma(t)) > 0 \tag{6.2.5}$$

for all $t \in [t_1, T)$. From (6.2.4), (6.2.5) and (i) the estimate (6.2.2) follows for all $t \in (t_1, T)$. Using this and conditions (iii), (iv) we get the following relation, where the argument $\sigma(t)$ is omitted in f, Ψ, F and F'

$$\lambda F^2 + [\Psi + F'F][\dot{\sigma} + f] - F'Ff \leq \left(\sqrt{2\nu}\right)^{-1} F\left[F'F + \lambda\sqrt{2\nu}\, F - \sqrt{2\nu}\, fF' + \Psi\right] = 0 \tag{6.2.6}$$

for all $t \in (t_1, T)$.

From (6.2.6) and condition (vi) we have

$$\dot{v} + 2\lambda v \leq 0 \tag{6.2.7}$$

for all $t \in (t_1, T)$. Suppose now that $v(t) < 0$ and $\sigma(t) \in (\vartheta_1, \vartheta_2)$ for all $t \in (t_1, T)$ and one of the following equalities is true: $v(T) = 0$ or $\sigma(T) = \vartheta_1$. But in this case we have already shown that $\dot{\sigma}(t) > 0$ on $[t_1, T)$. So we have $\sigma(T) > \sigma(t_1) \geq \vartheta_1$ and it is impossible that $\sigma(T) = \vartheta_1$. But then from (6.2.7) it follows that

$$v(T) \leq v(t_1)e^{-2\lambda(T-t_1)} < 0$$

and, consequently, $v(T) < 0$. Thus $v(t)$ is defined for all $t \in [t_1, t_2]$ and (6.2.3) is fulfilled. Then the estimate (6.2.2) is also true on (t_1, t_2). $\blacksquare$

In the next theorem, which was proved in [89], we consider the system (6.1.1) using the notation

$$D(s) := c^*(A - sI)^{-1}b \quad \text{and} \quad \Gamma := -c^*b = -\lim_{s \to \infty} sD(s).$$

Suppose that

$$\Gamma > 0 \quad \text{and} \quad \varrho \leq 0.$$

Theorem 6.2.1 *Suppose that there exists a non-negative number λ such that the following conditions are fulfilled:*

(i) *the system*

$$\dot{\vartheta} = \eta + \frac{\varrho}{\sqrt{\Gamma}}\varphi(\vartheta), \quad \dot{\eta} = -\frac{\lambda}{\sqrt{\Gamma}}\eta - \varphi(\vartheta) \tag{6.2.8}$$

has a circular solution;

(ii) $\operatorname{Re} D(i\omega - \lambda) < 0$ *for all* $\omega \in \mathbf{R}$ *and* $\lim\limits_{\omega \to \infty} \omega^2 D(i\omega - \lambda) < 0$;

(iii) *one of the following properties is true:*

    (a) *matrix $A + \lambda I$ has one positive eigenvalue and $n - 1$ eigenvalues with negative real part;*

    (b) *matrix $A + \lambda I$ has two zero eigenvalues and $n - 2$ eigenvalues with negative real part, and $\lim\limits_{s \to \infty} s^2 D(s) = 0$.*

Then system (6.1.1) has a circular solution.
Suppose in addition that the following hypothesis is valid:

(iv) *for a certain circular solution $(\vartheta_0(\cdot), \eta_0(\cdot))$ of the reduced system (6.2.8) we have*

$$\eta_0(0) = 0, \quad \vartheta_0(0) =: \vartheta^0 \quad and \quad \dot{\vartheta}_0(t) > 0 \quad for \; all \;\; t > 0. \tag{6.2.9}$$

Then for any $\delta > 0$ there exists a circular solution $(z(\cdot), \sigma(\cdot))$ of system (6.1.1), satisfying the conditions $|z(0)| < \delta$ and $\sigma(0) = \vartheta^0$.

In preparing the proof of Theorem 6.2.1 we state and prove the following lemma. Consider the linear system

$$\dot{x} = Bx, \tag{6.2.10}$$

in which the $n \times n$ matrix B has the form $B = \begin{bmatrix} \begin{array}{cc} 0 & 1 \\ 0 & 0 \end{array} & \vdots & 0 \\ \cdots & \vdots & \cdots \\ 0 & \vdots & B_{22} \end{bmatrix}$ where B_{22} is an $(n-2) \times (n-2)$

Hurwitz matrix.

Lemma 6.2.2 *Suppose that for a certain $n \times n$ matrix $H = H^*$ we have*

$$x^* H B x < 0 \qquad (6.2.11)$$

for all $x \notin M := \{x = (x_1, x_2, x_3) \in \mathbf{R} \times \mathbf{R} \times \mathbf{R}^{n-2} : x_2 = 0, x_3 = 0\}$. Then H has one negative and $n-1$ positive eigenvalues.

Proof Consider the set $N := \{(x_1, x_2, x_3) \in \mathbf{R} \times \mathbf{R} \times \mathbf{R}^{n-2} : x_3 = 0\}$. It is easy to check, that $x^* H x = h_{11} x_1^2 + 2 h_{12} x_1 x_2 + h_{22} x_2^2$ and $x^* H B x = h_{11} x_1 x_2 + h_{12} x_2^2$ for $x \in N$ where h_{ij} denotes the element of H with the coordinates (i, j).

From (6.2.11) it follows that $h_{11} = 0$ and $h_{12} < 0$. Thus H has at least one negative eigenvalue. Consider now the solution $x(t) = e^{Bt} x_0$ of (6.2.10) with

$$x_0 \in P := \left\{x = (x_1, x_2, x_3) \in \mathbf{R} \times \mathbf{R} \times \mathbf{R}^{n-2} : x_2 = 0\right\}.$$

Because of (6.2.11) we have $\lim_{t \to +\infty} x(t) \in M$. It follows also from (6.2.11) that the function $t \mapsto x(t)^* H x(t)$ is not increasing. As a consequence we have

$$x_0^* H x_0 \geq 0 \qquad \text{for all} \qquad x_0 \in P. \qquad (6.2.12)$$

Let us demonstrate now that $\det H \neq 0$. Indeed, suppose that $Hd = 0$ for a certain vector $d \neq 0$. Then $d^* H B d = 0$ and it follows from (6.2.11) that $d \in M$. Furthermore we see that the second component of vector Hd is $h_{12} d_1$ where d_1 is the first component of d. Hence it is clear that $Hd \neq 0$. This contradiction proves that $\det H \neq 0$. Hence from (6.2.12) and from the fact that H has one negative eigenvalue the conclusion of the lemma follows. $\blacksquare$

Proof of Theorem 6.2.1 Using assumption (ii) we see by Theorem 1.4.2, p. 9, that there exists a matrix $H = H^*$ satisfying

$$(A + \lambda I)^* H + H(A + \lambda I) < 0 \qquad (6.2.13)$$

$$2Hb = c. \qquad (6.2.14)$$

From assumption (iii) and (6.2.13) it follows that H has one negative eigenvalue and $n-1$ positive ones. In the case (a) this fact is guaranteed by Lemma 1.2.1, p. 6, and in case (b) it follows from Lemma 6.2.1.

Proceeding as in the proof of Theorem 5.5.1, p. 109, we can establish that

$$\det[H - (2c^* b)^{-1} cc^*] = \det[1 - (2c^* b)^{-1} c^* H^{-1} c].$$

From this and (6.2.14) it follows that

$$\det[H - (2c^* b)^{-1} cc^*] = 0.$$

Hence using the facts that H has only one negative eigenvalue, $c^* b < 0$ and the matrix cc^* is positive semi-definite we obtain that

$$H - (2c^* b)^{-1} cc^* \geq 0. \qquad (6.2.15)$$

In order to use Lemma 6.2.1, we define for a solution $(z(\cdot), \sigma(\cdot))$ of (6.1.1) the following functions:

$$\begin{aligned}
w(t) &= z(t)^* H z(t) \quad (t \geq 0), \\
\Psi(\sigma) &= \varphi(\sigma), \\
f(\sigma) &= -\varrho \varphi(\sigma) \quad (\sigma \in \mathbf{R}).
\end{aligned}$$

Determine also the constant

$$\nu = -(2c^*b)^{-1} = (2\Gamma)^{-1}.$$

Inequality (6.2.15) guarantees that condition (v) of Lemma 6.2.1 is fulfilled. Condition (vi) of this lemma is satisfied by virtue of (6.2.13) and (6.2.14). From condition (i) of the present theorem we find that there exists a solution $(\vartheta_0(\cdot), \eta_0(\cdot))$ of system (6.2.8) and numbers τ and $\varepsilon > 0$ such that

$$\varphi(\vartheta(\tau)) = 0 \quad \text{and} \quad \dot{\vartheta}_0(t) \geq \varepsilon \quad \text{for all} \quad t \geq \tau.$$

Let us consider the first-order equation

$$\frac{dF}{d\vartheta} = \frac{-\frac{\lambda}{\sqrt{\Gamma}}F - \varphi(\vartheta)}{\Gamma + \frac{\varrho}{\sqrt{\Gamma}}\varphi(\vartheta)}, \tag{6.2.16}$$

which corresponds to (6.2.8), and denote by $F(\vartheta)$ the solution of (6.2.16) corresponding to $(\vartheta_0(\cdot), \eta_0(\cdot))$. Then function F satisfies conditions (ii) and (iv) of Lemma 6.2.1 with $\vartheta_1 = \vartheta_0(\tau)$ and $\vartheta_2 = +\infty$. Furthermore we have

$$F(\vartheta) + \frac{\varrho}{\sqrt{\Gamma}}\varphi(\vartheta) \geq \varepsilon \tag{6.2.17}$$

for all $\vartheta \in [\vartheta_1, \vartheta_2]$. From (6.2.16) and Lemmas 5.5.3, p. 107, and 5.5.4, p. 107, it follows that conditions (i) and (iii) of Lemma 6.2.1 are also satisfied.

Since matrix H has one negative and $n - 1$ positive ones there exists a vector z_0 such that

$$2z_0^* H z_0 + F(\vartheta_1)^2 < 0, \; c^* z_0 > 0 \quad \text{and} \quad c^* z_0 + \varphi(\vartheta_1) > 0. \tag{6.2.18}$$

These inequalities guarantee that condition (vii) of Lemma 6.2.1 is also fulfilled.

Thus for the solution of (6.1.1) with the initial points $z(0) = z_0$ and $\sigma(0) = \vartheta_1$ all the assumptions of Lemma 6.2.1 are fulfilled. We conclude then by (6.2.2) and (6.2.17) that this solution is a circular one. Suppose now that for the circular solution $(\vartheta_0(\cdot), \eta_0(\cdot))$ conditions (6.2.9) hold. Then the corresponding solution F of equation (6.2.16) satisfies the relations

$$F(\vartheta^0) = 0 \quad \text{and} \quad F(\vartheta) + \frac{\varrho}{\sqrt{\Gamma}}\varphi(\vartheta) > 0 \tag{6.2.19}$$

for all $\vartheta > \vartheta^0$. More than that, there exists for any $\vartheta_1 > \vartheta_0$ an $\varepsilon > 0$ such that (6.2.17) is valid. From (6.2.17) it follows directly that $\varrho\varphi(\vartheta^0) \geq 0$. So for any $\delta > 0$ there exists a vector z_0 which satisfies (6.2.18) with $\vartheta_1 = \vartheta^0$ and $|z_0| < \delta$. It is not difficult to see now that for a solution $(z(\cdot), \sigma(\cdot))$ of (6.1.1) with initial points $z(0) = z_0$, $\sigma(0) = \vartheta^0$ all conditions are satisfied. Then from estimates (6.2.2) and (6.2.19) and the inequality (6.2.17) (with any $\vartheta_1 > \vartheta^0$) it follows that this solution is a circular one. Theorem 6.2.1 is proved. ∎

In the next theorem, borrowed from [103], the existence of a circular solution is used to derive the existence of cycles.

Theorem 6.2.2 *Suppose there exists a non-negative λ such that conditions (ii) and (iii-a) of Theorem 6.2.1 are fulfilled. Suppose also that matrix A is Hurwitzian and the reduced system (6.2.8) has a circular cycle of the second kind. Then system (6.1.1) has a circular cycle of the second kind.*

Proof Note that under the conditions of Theorem 6.2.2 the function F, which has been introduced in the proof of Theorem 6.2.1, may be chosen as Δ-periodic function. From the proof of Theorem 6.2.1 it follows also that for solutions $(z(\cdot), \sigma(\cdot))$ of (6.1.1) satisfying

$$c^* z(0) > 0, \quad c^* z(0) + \varrho\varphi(\sigma(0)) > 0 \quad \text{and}$$

$$2z(0)^*Hz(0) + F(\sigma(0)) < 0$$

the inequalities

$$z(t)^*Hz(t) + \frac{1}{2}F(\sigma(t))^2 < 0 \qquad (6.2.20)$$

and

$$\frac{d\sigma(t)}{dt} \geq \sqrt{\Gamma}F(\sigma(t)) + \varrho\varphi(\sigma(t)) \geq \epsilon > 0 \qquad (6.2.21)$$

are fulfilled for all $t \geq 0$.

So the set

$$\Omega := \left\{ (z,\sigma) \in \mathbf{R}^n \times \mathbf{R} \ : \ z^*Hz + \frac{1}{2}F(\sigma)^2 < 0, c^*z > 0, c^*z + \varrho\varphi(\sigma) > 0 \right\}$$

is positively invariant for (6.1.1). Because of the continuous dependence of the solutions on the initial points the closure $\overline{\Omega}$ is also positively invariant. Let us choose now a number $\gamma > 0$ such that the matrix $A + \gamma I$ is Hurwitzian. The well known Lyapunov lemma says that the equation

$$(A + \gamma I)^*L + L(A + \gamma I) = -I$$

possesses a solution $L = L^* > 0$.

Let us introduce the function $U : \mathbf{R}^n \times \mathbf{R}$ by

$$U(z) = z^*Lz - \nu$$

where

$$\nu > \frac{1}{\gamma}|Lb|^2 \max_{\sigma \in \mathbf{R}} |\varphi(\sigma)|^2.$$

Then for all $(z,\sigma) \in \mathbf{R}^n \times \mathbf{R}$ we have

$$\begin{aligned}
\dot{U} + 2\gamma U &= 2z^*L[Az + b\varphi(\sigma)] + 2\gamma z^*Lz - 2\gamma\nu \\
&= z^*[(A + \gamma I)^*L + L(A + \gamma I)]z + 2z^*Lb\varphi(\sigma) - 2\gamma\nu \\
&= -z^*z + 2z^*L\varphi(\sigma) - 2\gamma\nu \\
&= -[z - Lb\varphi(\sigma)]^*[z - Lb\varphi(\sigma)] + |Lb|^2\varphi(\sigma)^2 - 2\gamma\nu < 0,
\end{aligned}$$

where $\dot{U}$ denotes the derivative of U with respect to system (6.1.1). From this inequality it follows by Lemma 3.1.1, p. 48, that.the set

$$D := \{(z,\sigma) \ : \ z^*Lz \leq \nu\}$$

is positively invariant for (6.1.1). Then the set $\overline{\Omega} \cap D$ is also positively invariant.

Now let σ_1 be a certain zero of φ. From (6.2.21) it follows that for any initial point

$$(z_0,\sigma_0) \in \Omega_1 := \overline{\Omega} \cap D \cap \{\sigma = \sigma_1\}$$

there exists a time $\tau = \tau(z_0)$ such that for the solution $(z(\cdot),\sigma(\cdot))$ of (6.1.1) starting in (z_0,σ_0) we have $(z(\tau),\sigma(\tau)) \in \Omega_2$ and $(z(t),\sigma(t)) \notin \Omega_2$ for $t \in [0,\tau)$, where $\Omega_2 := \overline{\Omega} \cap D \cap \{\sigma = \sigma_1 + \Delta\}$. Let us show now that Ω_1 is convex. Because of $L > 0$ the set D is convex. Denote $\beta = -\frac{1}{2}F(\sigma_1)^2$. Thus for $(z,\sigma) \in \Omega \cap \{\sigma = \sigma_1\}$ we have $z^*Hz < -\beta$ and we can write

$$\Omega \cap \{\sigma = \sigma_1\} = \{(z,\sigma) \ : \ z^*Hz < -\beta, c^*z > 0, \sigma = \sigma_1\}.$$

By the same argument as in the proof of Theorem 6.2.1 we see that

$$H + lcc^* > 0 \quad \text{for all} \quad l > (2\Gamma)^{-1}.$$

Let us suppose w.l.o.g. that $H + lcc^* = I$. Then for any two vectors (z_1, σ_1) and (z_2, σ_2) from $\Omega \cap \{\sigma = \sigma_1\}$ we have $c^* z_1 > 0, c^* z_2 > 0, l(c^* z_1)^2 \geq \beta + |z_1|^2$ and $l(c^* z_2)^2 \geq \beta + |z_2|^2$. From this and the obvious inequality

$$|z_1||z_2| + \beta \leq \sqrt{(\beta + |z_1|^2)(\beta + |z_2|^2)}$$

we get that

$$z_1^* z_2 + \beta \leq l(c^* z_1)(c^* z_2).$$

The latter inequality we can represent as

$$z_1^* H z_2 \leq -\beta.$$

So for any pair $(z_1, \sigma_1), (z_2, \sigma_2) \in \Omega \cap \{\sigma = \sigma_1\}$ the relation

$$[(1 - \tau)z_1 + \tau z_2]^* H[(1 - \tau)z_1 + \tau z_2] \leq \beta$$

is valid for all $\tau \in [0, 1]$ which gives the convexity of $\Omega \cap \{\sigma = \sigma_1\}$ and, consequently, of Ω_1.

Let us define a transformation $T : \Omega_1 \to \Omega_2$ by $T(z_0, \sigma_1) = (z(\tau(z_0), z_0, \sigma_1), \sigma_1 + \Delta)$ (with $\tau = \tau(z_0)$ defined above) and a transformation $Q : \Omega_2 \to \Omega_1$ given by $Q(z, \sigma_1 + \Delta) = (z, \sigma_1)$. Since $\overline{\Omega} \cap D$ is invariant we have $T\Omega_1 \subset \Omega_2$ and from the equality $F(\sigma_1 + \Delta) = F(\sigma_1)$ it follows that $Q\Omega_2 \subset \Omega_1$, consequently, $(Q \circ T)\Omega_1 \subset \Omega_1$.

We see by (6.2.21) that the set Ω_2 is without contact with the vector field (6.1.1). From this fact and from the theorem on continuous dependence of solutions on initial data we deduce that transformation T is continuous. Hence $Q \circ T$ is also continuous and by the Brouwer theorem (Theorem 2.1.2, p. 16) $Q \circ T$ has a fixed point in the convex and compact set Ω_1 which we denote by $(\overline{z}, \sigma_1)$. Then for the solution $(z(\cdot), \sigma(\cdot))$ of (6.1.1) starting in $z(0) = \overline{z}$ and $\sigma(0) = \sigma_1$ it follows that $z(\tau) = \overline{z}$ and $\sigma(\tau) = \sigma_1 + \Delta$, where $\tau = \tau(\overline{z})$ is defined above. Thus this solution is a cycle of the second kind. $\blacksquare$

We note in passing that in the proof of Theorem 6.2.2 the following result was obtained.

Lemma 6.2.3 *Let L be an $n \times n$ matrix and h be an n-vector such that the relation*

$$\{z \ : \ z^* H z \leq 0\} \cap \{z \ : \ h^* z = 0\} = \{0\}$$

is satisfied. Then for any positive α the set

$$\{z \ : \ z^* L z \leq -\alpha, h^* z \geq 0\}$$

is convex.

Remark 6.2.1 Let us consider separately the case $\varrho = 0$ in (6.1.1). Then the reduced system (6.2.8) can be transformed into the second-order equation

$$\ddot{\vartheta} + \frac{\lambda}{\sqrt{\Gamma}}\dot{\vartheta} + \varphi(\vartheta) = 0. \tag{6.2.22}$$

Suppose that for a certain $\lambda > 0$ the hypotheses (ii) and (iii-a) of Theorem 6.2.1 are fulfilled and the solution ϑ of (6.2.22) originating in $\vartheta(0) = \vartheta^0, \dot{\vartheta}(0) = 0$ is a circular one satisfying

$$\dot{\vartheta}(t) > 0 \quad \text{for all} \quad t > 0,$$

(i.e. hypotheses (i) and (iv) of Theorem 6.2.1 are also fulfilled). Then if A is Hurwitzian then system (6.1.1) has a cycle of the second kind.

Note that the proof of this assertion is similar to Theorem 6.2.2. It differs by the choice of σ_1 in Ω_1. We have to take this σ_1 near ϑ^0 such that $\sigma_1 > \vartheta^0$ and $F(\sigma_1 + \Delta) > F(\sigma_1)$. The existence of such σ_1 is guaranteed by the continuity of F on $(\vartheta^0, +\infty)$, by $F(\vartheta^0) = 0$ and the fact that $F(\sigma) > 0$ for all $\sigma > \vartheta^0$. The invariant set Ω is taken as

$$\Omega = \left\{ (z, \sigma) \; : \; z^* H z + \frac{1}{2} F(\sigma)^2 < 0, c^* z > 0, \sigma > \vartheta^0 \right\}.$$

The sets Ω_1 and Ω_2 are just the same as in the text of Theorem 6.2.2. The inequality $F(\sigma_1 + \Delta) > F(\sigma_1)$ guarantees the inclusion $(Q \circ T)\Omega_1 \subset \Omega_1$. Applying Theorems 6.2.1 and 6.2.2 to $(n + 1)$-dimensional systems (6.1.1) it is possible to extend the conditions on the existence of limit cycles for certain second-order systems to system (6.1.1). Consider as an example the equation

$$\ddot{\vartheta} + a\dot{\vartheta} + \varphi(\vartheta) = 0 \tag{6.2.23}$$

with

$$\varphi(\vartheta) = \sin(\vartheta + \Sigma_0) - \sin \Sigma_0 \qquad (\vartheta \in \mathbf{R}), \tag{6.2.24}$$

where $a > 0$ and $\Sigma_0 \in (0, \frac{\pi}{2})$ are constants.

Recall that a circular solution $\vartheta(\cdot)$ is called *irregular* if there exists a time $\bar{t} \geq 0$ with $\dot{\sigma}(\bar{t}) = 0$. If φ in (6.2.23) is of class C^1 it is known that if (6.2.23) has a circular solution it has also an irregular circular solution ([20]). A classical result is due to W.D. Hayes [58].

Theorem 6.2.3 (Hayes) *If*

$$a^2 \leq \sqrt{3 \cos^2 \Sigma_0 + 1} - 2 \cos \Sigma_0, \tag{6.2.25}$$

then equation (6.2.23), (6.2.24) has an irregular circular solution.

Let us consider now the system

$$\begin{aligned} \dot{z} &= Az + b\varphi(\sigma), \\ \dot{\sigma} &= c^* z, \end{aligned} \tag{6.2.26}$$

where A, b and c are matrices of order $n \times n$, $n \times 1$ resp. $n \times 1$ and φ is defined by (6.2.24).

Theorem 6.2.1 together with Remark 6.2.1 imply the following assertion.

Theorem 6.2.4 (Generalized Hayes theorem) *Suppose that the transfer-function of (6.2.26) is non-degenerate, $c^* b < 0$ and there exists a number λ such that the hypotheses (ii) and (iii-a) of Theorem 6.2.1 and the inequality*

$$\lambda^2 \leq (-c^* b) \left(\sqrt{3 \cos^2 \Sigma_0 + 1} - 2 \cos \Sigma_0 \right),$$

are satisfied. Then for any small $\delta > 0$ there exists a circular solution $(z(\cdot), \sigma(\cdot))$ of (6.2.26), (6.2.24) with $|z(0)| < \delta$. If in addition matrix A is Hurwitzian then system (6.2.26), (6.2.24) has a cycle of the second kind.

Let us now state some modifications of Theorem 6.2.2, taken from [86], which are of considerable use for investigating concrete electro-mechanical systems.

Theorem 6.2.5 *Suppose in (6.1.1) that $\varrho < 0$, $|\varphi'(\sigma)| < \mu$ on $\mathbf{R}$ and that there exists a number $\lambda > 0$ with $a > \frac{\lambda}{\sqrt{\mu}}$ such that the conditions (ii), (iii-a) of Theorem 6.2.1 as well, as the following conditions, are fulfilled:*

(i) *equation (6.2.23) has a cycle of the second kind;*

(ii) $\frac{1}{2\mu}\left[\sqrt{a^2 + 4\mu} - a\right] \geq \frac{-\varrho a}{2(a\sqrt{\Gamma} - \lambda)} + \sqrt{\frac{\varrho^2 a^2}{4(a\sqrt{\Gamma} - \lambda)^2} - \frac{\varrho}{a\sqrt{\Gamma} - \lambda}}.$

Then the system (6.1.1) has a circular cycle of the second kind.

We have been considering up to now the case $\Gamma = -c^*b > 0$. But there exist important systems in applications for which this condition is not fulfilled. For that reason the following result is specially devoted to systems (6.1.1) with $\Gamma = 0$.

Introduce for (6.1.1) the number

$$\Gamma_1 := \lim_{s \to \infty} s(sD(s) - \Gamma).$$

Theorem 6.2.6 *Suppose that in (6.1.1) we have $|\varphi'(\sigma)| < \mu$ on* **R**, *$\Gamma = \varrho = 0$ and there exist numbers $\lambda > 0, a > 0, \xi \in (0, 2\mu), \delta \in (0, 1)$ and $l \geq \xi/(\mu + 2\lambda)$ such that the following conditions are fulfilled:*

(i) *the second-order equation*

$$\ddot{\vartheta} + a\dot{\vartheta} + (1 - \frac{1}{2}\mu\xi)\varphi(\vartheta) = 0$$

has a circular cycle of the second kind;

(ii) *the matrix $A + \lambda I$ has one positive eigenvalue and $n - 1$ eigenvalues with negative real part;*

(iii) $\text{Re}\left[(1 + \lambda\xi + i\omega\xi)D(i\omega - \lambda)\right] + \xi\mu|D(i\omega - \lambda)|^2 + l(\omega^2 + \lambda^2)|D(i\omega - \lambda)|^2 \leq 0$ *for all $\omega \geq 0$;*

(iv) $\dfrac{\delta^2\xi\sqrt{\Gamma_2\xi}(1 - \frac{1}{2}\mu\xi)}{(1 - \delta)(1 + 2\lambda\xi)^2 + a\delta^2\xi\sqrt{\Gamma_2}} \leq \dfrac{1}{2\mu}\left[\sqrt{a^2 + 4\mu(1 - \frac{1}{2}\mu\xi)} - a\right];$

(v) $a \geq \dfrac{\lambda(1 + 2\lambda\xi)}{\delta\sqrt{\Gamma_2\xi}}.$

Then system (6.1.1) has a circular cycle of the second kind.

Remark 6.2.2 Condition (iv) of Theorem 6.2.6 may be substituted by the simpler verficable requirement

$$\delta = 1 - \sqrt{\mu\xi^3\Gamma_2 + a\xi\sqrt{\Gamma_2}}. \tag{6.2.27}$$

Indeed, since $\delta < 1, 1 + 2\lambda\xi < 1, 1 - \frac{1}{2}\mu\xi < 1$ and $\mu\xi^3\Gamma_2 + a\xi\sqrt{\Gamma_2\xi} < 1$ it is easy to demonstrate that (6.2.27) implies

$$\left[\frac{\delta^2\xi\sqrt{\Gamma_2\xi}(1 - \frac{1}{2}\mu\xi)}{(1 - \delta)(1 + 2\lambda\xi)^2}\right]^2 + \frac{a}{\mu}\left[\frac{\delta^2\xi\sqrt{\Gamma_2\xi}(1 - \frac{1}{2}\mu\xi)}{(1 - \delta)(1 + 2\lambda\xi)^2}\right] \leq \frac{1 - \frac{1}{2}\mu\xi}{\mu}.$$

Hence it follows that condition (iv) of the theorem is true.

6.3 Circular Solutions and Cycles of the Second Kind in Concrete Systems of Phase Synchronization

In this section we shall use theorems proved in the previous section in order to approximate the boundaries of the parameter domains of existence of cycles of the second kind and of circular solutions for some concrete systems.

Example 6.3.1 Let us consider system (6.1.1) for n=2 described already in Example 5.5.1, p. 110, for $\varrho = 0$ and

$$K(s) = D(s) = \frac{s+1}{0.8s^2 + s + \beta} \tag{6.3.1}$$

where $\beta > 0$ is a parameter. Note that $\Gamma = \frac{1}{0.8}$. We use the frequency-response function $\omega \to D(i\omega - \lambda)$ calculated in Example 5.5.1, p. 110, to see that the conditions (ii) and (iii-a) of Theorem 6.2.1 are fulfilled if

$$\lambda \in (\frac{1}{4}, 1) \quad \text{and} \quad 0.8\lambda^2 - \lambda + \beta < 0. \tag{6.3.2}$$

Note that the latter system of inequalities has the following solutions: either $\beta < \frac{1}{5}$ and $\frac{1}{4} < \lambda < 1$ or $\frac{1}{5} < \beta < \frac{5}{16}$ and

$$\frac{1 - \sqrt{1 - 3.2\beta}}{1.6} < \lambda < \frac{1 + \sqrt{1 - 3.2\beta}}{1.6}.$$

As a reduced equation we use

$$\ddot{\vartheta} + \sqrt{0.8}\lambda\dot{\vartheta} + \varphi(\vartheta) = 0. \tag{6.3.3}$$

It follows from Section 2.2 that (6.3.3) has a cycle of the second kind if $\sqrt{0.8}\lambda \le a_{cr}$, where a_{cr} depends on the function φ. So it is clear that we must make λ as small as possible. Thus in the case $\beta < \frac{1}{5}$ the system (6.1.1), (6.3.1) has cycles of the second kind if the equation

$$\ddot{\vartheta} + \frac{\sqrt{0.8}}{4}\dot{\vartheta} + \varphi(\vartheta) = 0$$

has them.

In the case $\frac{1}{5} < \beta < \frac{5}{16}$ the system (6.1.1), (6.3.1) has a cycle of the second kind supposed that the equation

$$\ddot{\vartheta} + \frac{\sqrt{0.8}}{1.6}\left[1 - \sqrt{1 - 3.2\beta}\right]\dot{\vartheta} + \varphi(\vartheta) = 0$$

has such a cycle.

Suppose now φ in the form $\varphi(\vartheta) = \sin\vartheta - \gamma$ with $\gamma \in (0,1)$. Consider in the (β^{-2}, γ)-plane of parameters the cycle problem. The boundary below of the region of parameters for which Theorem 6.2.1 guarantees the existence of cycles of the second kind is indicated by a solid curve in Figure 6.3.1. Here we have used the curve 0 from Figure 5.4.1, p. 101. The thick dotted line 1 in Figure 6.3.1 indicates the boundary from below of the region of cycles of the second kind received in [139] by means of the two-dimensional comparision systems method. By thin dotted lines 2 is shown the boundary from above, of the region of global convergence received in Figure 5.5.1, p. 111.

Example 6.3.2 We consider again Example 5.5.2, p. 111, i.e. System (6.1.1) with $n = 2$ and

$$D(s) = \frac{\beta_1}{s + \alpha_1} + \frac{\beta_2}{s + \alpha_2}, \tag{6.3.4}$$

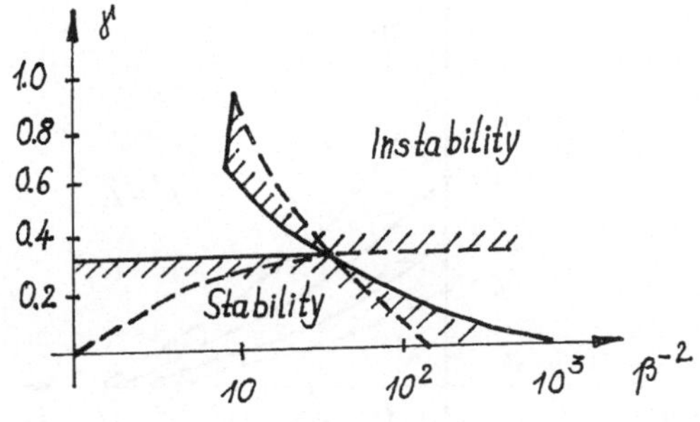

Figure 6.3.1

Figure 6.3.2

where for the parameters we have $0 < \alpha_1 < \alpha_2$ and $\beta_1 > \beta_2 > 0$. It is easy to calculate that for $\omega \in \mathbf{R}$

$$\operatorname{Re} D(i\omega - \lambda) = \frac{\beta_1(\alpha_1 - \lambda)}{(\alpha_1 - \lambda)^2 + \omega^2} + \frac{\beta_2(\alpha_2 - \lambda)}{(\alpha_2 - \lambda)^2 + \omega^2}.$$

Because of the parameter conditions we have

$$\alpha_1 < \gamma_1 := \frac{\alpha_1\beta_1 + \alpha_2\beta_2}{\beta_1 + \beta_2} < \frac{\beta_1\beta_2 + \alpha_1\beta_2}{\beta_1 + \beta_2} =: \gamma_2 < \alpha_2.$$

Now it easy to establish that for $\lambda \in (\gamma_1, \gamma_2)$ the conditions (ii) and (iii-a) of Theorem 6.2.1 are fulfilled. It ist clear that matrix A in (6.1.1) is Hurwitzian. Consider now the case $\varphi(\sigma) = \sin \sigma - \gamma$ with $\gamma \in (0,1)$. Condition (i) of Theorem 6.2.1 can be verified with the help of paper [33] (see curve 0, Figure 5.4.1). In Figure 6.3.2 the solid curve 2 shows the boundary from below of the parameter region for cycles of the second kind received on the basis of Theorem 6.2.1 for $\alpha_1 = 0.1, \alpha_2 = 0.2, \beta_1 = 2$ and $\beta_2 = 1$. By the dotted line is denoted the boundary of the region of global convergence computed by the Shakhgil'dyan-Lyakhovkin formula ([140]), which is based on the harmonic balance method. By 1 we have indicated the curve from Figure 6.3.2 which is the boundary from above of the region of global convergence received in Section 5.

Example 6.3.3 Let us consider system (6.1.1) with $n = 2$,

$$D(s) = \frac{\beta_1}{s} + \frac{\beta_2}{s^2} \quad \text{and} \quad \varphi(\sigma) = \sin \sigma - \gamma, \tag{6.3.5}$$

where for the parameters we have $\beta_1, \beta_2 > 0$ and $\gamma \in (0,1)$. This system describes the dynamics of a phase-locked loop with ideal filter of second order and sinusoidal characteristic of the phase detector ([149]). It is not difficult to see that conditions (ii) and (iii-b) of Theorem 6.2.1 are fulfilled with $\lambda = 0$. So if the reduced system (6.2.8) has for $\lambda = 0, \Gamma = \beta_1$ and $\varphi(\sigma) = \sin \sigma - \gamma$ a circular solution which satisfies (6.2.9) then Theorem 6.2.1 is applicable. In [149] it is shown that for $\varrho = -\sqrt{2\beta_2}$ and $\gamma > 0.5$ system (6.2.8) has a circular solution satisfying (6.2.9). Thus on the basis of Theorem 6.1.1 we can conclude that for $\beta_2 > 0$, $\varrho = -\sqrt{2\beta_1}$, $\gamma > 0.5$ and for any $\delta > 0$ system (6.1.1), (6.3.5) has a circular solution $z(\cdot), \sigma(\cdot)$ with $|z(o)| < \delta$.

Example 6.3.4 Let us consider system (6.1.1) with $n = 2$, $\varrho = 0$

$$K(s) = \frac{\Gamma s + 1}{s^2 + \alpha s + \beta} \tag{6.3.6}$$

139

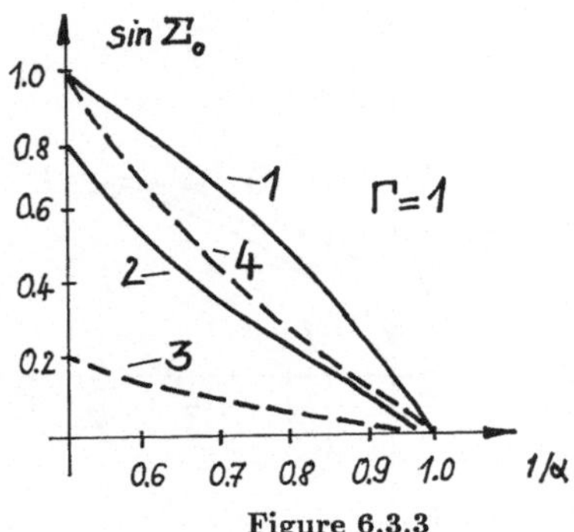

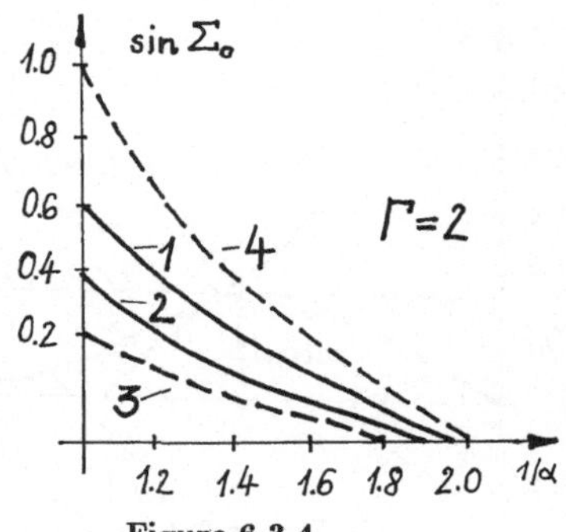

Figure 6.3.3   Figure 6.3.4

and

$$\varphi(\sigma) = \sin(\sigma + \Sigma_0) - \sin \Sigma_0. \tag{6.3.7}$$

We suppose that $\alpha > 0, \beta \geq 0, \Gamma > 0$ and $\Sigma_0 \in (0, \frac{\pi}{2})$. It is easy to see that $D(s) \equiv K(s), \Gamma = -c^*b = \lim_{s \to \infty} K(s)$ and

$$\det[(s - \lambda)I - A] = s^2 + (\alpha - 2\lambda)s + \lambda^2 - \alpha\lambda - \beta.$$

At first let $\beta = 0$ and $\alpha\Gamma \neq 1$. In this case system (6.1.1) describes the dynamics of certain search phase locked-loops ([28], [138]). A direct computation shows that $\chi(p) \equiv \frac{1}{p}K(p)$ is non-degenerate. It is clear that for $\lambda \in (0, \alpha)$ the condition (iii-a) of Theorem 6.2.1 is fulfilled. Since

$$\operatorname{Re} K(i\omega - \lambda) = \frac{(\alpha\Gamma - 1 - \lambda\Gamma)\omega^2 + (\lambda - \alpha)(1 - \Gamma\lambda)\lambda}{|(i\omega - \lambda)(i\omega + \alpha - \lambda)|^2}$$

we conclude that for

$$\lambda \in (0, \alpha) \quad \text{and} \quad \lambda\Gamma \in (\alpha\Gamma - 1, 1) \tag{6.3.8}$$

the condition (ii) of Theorem 6.2.1 is also fulfilled. If in addition the free parameter λ satisfies

$$\lambda^2 \leq \Gamma \left(\sqrt{3\cos^2 \Sigma_0 + 1} - 2\cos \Sigma_0 \right) \tag{6.3.9}$$

then all the conditions of Theorem 6.2.4 are fulfilled. Thus if $\beta = 0$, $\alpha\Gamma \neq 1$, $\alpha\Gamma < 2$ and if there exists a $\lambda > 0$ satisfying (6.3.8) and (6.3.9) then for any $\delta > 0$ system (6.1.1), (6.3.6), (6.3.7) has a circular solution $z(\cdot), \sigma(\cdot)$ with $|z(0)| < \delta$.

It is evident that instead of (6.3.9) we can use any other sufficient existence condition of irregular solutions of the equation

$$\ddot{\vartheta} + \frac{\lambda}{\sqrt{\Gamma}}\dot{\vartheta} + \sin(\vartheta + \Sigma_0) - \sin \Sigma_0 = 0. \tag{6.3.10}$$

Thus, for example, the domain which is situated in Figure 5.4.1, p. 101, above the curve 0 is the domain of existence of irregular circular solutions of (6.3.10) in the plane of parameters (T, γ) defined by $\gamma = \sin \Sigma_0$ and $T = \sqrt{\Gamma}/\lambda$. This domain was obtained in [138] by numerical methods.

In Figures 6.3.3, 6.3.4, 6.3.5 and 6.3.6 various approximations from below of the boundaries of the circular solutions region for $\Gamma = 1, 2, 3, 5$, respectively, are shown. The curves 1 denote the approximations of this boundaries which are obtained on the basis of conditions (6.3.8), (6.3.9). Curves 2 are constructed on the basis of qualitative numerical estimates obtained in

140

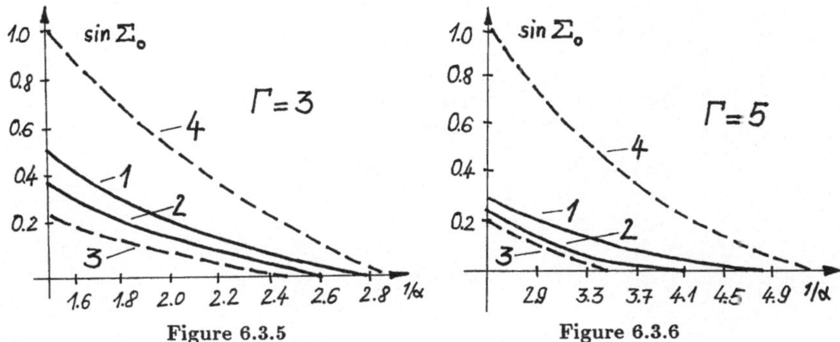

Figure 6.3.5   **Figure 6.3.6**

[33] (see curve 0, Figure 5.4.1) for equation (6.3.10). Curve 3 represents the exact boundaries received in [138] on the basis of qualitative-numerical methods. Curves 4 correspond to the boundaries of parameter regions associated with circular solutions and received by the method of two-dimensional comparison systems of Belykh and Nekorkin [28], [138]. It follows from Figs. 3.3 to 3.6 that Theorem 6.2.4, in common with the estimates from [33], always give better approximation than the two-dimensional systems comparison method (cf. curves 2 and 4). Let us now consider the case $\beta > 0$. Suppose also that

$$\alpha^2 > 4\beta \quad \text{and} \quad \alpha\Gamma + \Gamma\sqrt{\alpha^2 - 4\beta} \neq 2. \tag{6.3.11}$$

Employing the fact that

$$\operatorname{Re} K(i\omega - \lambda) = \frac{(-\Gamma\alpha - 1 + \alpha\Gamma)\omega^2 + (\lambda^2 - \Gamma\lambda^3 + \alpha\Gamma\lambda^2 - \alpha\lambda - \Gamma\lambda\beta + \beta)}{(\lambda^2 - \omega^2 - \alpha\lambda + \beta)^2 + (\alpha\omega - 2\lambda\omega)^2}$$

we see that with λ satisfying

$$\Gamma > \alpha\Gamma - 1 \quad \text{and} \quad \lambda^2 - \Gamma\lambda^3 + \alpha\Gamma\lambda^2 - \alpha\lambda - \Gamma\lambda\beta + \beta < 0 \tag{6.3.12}$$

the condition (ii) of Theorem 6.2.1 is fulfilled. It is also clear that for

$$\lambda \in \left(-\frac{\alpha}{2} - \sqrt{\frac{\alpha^2}{4} - \beta}, -\frac{\alpha}{2} + \sqrt{\frac{\alpha^2}{4} - \beta} \right) \tag{6.3.13}$$

the condition (iii-a) is valid.

So if $\beta > 0$, inequalities (6.3.11) hold and the free parameter satisfies (6.3.9), (6.3.12) and (6.3.13) the assumptions of Theorem 6.2.1 are fulfilled. Note also that matrix A from (6.1.1) is Hurwitzian. It follows that system (6.1.1), (6.3.6), (6.3.7) has under these conditions a cycle of the second kind. Instead of (6.3.9) one can also use numerical estimates from [33]. Figure 6.3.7 shows the case $\Gamma = 0.5$ and $\sin \Sigma_0 = 0.5$. Curve 2 approximates from below the boundary of the region with cycles of the second kind obtained on the basis of (6.3.11) to (6.3.13) and on the estimates of paper [33]. The dotted line 1 and the line 3 taken from Figure 5.4.5, p. 103, approximate from above the boundary of global convergence regions.

Example 6.3.5 Let us consider the system (6.1.1) with $n = 1$ and

$$K(s) = T\frac{1 + mTs}{1 + Ts} \tag{6.3.14}$$

141

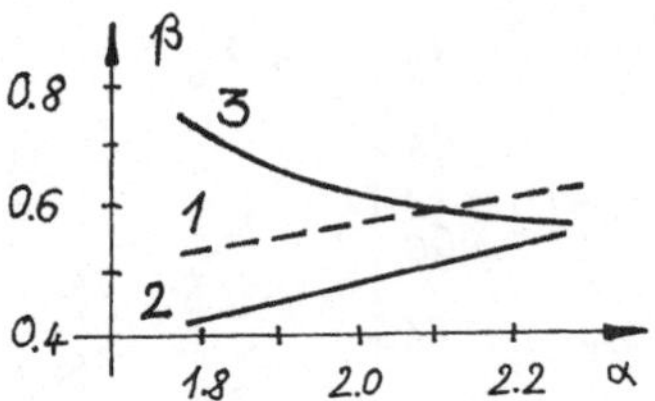

Figure 6.3.7

which was already discussed in Examples 5.4.1, p. 100 and 4.3.2, p. 75. In the notation of system (6.1.1) we have $\varrho = -mT$, $\Gamma = 1 - m$ and $D(s) = -T(1 - m)(Ts + 1)^{-1}$. Remember that m and T are non-zero parameters. It is clear that conditions (ii) and (iii-a) of Theorem 6.2.1 are satisfied if we choose $\lambda > T^{-1}$. Thus conditions of Theorem 6.2.5 are fulfilled if we require $a > T(\sqrt{1-m})^{-1}$ and

$$\frac{1}{2\mu}\left[\sqrt{a^2 + 4\mu} - a\right] > \frac{amT}{2(a\sqrt{1-m} - \frac{1}{T})} + \left[\frac{(amT)^2}{4(a\sqrt{1-m} - \frac{1}{T})^2} + \frac{mT}{a\sqrt{1-m} - \frac{1}{T}}\right]^2.$$

For $T \gg 1$ and $m \ll \frac{1}{T}$ the latter inequalities are fulfilled if $a \approx \frac{1}{T} + \mu mT$.

It follows on the basis of Theorem 5.3.1, p. 96, that system (6.1.1), (6.3.14) does not have cycles of the second kind supposed that the second-order comparison equation

$$\ddot{\vartheta} + \left[\frac{1}{T} + \mu mT\right]\dot{\vartheta} + \varphi(\vartheta) = 0 \tag{6.3.15}$$

does not have such cycles.

It is known from [140] that for equation (6.3.15) with $\varphi(\vartheta) = \sin\vartheta - \gamma$, $T \gg 1$ and $m \ll \frac{1}{T}$ the condition

$$\gamma \le 1.27\left[\frac{1}{T} + mT\right] \tag{6.3.16}$$

guarantees the absence of cycles of the second kind. Hence the estimate (6.3.16) enlarges the parameter region for stability of system (6.1.1), (6.3.14) which are obtained in Chapters 4 and 5. Note that estimate (6.3.16) is near to the stability condition

$$\gamma \le \frac{4}{\pi}\left[\frac{1}{T} + T\frac{m}{3}\right]$$

received in [10] by the averaging method.

Example 6.3.6 Let us consider system (6.1.1) with $n = 3$, $\varrho = 0$ and

$$K(s) \equiv D(s) = \frac{1}{s^2 + \alpha s + \beta}, \tag{6.3.17}$$

already discussed in Example 5.4.2, p. 101. The parameters α and β are assumed to be positive.

142

In order to apply Theorem 6.2.6 we calculate

$$\Gamma = \lim_{s\to\infty} sD(s) = 0 \quad \text{and} \quad \Gamma_2 = \lim_{s\to\infty} s^2 D(s) = 1.$$

Let us choose with respect to Theorem 6.2.6

$$\delta = 1 - \sqrt{\mu\xi^3 + a\xi\sqrt{\xi}}, \qquad \lambda = \frac{\beta}{\alpha} + \frac{3\mu}{\alpha^2},$$

$$\xi = \frac{1}{\alpha + (\mu + 2\lambda)^{-1}} \quad \text{and} \quad l = \frac{\xi}{\mu + 2\lambda}.$$

We suppose

$$2\alpha > \mu \tag{6.3.18}$$

and get $\xi \in (0, \frac{2}{\mu})$. The characteristic polynomial of the matrix A in (6.1.1) is

$$\det[(s - \lambda)I - A] = (s - \lambda)^2 + \alpha(s - \lambda) + \beta = s^2 + (\alpha - 2\lambda)s + (\lambda^2 - \alpha\lambda + \beta).$$

Thus the condition (ii) of Theorem 6.2.6 is fulfilled if

$$\lambda^2 - \alpha\lambda + \beta < 0. \tag{6.3.19}$$

The frequency-domain condition (iii) is satisfied if

$$\omega^2[-(1 + 2\lambda\xi) - \lambda\xi + \alpha\xi + 1] + (1 + \lambda\xi)(\beta - \alpha\lambda + \lambda^2) + \xi\mu + l\lambda^2 \leq 0$$

for all $\omega \geq 0$.

This relation is true if

$$1 + \lambda\xi \geq (\alpha - 2\lambda)\xi + l \text{ and } (1 + \lambda\xi)(\beta - \alpha\lambda + \lambda^2) + \xi\mu + l\lambda^2 \leq 0. \tag{6.3.20}$$

Let us choose now the parameters of the system by the condition

$$(\alpha\beta + 3\mu)^2 \leq \mu\alpha^3. \tag{6.3.21}$$

Then we have

$$\lambda^2 - \alpha\lambda + \beta = -\frac{3\mu}{\alpha} + \left(\frac{\beta}{\alpha} + \frac{3\mu}{\alpha^2}\right)^2 \leq -\frac{2\mu}{\alpha} \leq 0,$$

$$1 + \lambda\xi - (\alpha - 2\lambda)\xi - \frac{\xi}{\mu + 2\lambda} \geq 1 - \left(\alpha + \frac{1}{\mu + 2\lambda}\right) = 0$$

and

$$(1 + \lambda\xi)(\beta - \alpha\lambda + \lambda^2) + \mu + \frac{\lambda^2\xi}{\mu + 2\lambda} \leq -\frac{2\mu}{\alpha} + \xi\mu + \lambda^2 \leq -\frac{\mu}{\alpha} + \xi\mu \leq 0.$$

It follows that the conditions (6.3.19) and (6.3.20) are fulfilled. Thus if the inequalities (6.3.18) and (6.3.21) are true and if condition (i) of Theorem 6.2.6 with parameter a, which satisfies (v), holds, then all the assumptions of Theorem 6.2.6 may be satisfied by the choice of δ, λ, ξ and l as above.

Note that if μ and β are fixed and $\alpha \to \beta$ we may put $\lambda \approx \beta/\alpha$, $\xi \approx 1/\alpha$, $\delta \approx 1$ and $\alpha \approx \beta/\sqrt{\alpha}$. Then all the conditions of Theorem 6.2.6 are reduced to the requirement that the equation

$$\ddot{\vartheta} + \frac{\beta}{\sqrt{\alpha}}\dot{\vartheta} + \varphi(\vartheta) = 0 \tag{6.3.22}$$

143

has a cycle of the second kind. Thus if μ and β are fixed and $\alpha \to +\infty$ the existence of a cycle of the second kind in (6.3.22) implies that system (6.1.1), (6.3.17) has also a cycle of the second kind. According to the results of Example 5.4.2, p. 101, it is also a necessary condition for (6.1.1), (6.3.17) to have cycles of the second kind. Indeed, if we take in Example 5.4.2

$$\lambda = \frac{1}{2}\mathrm{Re}\,(\alpha - \sqrt{\alpha^2 - 4\beta})$$

then the reduced equation (5.4.11) from Chapter 5 will take the form (6.3.22).

6.4   Cycles of the First Kind

In this section we want to prove a result for pendulum-like systems concerning the existence of non-trivial periodic solutions, i.e. of cycles of the first kind.

Let us consider the system

$$\dot{x} = Px + q\varphi(\sigma), \qquad \sigma = r^*x, \tag{6.4.1}$$

in which P is an $n \times n$ matrix, q and r are n-vectors and $\varphi : \mathbf{R} \to \mathbf{R}$ is Δ-periodic and belongs to C^2. Suppose also that φ has exactly the two zeros 0 and $\overline{\sigma}$ on $[0, \Delta)$ and $\varphi'(\overline{\sigma}) > 0$.

Let us assume that the transfer function $\chi(s) = r^*(P - sI)^{-1}q$ is non-degenerate and satisfies the condition

$$\lim_{s \to \infty} s\chi(s) = 0. \tag{6.4.2}$$

This requirement is equivalent to the equality $r^*q = 0$. It is not difficult to demonstrate that system (6.4.1), having the property (6.4.2), can be transformed into the system

$$\dot{z} = Ax + b\varphi(\sigma), \qquad \dot{\sigma} = c^*x, \tag{6.4.3}$$

with an $(n-1) \times (n-1)$ matrix A and $(n-1)$-vectors b and c. Hence we assume in (6.4.1) w.l.o.g. that

$$P = \begin{bmatrix} A & o \\ c^* & 0 \end{bmatrix}, \quad q = \begin{bmatrix} b \\ 0 \end{bmatrix} \quad \text{and} \quad r = \begin{bmatrix} o \\ 1 \end{bmatrix},$$

where o denotes the zero element in $\mathbf{R}^{n-1}$.

In the sequel we need the functions $K(\cdot)$, $n(\cdot)$ and $m(\cdot)$ defined for suitable complex s by

$$K(s) = s\chi(s), \quad n(s) = \det(sI - A) \quad \text{and} \quad m(s) = n(s)K(s).$$

Let us introduce for arbitrary $\lambda > 0$ the value

$$\varepsilon(\lambda) = \inf_{\omega \geq 0}[-2\lambda\mathrm{Re}\,\chi(i\omega - \lambda)|K(i\omega - \lambda)|^2].$$

Assume also that

$$\lim_{s \to \infty} sK(s) = -c^*b = 0 \tag{6.4.4}$$

and that there exists a constant $\mu > 0$ such that

$$\varphi(\sigma)\sigma \leq \mu\sigma^2, \qquad \sigma \in \mathbf{R}. \tag{6.4.5}$$

Theorem 6.4.1 *Suppose that for system (6.4.1) the conditions of Theorem 3.4.1, p. 59, are satisfied. Suppose also that there exists a positive number λ_1 such that the following requirements are true:*

(i) $\operatorname{Re} K(i\omega - \lambda_1) < 0$ *for all* $\omega \in \mathbf{R}$ *and* $\lim\limits_{\omega \to \infty} \omega^2 \operatorname{Re} K(i\omega - \lambda_1) < 0$;

(ii) *the polynomial $n(s - \lambda_1)$ has one positive zero and $n - 2$ zeros with negative real part;*

(iii) *the polynomial $m(s)\varphi'(\sigma_1) + sn(s)$ has two positive zeros with positive real parts and $n - 2$ zeros with negative real part;*

Then system (6.4.1) is Bakaev stable and in every strip $\Pi_k := \{x \in \mathbf{R}^n : k\Delta \leq r^*x \leq (k+1)\Delta\}$, $(k \in \mathbf{Z})$ this system has a non-trivial periodic solution, i.e., system (6.4.1) has at least one cycle of the first kind.

In order to prove this theorem we need the following auxiliary propositions.

Lemma 6.4.1 *Consider the system*

$$\dot{x} = Bx, \quad \sigma = C^*x,$$

where B and C are $n \times n$ and $n \times m$ matrices, respectively. We suppose that the pair (B, C) is observable and there exist a quadratic form $V(x) = x^*Hx$ and numbers $\varepsilon > 0$ and $\kappa > 0$ such that the derivative $\dot{V}$ with respect to the system satisfies

$$\dot{V}(x) + 2\varepsilon V(x) \leq -\kappa |C^*x|^2$$

for all $x \in \mathbf{R}^n$.

Then, if matrix B has no eigenvalues in the strip $-\varepsilon \leq \operatorname{Re}\lambda \leq 0$, there exist a positive number $\delta > 0$ and a positive definite form $U(x) = x^*Rx$ such that the derivative of U with respect to $\dot{x} = Bx$ satisfies

$$\dot{U}(x) + V(x) \geq \delta |x|^2 \quad \text{for all} \quad x \in \mathbf{R}^n.$$

The proof of this lemma can be found in [155].

Lemma 6.4.2 *Suppose that $(z(\cdot), \sigma(\cdot))$ is a bounded on $\mathbf{R}_+$ solution of (6.4.3) satisfying the inclusion $\sigma(t) \in [0, \Delta - \delta]$ for all $t > 0$, where $\delta > 0$ is some number.*

Then either $\lim\limits_{t \to +\infty} \sigma(t) = \overline{\sigma}$ ($\overline{\sigma}$ the positive zero of σ) or the expression $\sigma(t) - \overline{\sigma}$ changes infinitely times the sign for $t \to +\infty$.

Proof From (6.4.3) we get immediately for the solution $(z(\cdot), \sigma(\cdot))$

$$c^*A^{-1}\dot{z}(t) - \dot{\sigma}(t) = c^*A^{-1}b\varphi(\sigma(t)). \tag{6.4.6}$$

Suppose that $\sigma(t) - \overline{\sigma}$ does not change the sign for $t \geq \overline{t} > 0$ and assume w.l.o.g. that $\sigma(t) - \overline{\sigma} > 0$ for $t \geq \overline{t}$. Then $\varphi(\sigma(t))$ also does not change the sign on $[\overline{t}, +\infty)$. Integrating both sides of (6.4.6) from $\overline{t}$ to t we have

$$c^*A^{-1}b \int_{\overline{t}}^{t} \varphi(\sigma(\tau))\,d\tau = \left[c^*A^{-1}z(\tau) - \sigma(\tau) \right]_{\tau=\overline{t}}^{\tau=t}. \tag{6.4.7}$$

Since χ is non-degenerate it follows that $K(0) = c^*A^{-1}b \neq 0$. Therefore (6.4.7) implies for the bounded solution $(z(\cdot), \sigma(\cdot))$ that

$$\int_{\overline{t}}^{\infty} \varphi(\sigma(t))\,dt \leq \text{const.} \tag{6.4.8}$$

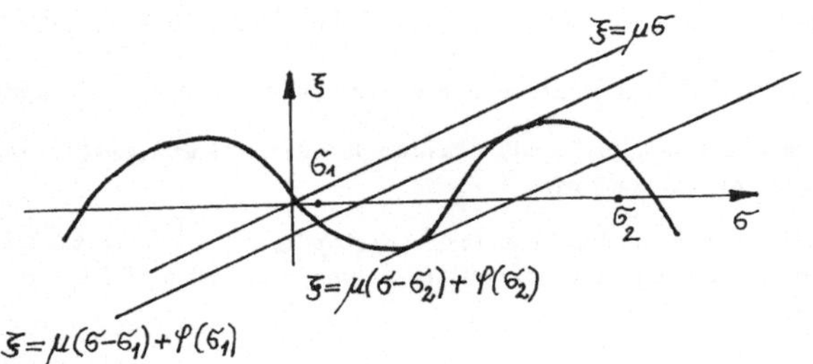

Figure 6.4.1

Furthermore we have by the boundedness of φ and σ

$$\left|\frac{d}{dt}\varphi(\sigma(t))\right| = |\varphi'(\sigma(t))\dot{\sigma}(t)| \leq \text{const.}$$

Using the last estimate and inequality (6.4.8) we obtain by the Barbalat lemma (Theorem 2.1.3, p. 16) that $\lim_{t\to\infty} \varphi(\sigma(t)) = 0$. Therefore $\lim_{t\to\infty} \sigma(t) = \overline{\sigma}$ and the lemma is proved. ∎

Proof of Theorem 6.4.1 By Theorem 3.4.1, p. 59, we get immediately that system (6.4.1) is Bakaev stable. We will consider in the sequel only the strip $\Pi_0 = \{x : 0 \leq r^*x \leq \Delta\}$. Using the equivariance property of (6.4.1) we can extend our result from Π_0 to any Π_k, $k \in \mathbf{Z}$. We will show that in Π_0 there exists a compact and positively invariant with respect to (6.4.1) set which contains exactly one equilibrium of (6.4.1). From (6.4.5) and the Δ-periodicity of φ it follows that there exist numbers $\sigma_1, \sigma_2 \in (0, \Delta)$, $\sigma_1 < \sigma_2$, and such that for j=1,2

$$[\varphi(\sigma) - \varphi(\sigma_j)](\sigma - \sigma_j) \leq \mu(\sigma - \sigma_j)^2 \tag{6.4.9}$$

for all $\sigma \in \mathbf{R}$ (see Figure 6.4.1).

Let us take now the matrix H from the proof of Theorem 3.4.1, p. 59, satisfying (3.4.6) p. 60. We choose two vectors $y_1, y_2 \in \mathbf{R}^n$ with $Py_j = 0$ and $r^*y_j = \sigma_j$, $(j = 1, 2)$ and consider the functions $U_j : \mathbf{R}^n \to \mathbf{R}$ defined by

$$U_j(x) = (x - y_j)^* H(x - y_j), \qquad j = 1, 2.$$

Putting $K_j := \{x : (x - y_j)^* H(x - y_j) < 0\}$ (i.e. an open quadratic cone) we find by Lemma 3.4.1, p. 59, that

$$K_j \cap \{x : r^*x = \sigma_j\} = \emptyset, \qquad j = 1, 2.$$

The conditions of Theorem 3.4.1, p. 59, guarantee also that with a certain $\delta > 0$ the relation

$$2x^*H[(P + \lambda I)x + q\xi] + r^*x[(\mu + \delta)r^*x - \xi] \leq 0 \tag{6.4.10}$$

is satisfied for all $(x, \xi) \in \mathbf{R}^n \times \mathbf{R}$. With the help of (6.4.10) we obtain for j= 1, 2 and $x(\cdot) := (z(\cdot), \sigma(\cdot))$ that

$$\dot{U}_j(x(t)) + 2\lambda U_j(x(t)) \leq - \{(\mu + \delta)[\sigma(t) - \sigma_j] - [\varphi(\sigma(t)) - \varphi(\sigma_1)]\} [\sigma(t) - \sigma_j] + \varphi(\sigma_j)[\sigma(t) - \sigma_j] \tag{6.4.11}$$

146

for all $t > 0$. From (6.4.9) and (6.4.11) and the properties of K_1, K_2 it follows according to Lemma 3.1.1, p. 48, that the sets

$$W_j = \left\{ x \ : \ U_j(x) < 0, \ (-1)^j r^* x < (-1)^j \sigma_j \right\}, \qquad j = 1,2$$

are positively invariant for the solutions of (6.4.1). By the continuous dependence of solutions on the initial data the set

$$R := \overline{W}_1 \cap \overline{W}_2 = \{ x \in \mathbf{R}^n \ : \ U_j(x) \leq 0, \ j = 1,2, \ \sigma_1 \leq r^* x \leq \sigma_2 \}$$

is also positively invariant. Arguing as in the proof of Theorem 3.1.1, p. 48, we see that R is bounded. It is evident that R contains the only equilibrium $\overline{x}$, given by $P\overline{x} = 0$ and $r^* \overline{x} = \overline{\sigma}$.

Let us construct now a two-dimensional quadratic cone with the top in $\overline{x}$ which is positively invariant for (6.3.1). Because of the assumption (i) the Yakubovich-Kalman theorem (Theorem 1.4.2, p. 9) says that there exist an $(n-1) \times (n-1)$ matrix $H_1 = H_1^*$ and a number $\delta_1 > 0$ such that

$$2z^* H_1[(A + \lambda_1 I)z + b\xi] - c^* z\xi \leq -\delta_1 |z|^2 \tag{6.4.12}$$

for all $(z, \xi) \in \mathbf{R}^n \times \mathbf{R}$. It follows that

$$2H_1 b = c. \tag{6.4.13}$$

For $\xi = 0$ we get from (6.4.12)

$$H_1(A + \lambda_1 I) + (A + \lambda_1 I)^* H_1 \leq -\delta I.$$

Using this and condition (ii) of the theorem we see that H_1 has one negative eigenvalue and $n-2$ positive ones. It follows from (6.4.4) and (6.4.13) that $c^* H_1^{-1} c = 2c^* b = 0$. Then, according to Lemma 3.4.1, p. 59, we have

$$\{z \ : \ z^* H_1 z < 0\} \cap \{z \ : \ c^* z = 0\} = \emptyset. \tag{6.4.14}$$

Note that from conditions $\varphi(\overline{\sigma}) = 0$ and $\varphi'(\overline{\sigma}) > 0$ it follows that there exists an $l > 0$ such that

$$|\varphi(\sigma)(\sigma - \overline{\sigma})^{-1}| > l \quad \text{for} \quad \sigma \in [\sigma_1, \sigma_2].$$

Then

$$-\int_{\overline{\sigma}}^{\sigma} \varphi(\vartheta) d\vartheta \leq -\frac{l}{2}(\sigma - \overline{\sigma})^2 \tag{6.4.15}$$

for all $\sigma \in [\sigma_1, \sigma_2)$. Define now the function $W : \mathbf{R}^{n+1} \to \mathbf{R}$ by

$$W(x) = z^* H_1 z - \int_{\overline{\sigma}}^{\sigma} \varphi(\vartheta)\, d\vartheta \qquad (x = (z, \sigma))$$

and denote by $\dot{W} : \mathbf{R}^{n+1} \to \mathbf{R}$ its derivative with respect to system (6.4.3). It follows from (6.4.12) and (6.4.13) that the estimate

$$\dot{W} + 2\lambda_1 W \leq -\delta_2 \left[|z|^2 + (\sigma - \overline{\sigma})^2 \right] \tag{6.4.16}$$

with $\delta_2 = \min\{\delta_1, \lambda_1 l\}$ is satisfied on R. Since R is positively invariant the estimate (6.4.16) implies that the set

$$D_1 = \{ x \ : \ W(x) \leq 0 \} \cap R$$

is also positively invariant.

Let us consider now system (6.4.1) with a nonlinearity φ_1 defined by $\varphi_1(\sigma) = \varphi'(\overline{\sigma})(\sigma - \overline{\sigma})$ which has all the properties of φ except the Δ-periodicity. Thus we get the linear system $\dot{x} = P_1 x - q\varphi'(\overline{\sigma})\overline{\sigma}$ with $P_1 = P + qr^*\varphi'(\overline{\sigma})$. According to condition (iii) of the theorem matrix P_1 has exactly two eigenvalues with positive real parts and $n - 2$ eigenvalues with negative real part. Indeed, using the well-known property $\det(I - bc^*) = 1 + c^*b$ $(b, c \in \mathbf{R}^n)$ we get

$$\begin{aligned}
\det(P_1 - sI) &= \det[I + qr^*(P - sI)^{-1}\varphi'(\overline{\sigma})]\det(P - sI) \\
&= [1 + r^*(P - sI)^{-1}q\varphi'(\overline{\sigma})]\det(P - sI) \\
&= -\left[1 + \frac{m(s)}{sn(s)}\varphi'(\overline{\sigma})\right] sm(s).
\end{aligned}$$

The function W_1 from above is defined now by

$$W_1(x) = z^* H_1 z - \frac{1}{2}\varphi'(\overline{\sigma})(\sigma - \overline{\sigma})^2.$$

For the derivative of W_1 with respect to (6.4.1) with φ_1 we get the estimate

$$\dot{W}_1 + 2\lambda_1 W_1 \leq -\delta_1|z|^2 - \lambda_1\varphi'(\overline{\sigma})(\sigma - \overline{\sigma})^2 \tag{6.4.17}$$

for all $(z, \sigma) \in \mathbf{R}^n \times \mathbf{R}$.

Since the matrix $\widetilde{H} := \mathrm{diag}\left\{H_1, -\frac{1}{2}\varphi'(\overline{\sigma})\right\}$ has two negative and $n - 2$ positive eigenvalues it follows from (6.4.17) that the matrix $P_1 + \lambda_1 I$ has two eigenvalues with positive and $n - 2$ eigenvalues with negative real part. Because two eigenvalues of P_1 have positive real parts we may conclude that P_1 has no eigenvalues in the strip $-\lambda_1 \leq \mathrm{Re}\, p \leq 0$. From this and (6.4.17) it follows by Lemma 6.4.1 that there exists a positive definite matrix M and a positive number δ_0 such that the derivative of $W_2(x) := (x - \overline{x})^* M(x - \overline{x})$ with respect to the system (6.4.1) with φ_1 satisfies

$$\dot{W}_2(x) + W_1(x) \geq \delta_0|x - \overline{x}|^2 \tag{6.4.18}$$

for all $x \in \mathbf{R}^n$. Inequality (6.4.18) may also be written in the form

$$\dot{W}_2(x) + W(x) \geq \delta_0|x - \overline{x}|^2 + \int\limits_{\overline{\sigma}}^{\sigma} [\varphi'(\overline{\sigma})(\vartheta - \overline{\sigma}) - \varphi(\vartheta)]\, d\vartheta, \tag{6.4.19}$$

where $x = (z, \sigma) \in \mathbf{R}^n \times \mathbf{R}$.

Note that

$$\int\limits_{\overline{\sigma}}^{\sigma} [\varphi'(\overline{\sigma})(\vartheta - \overline{\sigma}) - \varphi(\vartheta)]\, d\vartheta = o(|x - \overline{x}|^2).$$

Note that the difference between the derivative of W_2 with respect to (6.4.1) with φ_1 and the derivative of W_2 with respect to (6.4.1) with the original function φ is also $o(|x - \overline{x}|^2)$. Then we obtain from (6.4.19) for the derivative of W_2 with respect to the original system

$$\dot{W}_2(x) \geq -W(x) + \delta_0|x - \overline{x}|^2 + o(|x - \overline{x}|^2). \tag{6.4.20}$$

Let the numbers $\kappa_1 > 0$ and $\delta_3 > 0$ be such that

$$\delta_0|x - \overline{x}|^2 + o(|x - \overline{x}|^2) \geq \delta_3|x - \overline{x}|^2$$

if $|x - \overline{x}|^2 \leq \kappa_1$. Then it follows from (6.4.20) that

$$\dot{W}_2(x) \geq \delta_3|x - \overline{x}|^2 \tag{6.4.21}$$

148

whenever $W(x) \leq 0$ and $|x - \overline{x}|^2 \leq \kappa_1$. Define for $\kappa_2 > 0$ the set

$$D_2 = \left\{ x \; : \; (x - \overline{x})^* M(x - \overline{x}) \geq \kappa_2 \right\}.$$

Since M is positive definite it follows from (6.4.21) that for a certain κ_2 the set $D_1 \cap D_2$ is positively invariant for solutions of (6.4.2). It is clear that for a certain $\kappa_3 > 0$

$$D_2 \subset \left\{ x \; : \; |x - \overline{x}|^2 \geq \kappa_3 \right\}. \tag{6.4.22}$$

Let us demonstrate now that from this inclusion and inequality (6.4.16) it follows that for $\gamma \in (0, \frac{1}{2}\delta_2\kappa_3\lambda_1^{-1})$ the set

$$D_3 = \left\{ x \; : \; W(x) \leq -\gamma \right\} \cap D_2$$

is positively invariant for (6.4.1). Indeed, suppose that for a solution $x(\cdot)$ from (6.4.1) we have $W(x(\tau)) = -\gamma$ and $x(\tau) \in D_2$ for some $\tau \geq 0$. From (6.4.19) and (6.4.13) it follows that

$$\dot{W}(x(\tau)) \leq 2\lambda_1\gamma - \delta_2\kappa_3 < 0.$$

Thus the invariance of D_3 is proved.

Let us now introduce the set

$$\Omega = D_1 \cap D_3 \cap \left\{ x = (z, \sigma) \; : \; \sigma = \overline{\sigma} \right\} \cap \left\{ x = (z, \sigma) \; : \; c^* z \geq 0 \right\}.$$

We conclude from (6.4.11) that

$$D_3 \cap \left\{ x = (z, \sigma) \; : \; \sigma = \overline{\sigma} \right\} \cap \left\{ x = (z, \sigma) : c^* z = 0 \right\} = \emptyset.$$

Thus $\partial\Omega$ is is without contact with the vector field of (6.4.1). It is clear that Ω is closed, bounded and convex. We consider now solutions $x(\cdot)$ of (6.4.1) with $x(0) \in \Omega$. Since D_1 is positively invariant it follows that $x(t) \in D_1$ for $t \geq 0$ and, consequently, $x(t)$ is bounded on $\mathbf{R}_+$ with $\sigma(t) \in [\sigma_1, \sigma_2]$ for all $t \geq 0$. Then according to Lemma 6.4.2 either $\sigma(t) - \overline{\sigma}$ infinitely times changes the sign as $t \to \infty$ or $\lim_{t\to\infty} \sigma(t) = \overline{\sigma}$. But the latter situation cannot occur as $\partial\Omega$ is without contact with (6.4.1). Consequently there exists a number $\tau = \tau(x(0)) > 0$ such that $x(\tau) \in \Omega$ and $x(t) \notin \Omega$ for $t \in (0, \tau)$.

Let us define the transformation $T : \Omega \to \Omega$ by $T(x(0)) = x(\tau(x_0))$. Using now the continuous dependence of the solutions on the initial data and the fact that $\partial\Omega$ is without contact with the vector field (6.4.1) we conclude that T is continuous. According to Brouwer's theorem there exists a $\widetilde{x}_0$ with $T(\widetilde{x}_0) = \widetilde{x}_0$. It is clear that the solution $x(\cdot, \widetilde{x}_0)$ of (6.4.1) is a non-trivial periodic solution of (6.4.1) which belongs to the strip $\{x \; : \; 0 \leq r^* x \leq \Delta\}$ for all t. From the equivariance property of (6.4.1) it follows that (6.4.1) has a non-trivial periodic solution in every strip $\{x \; : \; k\Delta \leq r^* x \leq (k+1)\Delta\}$. $\blacksquare$

Let us state a further theorem on the existence of cycles of the first kind. The proof is similar to the previous one and omitted.

Theorem 6.4.2 *Suppose there exists a positive number λ such that the following conditions are true:*

(i) *$\operatorname{Re}\chi(i\omega - \lambda) < 0$ for all $\omega \geq 0$;*

(ii) *all zeros of $n(s - \lambda)$ have a negative real part;*

(iii) *there exists a number $a < 2\sqrt{\lambda\varepsilon(\lambda)}$ such that the solution of the Cauchy problem*

$$\frac{dF(\vartheta)}{d\vartheta}F(\vartheta) + aF(\vartheta) + \varphi(\vartheta) + \varphi'(\vartheta)\vartheta = 0, \quad F(0) = 0 \tag{6.4.23}$$

is defined on $(-\infty, +\infty)$ satisfying the properties $F(\vartheta) > 0$ for $\vartheta < 0$, $F(\vartheta) < 0$ for $\vartheta > 0$ and $\lim_{|\vartheta|\to\infty} F(\vartheta)^2 = +\infty$.

Suppose also that for a certain $\lambda_1 > 0$ conditions (i) to (iii) of Theorem 6.4.1 are fulfilled. Then system (6.4.1) is Bakaev stable and has at least one cycle of the first kind.

In order to verify condition (iii) of Theorem 6.4.2 the following assertion is useful.

Proposition 6.4.1 *Suppose for the function φ in (6.4.23) that $|\varphi(\sigma)| \leq 1$ on $\mathbf{R}$. Suppose also that a certain solution F of (6.4.23) satisfies $F(0) = 0$, $F(\vartheta) > 0$ for $\vartheta \in [-\Delta, 0)$ and $F(-\Delta) > \sqrt{2\Delta}$. Then $F(\vartheta) > 0$ for $\vartheta < 0$ and $\lim_{\vartheta \to -\infty} F(\vartheta) = \infty$.*

Example 6.4.1 (Barbashin's problem) In his papers [17], [20] E.A. Barbashin has posed the problem of finding conditions for the existence of cycles of various types for the equation

$$\frac{d^3}{dt^3}\sigma + \alpha\frac{d^2}{dt^2}\sigma + \beta\frac{d}{dt}\sigma + \varphi(\sigma) = 0, \tag{6.4.24}$$

where α and β are positive numbers, and φ is a Δ-periodic C^1 function that has zeros. Sufficient conditions for the existence of cycles of the second kind for (6.4.24) have been established in Example 6.3.6.

In order to obtain in (6.4.24) cycles of the first kind we apply Theorem 6.4.1 to this equation. A straightforward computation shows that $\chi(s) = \dfrac{1}{s(s^2 + \alpha s + \beta)}$, $m(s) = 1$ and $n(s) = s^2 + \alpha s + \beta$. The function π defined in the statement of Theorem 3.4.1, p. 59 is given by

$$\pi(\omega) = \frac{(3\lambda - \alpha)\omega^2 - \lambda(\lambda^2 - \alpha\lambda + \beta) + \mu}{|i\omega - \lambda|^2|(i\omega - \lambda)^2 + \alpha(i\omega - \lambda) + \beta|}.$$

If we suppose now that $\alpha^2 > 4\beta$ the frequency-domain condition 2) of Theorem 3.4.1, p. 59, is satisfied for any $\mu < \lambda(\lambda^2 - \alpha\lambda + \beta)$ if we take

$$\lambda = \frac{\alpha}{3} - \sqrt{\frac{\alpha^2}{9} - \frac{\beta}{3}}. \tag{6.4.25}$$

It is also clear that (6.4.25) implies that the polynomial $(s - \lambda)n(s - \lambda)$ has one positive zero and two negative ones. Condition (i) of Theorem 6.4.1 is fulfilled if for a certain $\lambda_1 > 0$

$$\lambda_1^2 - \alpha\lambda_1 + \beta < 0. \tag{6.4.26}$$

If $\alpha^2 > 4\beta$ the inequality (6.4.26) is satisfied for $\lambda_1 = \frac{\alpha}{2}$. In this case the polynomial $n(s - \lambda_1)$ has one positive and one negative zero. An analysis of the polynomial $sn(s) + \varphi'(\overline{\sigma})$ shows that $\varphi'(\overline{\sigma}) > \alpha\beta$ and condition (iii) of Theorem 6.4.1 is valid. Thus if $\alpha^2 > 4\beta$, the function φ satisfies (6.4.5) with $\mu < \lambda(\lambda^2 - \alpha\lambda + \beta)$, where λ is defined by (6.4.25) and $\varphi'(\overline{\sigma}) > \alpha\beta$, then equation (6.4.24) has a cycle of the first kind.

Example 6.4.2 Let us consider equation (6.4.24) with $\varphi(\sigma) = -\sin\sigma$. If $\alpha^2 \geq 4\beta$ the transfer function χ can be rewritten as

$$\chi(s) = \frac{1}{s(\gamma s + 1)(\delta s + 1)}, \tag{6.4.27}$$

where $0 < \delta \leq \gamma$ are parameters. System (6.1.1), (6.4.27) describes the dynamics of a synchronization system with two sequentially connected integrating filters. We want to use Theorem 6.4.2 and choose $\lambda = (2\gamma)^{-1}$. It follows that for this λ we have $\varepsilon(\lambda) = (1 - \frac{\delta}{2\gamma})$ and conditions (i) and (ii) of Theorem 6.4.2 are satisfied. Note that for nonlinearities φ with $\varphi(-\sigma) = -\varphi(\sigma) \quad (\sigma \in \mathbf{R})$

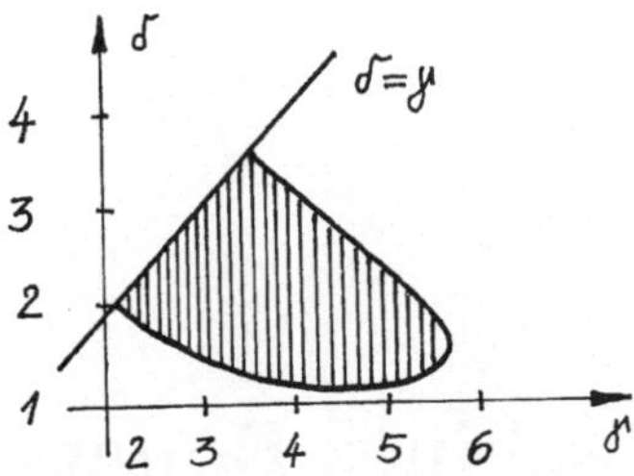

Figure 6.4.2

the considered solution F of (6.4.23) is also odd. Thus for $\varphi(\sigma) = -\sin\sigma$ it is sufficient according to Proposition 6.4.1 to have information about F on $[-2\pi, o)$. A computer calculation for $a = 1/\sqrt{3.3}$ shows that the assumptions of Proposition 6.4.1 are satisfied if

$$\frac{3.3}{\gamma}\left(2 - \frac{\delta}{\gamma}\right) > 1. \tag{6.4.28}$$

It follows from Example 6.4.1 that condition (i) to (iii) of Theorem 6.4.1 are fulfilled if $\varphi'(\overline{\sigma}) > \alpha\beta$, i.e., if

$$\varphi'(\overline{\sigma}) > \frac{1}{\gamma} + \frac{1}{\delta}. \tag{6.4.29}$$

Thus conditions (6.4.28) and (6.4.29) are sufficient conditions of the existence of cycles of the first kind in (6.4.24). This parameter domain is demonstrated in Figure 6.4.2 (shadowed region).

Remark 6.4.1 In papers [41], [90] certain extensions of Theorems 6.4.1 and 6.4.2 are given. In particular the cases $\lim_{s\to\infty} s\chi(s) \neq 0$ and $\varphi'(0) > 0$ are considered there.

Chapter 7

Synchronous Machines Equations

In this chapter some classes of pendulum-like systems describing the dynamics of synchronous machines are considered. For such systems there are given sufficient conditions for global convergence (synchronous behavior) or for the existence of cycles of the second kind (asynchronous behavior) which are mainly based on the non-local reduction principle allowed to relate the behavior of higher dimensional power systems to the one of associated second order equation of Tricomi type.

7.1 Special Properties of Synchronous Machines Equations

It is well-known [71] that in a certain approximation the synchronizing process in a synchronous motor may be described by the pendulum equation

$$\ddot{\vartheta} + \alpha\dot{\vartheta} + \sin\vartheta = \gamma, \tag{7.1.1}$$

where α and γ are positive parameters. With respect to the synchronous motor $\vartheta(t)$ is the actual phase difference, $\dot{\vartheta}(t)$ is the actual difference between the angular velocities of the rotating magnet field and the rotor. Furthermore, γ is the per unit value of the torque and α is the per unit value of the damping moment. It is important to note that in more complicated mathematical models of a synchronous machine, which take into consideration the influence of $\vartheta(t)$ and $\dot{\vartheta}(t)$ on the currents in the windings of the rotor, there appear nonlinearities of the type "inner product of the vector of the phase variables with a periodic with respect to angular coordinates vector function". In this connection the typical equations of a synchronous machine and a PLL system differ from each other, since the latter one contains only one scalar nonlinearity which is periodic. In the present chapter we investigate the global behavior of synchronous machines models with the described nonlinearities. More complicated models one can find in the monographs [52, 74, 138].

In analogy to PLL systems our aim is to give sufficient conditions for the gradient-like behavior of a machine equation. The practical importance of this question was already mentioned by F. Tricomi [148]. Gradient-like behavior of the machine equation guarantees the process of pull into synchronism of the machine independently on the initial states [74]. In contrast with a PLL system the stability in the sense of Yu.A. Bakaev is of little use for the machine since this type of stability does not exclude the presence of oscillations of the phase difference $\vartheta(t)$, which is not admissible for the machine. Note that with respect to a machine another solutions property of (7.1.1) is of great importance.

A typical situation is as follows. For a certain time the synchronous machine works without load ($\gamma = 0$). The stable stationary solution $\vartheta(t) \equiv 0$ of (7.1.1) corresponds with the working

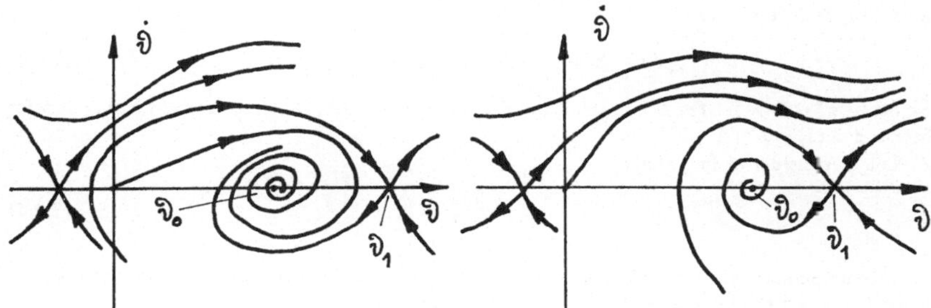

Figure 7.1.1

state of the machine. Suppose now that at a certain time $t = t_0$ the load increases abruptly from $\gamma = 0$ to $\gamma = \gamma_0 < 1$. Clearly, that for $t \geq t_0$ the working state of the machine is a non-stationary solution $\vartheta(t)$ of (7.1.1), which is a C^1-extension of the stationary solution for $t < t_0$. It was shown in Section 2.2 that for this solution $\vartheta(t)$ of (7.1.1) with $\gamma = \gamma_0$ satisfying $\vartheta(t_0) = 0$, $\dot\vartheta(t_0) = 0$ one of the following properties is possible: either

$$(A) \qquad \lim_{t\to+\infty} \vartheta(t) = \vartheta_0 \qquad \text{and} \qquad \lim_{t\to+\infty} \dot\vartheta(t) = 0$$

or

$$(B) \qquad \lim_{t\to+\infty} \vartheta(t) = +\infty$$

(see Figure 7.1.1).

The point $(\vartheta_0, 0)$ will be the new local stable equilibrium of (7.1.1). The supremum of all possible γ_0 with property (A), $\Gamma = \sup\{\gamma_0\}$, is called the *limit load* of the machine. Note that for γ near Γ equation (7.1.1) admits a basin of attraction of the local stable equilibrium point $(\vartheta_0, 0)$ and local stable cycle of the second kind. It follows that the limit load problem is not identical to the problem of defining the parameter region of the gradient-like behavior of the machine equation.

7.2 Gradient-Like Behavior of Machine Equations with a Zero Load

Some models of synchronous machine may be described by the system

$$\begin{aligned}
\dot\sigma &= \eta, \\
\dot\eta &= -g(\eta,\sigma) + z^* C f(\sigma) - \varphi(\sigma), \\
\dot z &= Az + Bf(\sigma)\eta.
\end{aligned} \qquad (7.2.1)$$

Note that the famous Lorenz equations may also be written in the form (7.2.1) ([96]). We suppose that A, B and C are constant $n \times n$, $n \times m$ resp. $n \times m$ matrices, $f : \mathbf{R} \to \mathbf{R}^m$ is a Δ-periodic function of class C^1, $\varphi : \mathbf{R} \to \mathbf{R}$ and $g : \mathbf{R} \times \mathbf{R} \to \mathbf{R}$ are C^1 functions satisfying with fixed μ_1, μ_2 the properties

$$\mu_1\eta^2 \leq g(\eta,\sigma)\eta \leq \mu_2\eta^2 \qquad (7.2.2)$$

153

for all $(\eta, \sigma) \in \mathbf{R} \times \mathbf{R}$ and

$$\varphi(\sigma + \Delta) = \varphi(\sigma) \qquad g(\eta, \sigma + \Delta) = g(\eta, \sigma) \tag{7.2.3}$$

for all $(\sigma, \eta) \in \mathbf{R} \times \mathbf{R}$. For the description of the linear part of (7.2.1) we introduce the transfer function $K(s) = C^*(A - sI)^{-1}B$.

Let us consider at first the case

$$\int_0^\Delta \varphi(\sigma)\, d\sigma = 0, \tag{7.2.4}$$

which corresponds with an unloaded synchronous machine. Suppose also that the matrix A is Hurwitzian and the nonlinearity φ admits a finite set of zeros on $[0, \Delta)$.

Theorem 7.2.1 *Suppose that $\mu_1 \geq 0$, (7.2.4) is satisfied and the following conditions hold:*

(i) $\operatorname{Re} K(i\omega) > 0$ *for all* $\omega \in \mathbf{R}$ *and* $\lim\limits_{\omega \to \infty} \omega^2 \operatorname{Re} K(i\omega) > 0$;

(ii) $\mid f(\sigma) \mid^2 + \mid f'(\sigma) \mid^2 \neq 0$ *for all* $\sigma \in \mathbf{R}$.

Then system (7.2.1) is gradient-like.

Proof Let us introduce the vector $x = (\sigma, \eta, z)$ and the function $V : \mathbf{R}^{n+2} \to \mathbf{R}$ given by

$$V(x) = z^*Hz + \frac{1}{2}\eta^2 + \int_0^\sigma \varphi(\vartheta)\, d\vartheta,$$

where the $n \times n$ matrix $H = H^* > 0$ satisfies with an $\varepsilon > 0$ the inequality

$$2z^*H(Az + B\xi) + z^*C\xi \leq -\varepsilon \mid z \mid^2 \tag{7.2.5}$$

for all $(z, \xi) \in \mathbf{R}^n \times \mathbf{R}^m$. The existence of a matrix H and a number ε follows from Theorem 1.4.1. Let us show that function V satisfies the asumptions of Theorem 4.2.1. Clearly, assumption (i) of Theorem 7.2.1 results from the equality (7.2.4) and assumption (ii) results from $H > 0$. Using (7.2.5) we immediately see that for any solution $x(\cdot) = (\sigma(\cdot), \eta(\cdot), z(\cdot))$ of (7.1.1) we have

$$\frac{d}{dt}V(x(t)) \leq -\varepsilon \mid z(t) \mid^2 \tag{7.2.6}$$

for all $t \geq 0$. This guarantees assumption (iii) of Theorem 7.2.1. Let $x(\cdot) = (\sigma(\cdot), \eta(\cdot), z(\cdot))$ be a solution of (7.2.1) with $V(x(0)) = V(x(\tau))$ for a $\tau > 0$. From (7.2.6) we get

$$z(t) = 0 \qquad \text{on} \qquad [0, \tau]. \tag{7.2.7}$$

The last equation of (7.2.1) gives the equality

$$Bf(\sigma(t))\eta(t) = 0 \qquad \text{on} \qquad [0, \tau].$$

Assumption (i) of the present theorem implies that rank $B = m$. Thus we have $f(\sigma(t))\eta(t) = 0$ on $[0, \tau]$.

Using the fact that $\eta(t) = \dot{\sigma}(t)$ we get

$$\int_{\sigma(0)}^{\sigma(t)} f(s)ds = 0 \qquad \text{for all} \qquad t \in [0, \tau].$$

154

From this and (7.2.7) it follows that the considered solution $x(\cdot)$ is a stationary one. $\blacksquare$

Consider now the following system

$$
\begin{aligned}
\dot{\sigma} &= \eta, \\
\dot{\eta} &= -g(\eta,\sigma) + y^* C f_1'(\sigma) - \varphi_1(\sigma), \\
\dot{y} &= Ay + B_1 f_1(\sigma) + a,
\end{aligned}
\tag{7.2.8}
$$

where f_1 and φ_1 functions satisfying the properties of f and φ in (7.2.1), B_1 is a $n \times m$ matrix and a is an n-vector. The other functions and matrices are as in system (7.2.1). By the Lienard transformation

$$
y = z - A^{-1}B_1 f_1(\sigma) - A^{-1}a
$$

system (7.2.8) can be transformed into system (7.2.1) with $B = A^{-1}B_1$, $f(\sigma) = f_1'(\sigma)$,

$$
\varphi(\sigma) = \varphi_1(\sigma) + [f_1'(\sigma)]^* C^* A^{-1}[B_1 f_1(\sigma) + a].
$$

Note that the transfer matrices $K(s)$ of (7.2.1) and $K_1(s) = C^*(A - sI)^{-1}B_1$ are connected by the property

$$
K(s) = \frac{1}{s}[K_1(s) - K_1(0)].
$$

Thus the following theorem results from Theorem 7.2.1.

Theorem 7.2.2 *Suppose that $\mu_1 \geq 0, f \in C^2$ and the following conditions are satisfied:*

(i) $K_1(0)$ *is a diagonal matrix;*

(ii) Re $\left[\frac{1}{i\omega}K_1(i\omega)\right] > 0$ *for all $\omega \neq 0$;*

(iii) $\displaystyle\lim_{\omega \to +\infty} \omega^2 \mathrm{Re}\left[\frac{1}{i\omega}K_1(i\omega)\right] > 0$;

(iv) $\displaystyle\int_0^{\Delta} \varphi_1(\sigma)\,d\sigma = 0$;

(v) $\mid f_1'(\sigma) \mid^2 + \mid f_1''(\sigma) \mid^2 \neq 0$ *on $\mathbf{R}$.*

Then system (7.2.2) is gradient-like.

Example 7.2.1 Consider the 5-dimensional system of differential equations describing the work of a synchronous motor with zero load [47]:

$$
\begin{aligned}
\dot{\sigma} &= \eta, \\
\dot{\eta} &= (-\alpha_1 y_1 - \alpha_2 y_2)\sin\sigma + (\alpha_3 y_3 \cos\sigma - \alpha_4 \sin\sigma\cos\sigma), \\
\dot{y}_1 &= \alpha_5 - \alpha_6 y_1 + \alpha_7 y_2 + \alpha_8 \cos\sigma, \\
\dot{y}_2 &= \alpha_9 y_1 - \alpha_{10} y_2 + \alpha_{11} \cos\sigma, \\
\dot{y}_3 &= -\alpha_{12} y_1 + \alpha_{13} \sin\sigma,
\end{aligned}
\tag{7.2.9}
$$

where $\alpha_i(i = 1,\ldots,13)$ are positive numbers satisfying $\alpha_6\alpha_{10} - \alpha_7\alpha_9 > 0$. Let us show that for (7.2.9) the assumptions of Theorem 7.2.2 are satisfied.

In this system we have

$$
g(\eta,\sigma) \equiv 0, \quad \varphi_1(\sigma) = -\alpha_4 \sin\sigma\cos\sigma, \quad f_1(\sigma) = \begin{bmatrix} \cos\sigma \\ -\sin\sigma \end{bmatrix},
$$

$$a = \begin{bmatrix} \alpha_5 \\ 0 \\ 0 \end{bmatrix}, \quad C = \begin{bmatrix} \alpha_1 & 0 \\ \alpha_2 & 0 \\ 0 & -\alpha_3 \end{bmatrix}, \quad B_1 = \begin{bmatrix} \alpha_8 & 0 \\ \alpha_{11} & 0 \\ 0 & -\alpha_{13} \end{bmatrix}, \quad A = \begin{bmatrix} -\alpha_6 & \alpha_7 & 0 \\ \alpha_9 & -\alpha_{10} & 0 \\ 0 & 0 & -\alpha_{12} \end{bmatrix}.$$

It is easy to see that $K_1(s) = \begin{bmatrix} K_{11}(s) & 0 \\ 0 & K_{22}(s) \end{bmatrix}$, where

$$K_{11}(s) = -[(\alpha_1\alpha_8 + \alpha_2\alpha_{11})s + \alpha_1\alpha_8\alpha_{10} + \alpha_1\alpha_{11}\alpha_7 + \alpha_2\alpha_{11}\alpha_6 + \alpha_2\alpha_9\alpha_8][(s+\alpha_6)(s+\alpha_{10}) - \alpha_9\alpha_7]^{-1},$$

$$K_{22}(s) = -\alpha_3\alpha_{13}[s + \alpha_{12}]^{-1}.$$

It is clear that the conditions (ii) and (iii) in Theorem 7.2.2 are equivalent to the inequalities

$$\mathrm{Re}\left[\frac{1}{i\omega}K_{jj}(i\omega)\right] > 0 \tag{7.2.10}$$

for all $\omega \in \mathbf{R}$ and $j = 1, 2$ and

$$\lim_{\omega\to\infty}\left(\omega^2\mathrm{Re}\left[\frac{1}{i\omega}K_{jj}(i\omega)\right]\right) > 0$$

for $j = 1, 2$. If we introduce the designations

$$\begin{aligned} \beta_1 &:= \alpha_1\alpha_8 + \alpha_2\alpha_{11}, \\ \beta_2 &:= \alpha_1\alpha_8\alpha_{10} + \alpha_{11}\alpha_1\alpha_7 + \alpha_2\alpha_{11}\alpha_6 + \alpha_2\alpha_8\alpha_9 \end{aligned}$$

we get

$$\mathrm{Re}\left[\frac{1}{i\omega}K_{11}(i\omega)\right] = \frac{\beta_2(\alpha_6 + \alpha_{10}) + \beta_1(\alpha_9\alpha_7 - \alpha_6\alpha_{10}) + \beta_1\omega^2}{|\,(i\omega + \alpha_6)(i\omega + \alpha_{10}) - \alpha_9\alpha_7\,|^2},$$

$$\mathrm{Re}\left[\frac{1}{i\omega}K_{22}(i\omega)\right] = \frac{\alpha_3\alpha_{13}}{|\,i\omega + \alpha_{12}\,|^2}.$$

It follows that the inequalities (7.2.10)) are satisfied if

$$\beta_2(\alpha_6 + \alpha_{10}) + \beta_1(\alpha_9\alpha_7 - \alpha_6\alpha_{10}) > 0. \tag{7.2.11}$$

The left-hand side of (7.2.11)) can be rewritten as

$$\alpha_1\alpha_{11}\alpha_7\alpha_6 + \alpha_2\alpha_{11}\alpha_6^2 + \alpha_2\alpha_9\alpha_8\alpha_6 + \alpha_1\alpha_8\alpha_{10}^2 + \alpha_1\alpha_{11}\alpha_{10}\alpha_7 + \alpha_2\alpha_9\alpha_8\alpha_{10} + \alpha_1\alpha_8\alpha_9\alpha_7 + \alpha_2\alpha_{11}\alpha_9\alpha_7.$$

This representation shows that (7.2.11) is satisfied. It is also obvious that the assumptions (iv) and (v) of Theorem 7.2.2 are fulfilled.

7.3   Application of Non-Local Reduction Method to some Classes of Synchronous Machines

As an associated second order equation for (7.1.1) we will consider for various a the equation

$$\ddot{\sigma} + a\dot{\sigma} + \varphi(\sigma) = 0,$$

i.e. the system

$$\begin{aligned} \dot{\sigma} &= \eta, \\ \dot{\eta} &= -a\eta - \varphi(\sigma), \end{aligned} \tag{7.3.1}$$

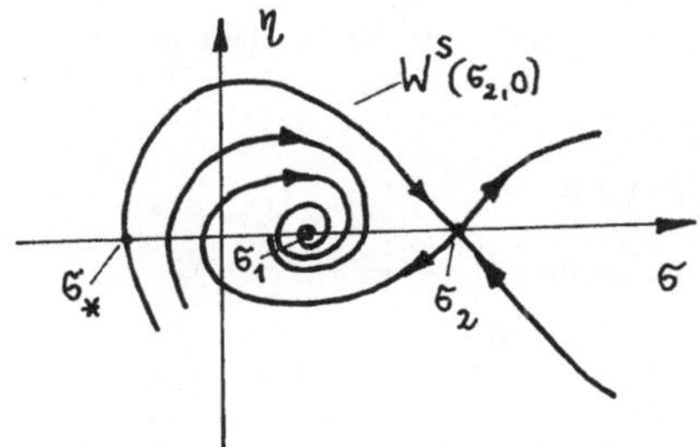

Figure 7.3.1

and the corresponding first order equation

$$F(\sigma)\frac{dF(\sigma)}{d\sigma} + aF(\sigma) + \varphi(\sigma) = 0. \qquad (7.3.2)$$

Suppose additionally that φ has exactly the two zeros $\sigma_1 < \sigma_2$ on the interval $[0, \Delta)$ satisfying $\varphi'(\sigma_1) > 0$ resp. $\varphi'(\sigma_2) < 0$. Suppose for the mean value of φ that

$$\frac{1}{\Delta}\int_0^\Delta \varphi(\vartheta)\, d\vartheta \le 0.$$

Let us consider now the parameter region $0 < a < a_{cr}$. Note that $(\sigma_2, 0)$ is by assumption a saddle point of (7.3.1). According to Section 2.2 we denote by $\widetilde{F}_0$ the solution of (7.3.2) representing the stable manifold $W^s(\sigma_2, 0)$ in $\eta > 0$. Let us denote by $\sigma_* = \sigma_*(a)$ the point of intersection of $\widetilde{F}_0$ with the σ axis (see Figure 7.3.1.

Any solution $(\sigma(\cdot), \eta(\cdot))$ of (7.3.1) with initial points

$$\sigma(0) \in (\sigma_*, \sigma_2), \qquad \eta(0) = 0 \qquad (7.3.3)$$

approaches the Lyapunov stable equilibrium point $(\sigma_1, 0)$ as $t \to +\infty$ and $\sigma(t) \in (\sigma_*, \sigma_2)$ for all $t \ge 0$. It is clear that $\sigma_* \in (\sigma_2 - \Delta, \sigma_1)$. We start our analysis of (7.2.1) with a reduction-type proposition. Suppose for this purpose that $w, \vartheta : [0, +\infty) \to \mathbf{R}$ are two arbitrary functions which are C^1 on $(0, +\infty)$.

Lemma 7.3.1 *Suppose that there exist numbers $\varepsilon \ge 0, \lambda \ge 0$ such that with $\sigma_* = \sigma_*(2\sqrt{\lambda\varepsilon})$ the following conditions are satisfied:*

(i) *$2\sqrt{\lambda\varepsilon} < a_{cr}$, i.e. any solution (σ, η) of (7.3.1) with the initial points $\sigma(0) \in (\sigma_*, \sigma_2)$, $\eta(0) = 0$ satisfies the inclusion $\sigma(t) \in (\sigma_*, \sigma_2)$, for all $t \ge 0$;*

(ii) *in a certain neighborhood of the set*

$$T_1 := \{t \in \mathbf{R}_+ \; : \; \vartheta(t) = \sigma_2, \; w(t) < 0\}$$

the function $\vartheta(t)$ is strictly decreasing and in a certain neighborhood of the set

$$T_2 := \{t \in \mathbf{R}_+ \; : \; \vartheta(t) = \sigma_*, \; w(t) < 0\}$$

the function $\vartheta(t)$ is strictly increasing;

(iii)
$$\dot{w}(t) + 2\lambda w(t) + \varepsilon\dot{\vartheta}(t)^2 + \varphi(\vartheta(t))\dot{\vartheta}(t) \leq 0, \ t \geq 0; \tag{7.3.4}$$

(iv) $\vartheta(0) \in (\sigma_*, \sigma_2), \quad w(0) \geq 0.$

Then $\vartheta(t) \in (\sigma_*, \sigma_2)$ for all $t \geq 0$.

Proof Let us consider the equation (7.3.2) with $a = 2\sqrt{\lambda\varepsilon}$

$$F\frac{dF}{d\sigma} + 2\sqrt{\lambda\varepsilon}F + \varphi(\sigma) = 0. \tag{7.3.5}$$

It follows from assumption (i) that (7.3.5) has a solution F such that $F(\sigma_*) = F(\sigma_2) = 0$ and $F(\sigma) > 0$ for all $\sigma \in (\sigma_*, \sigma_2)$. Define now the C^1-function

$$v(t) = w(t) - \frac{1}{2}F(\vartheta(t))^2, \quad t \geq 0.$$

We shall demonstrate that

$$\dot{v}(t) + 2\lambda v(t) \leq 0 \tag{7.3.6}$$

for all $t \geq 0$. Indeed, from (7.3.5) it follows that

$$\begin{aligned}
\dot{w}(t) + 2\lambda w(t) + \varepsilon\dot{\vartheta}(t)^2 + \varphi(\vartheta(t))\dot{\vartheta}(t) &= \dot{v}(t) + F(\vartheta(t))\frac{dF(\vartheta(t))}{d\sigma}\dot{\vartheta}(t) + 2\lambda v(t) + \lambda F(\vartheta(t))^2 + \\
&\quad + \varepsilon\dot{\vartheta}(t)^2 + \varphi(\vartheta(t))\dot{\vartheta}(t) \\
&= \dot{v}(t) + 2\lambda v(t) + [\sqrt{\lambda}F(\vartheta(t)) - \sqrt{\varepsilon}\dot{\vartheta}(t)]^2
\end{aligned}$$

for $t \geq 0$. Using this and assumption (iii) we get (7.3.6). According to Lemma 3.1.1 it follows from condition (iv) that $v(t) < 0$ for $t \geq 0$. Let us show now that $\vartheta(t) \in (\sigma_*, \sigma_2)$ for all $t \geq 0$. Suppose the opposite, i.e. suppose that there exists a number $t_1 > 0$ such that $\vartheta(t_1) \notin (\sigma_*, \sigma_2)$. Since $\vartheta(t)$ is continuous it follows from (iv) that there exists a number $t_2 \in [0, t_1]$ such that $[\sigma_2 - \vartheta(t_2)][\vartheta(t_2) - \sigma_*] = 0$ and

$$\vartheta(t) \in (\sigma_*, \sigma_2) \qquad \text{for} \qquad t \in [0, t_2]. \tag{7.3.7}$$

We see that either $t_2 \in T_1$ or $t_2 \in T_2$. Let for definiteness $t_2 \in T_2$ and $w(t_2) \geq 0$. Since $v(t_2) < 0$ we have then $\frac{1}{2}F(\vartheta(t_2))^2 > w(t_2) \geq 0$. Thus $F(\vartheta(t_2)) \neq 0$ and the equality $\vartheta(t_2) = \sigma_*$ is impossible. In the case $w(t_2) < 0$ the condition (ii) must be used. According to this condition there exists a $\delta > 0$ such that $\vartheta(t) < \sigma_*$ for $t \in (t_2 - \delta, t_2)$ a contradiction to (7.3.7). $\blacksquare$

Lemma 7.3.1 will be used in the proof of the next theorem guaranteeing the quasi-monostability and the convergence of solutions of (7.2.1) to $(\sigma_1, 0, 0)$ with initial points from a certain domain of attraction.

Theorem 7.3.1 *Suppose that in (7.2.2) $\mu_1 \geq 0$ and there exists a $\lambda \in [0, \mu_1]$ such that the following conditions are fulfilled:*

(i) $\operatorname{Re} K(i\omega - \lambda) > 0$ *for all $\omega \in \mathbf{R}$ and*

$$\lim_{\omega \to +\infty} \omega^2 \operatorname{Re} K(i\omega - \lambda) > 0;$$

(ii) *matrix $A + \lambda I$ is Hurwitzian;*

(iii) $2\sqrt{(\mu_1 - \lambda)\lambda} < a_{cr}$, *i.e. any solution of*

$$\ddot{\vartheta} + 2\sqrt{(\mu_1 - \lambda)\lambda}\dot{\vartheta} + \varphi(\vartheta) = 0 \tag{7.3.8}$$

with inclusion $\vartheta(t) \in (\sigma_*, \sigma_2)$ for $t \geq 0$. Here $\sigma_* = \sigma_* \left(2\sqrt{(\mu_1 - \lambda)\lambda}\right)$.

Then system (7.2.1) is quasi-monostable and any solution $(\sigma(\cdot), \eta(\cdot), z(\cdot))$ of (7.2.1) with initial points $\sigma(0) \in (\sigma_*, \sigma_2), \eta(0) = 0$ and $z(0) = 0$ tends to the equilibrium $x_1 = (\sigma_1, 0, 0)$ of (7.2.1) for $t \to +\infty$.

Proof First, let us show that there exists a positively invariant for (7.2.1) set Ω such that

$$\{x = (\sigma, \eta, z) \; : \; \sigma \in (\sigma_*, \sigma_2), \eta = 0, z = 0\} \subset \Omega \subset \{x = (\sigma, \eta, z) \; : \; \sigma \in (\sigma_*, \sigma_2)\}.$$

Let us introduce the quadratic form $G : \mathbf{R}^n \times \mathbf{R}^m \to \mathbf{R}$ by

$$G(z, \xi) = 2z^* H[(A + \lambda I)z + B\xi] + z^* C\xi,$$

where the matrix $H = H^*$ is to be determined. Condition (i) allows us to conclude by Theorem 1.4.1 that there exists a matrix $H = H^*$ and a number $\varepsilon > 0$ such that

$$G(z, \xi) \leq -\varepsilon \mid z \mid^2 \tag{7.3.9}$$

for all $(z, \xi) \in \mathbf{R}^n \times \mathbf{R}^m$. For $\xi = 0$ we get from (7.3.9) that

$$2z^* H(A + \lambda I)z < 0 \qquad \text{for} \qquad \mid z \mid \neq 0,$$

which implies by condition (ii) that $H > 0$. Consider now an arbitrary non-constant solution $x(\cdot) = (\sigma(\cdot), \eta(\cdot), z(\cdot))$ of (7.2.1) with initial data $\sigma(0) \in (\sigma_*, \sigma_2)$, $\eta(0) = 0$ and $z(0) = 0$. Define the functions

$$w(t) = z(t)^* H z(t) + \frac{1}{2}\eta(t)^2$$

and

$$\Phi(t) = \dot{w}(t) + 2\lambda w(t) + (\mu_1 - \lambda)\dot{\sigma}(t)^2 + \varphi(\sigma(t))\dot{\sigma}(t)$$

for $t > 0$. Note that for $t > 0$

$$\begin{aligned}
\Phi(t) &= z(t)^* H[(A + \lambda I)z(t) + Bf(\sigma(t))\eta(t)] + \lambda\eta(t)^2 + \eta(t)\dot{\eta}(t) + (\mu_1 - \lambda)\dot{\sigma}(t)^2 + \varphi(\sigma(t))\dot{\sigma}(t) \\
&= 2z(t)^* H[(A + \lambda I)z(t) + Bf(\sigma(t))\eta(t)] + \\
&\quad + \mu_1\eta(t)^2 + g(\eta(t), \sigma(t))\eta(t)^2 + z(t)^* Cf(\sigma(t))\eta(t).
\end{aligned}$$

Thus we have for $t > 0$

$$\Phi(t) \leq G(z(t), f(\sigma(t))\eta(t)).$$

Define now the mentioned set Ω by

$$\Omega = \left\{x = (\sigma, \eta, z) : \sigma \in (\sigma_*, \sigma_2), z^* H z + \frac{1}{2}\eta^2 \leq \frac{1}{2}F(\sigma)^2\right\},$$

where F is a solution of

$$F\frac{dF}{d\sigma} + 2\sqrt{\lambda(\mu_1 - \lambda)}F(\sigma) + \varphi(\sigma) = 0, \tag{7.3.10}$$

satisfying the conditions $F(\sigma_*) = F(\sigma_2) = 0$ and $F(\sigma) > 0$ for $\sigma \in (\sigma_*, \sigma_2)$. The set Ω is positively invariant for (7.2.1). To see this we use Lemma 5.2.1 with $\varepsilon = \mu_1 - \lambda$ and $\vartheta(t) \equiv \sigma(t)$.

Clearly that condition (i) of the lemma and condition (iii) of the theorem coincide. Assumption (iii) of the lemma follows from $\Phi(t) < 0$ on $\mathbf{R}_+$. Since $H > 0$ the sets T_1 and T_2 in assumption (ii) are empty. Thus Lemma 7.3.1 is applicable and $\sigma(t) \in (\sigma_*, \sigma)$ for $t \geq 0$. Let us demonstrate now that system (7.2.1) is quasi-monostable. In order to use Theorem 1.1.3 we consider the function $V : \mathbf{R}^{n+2} \to \mathbf{R}$ defined by

$$V(x) = z^* H z + \frac{1}{2}\eta^2 + \int\limits_0^\sigma \varphi(\vartheta)\, d\vartheta, \quad x = (\sigma, \eta, z),$$

where H is the matrix from (7.3.9). The derivative of V with respect to (7.2.1) is

$$V(x) = 2z^* H[Az + Bf(\sigma)\eta] - g(\eta, \sigma)\eta + z^* C f(\sigma)\eta.$$

Thus we have

$$\dot{V}(x) + 2\lambda z^* H z \leq G(z, f(\sigma)\eta) - \eta_1 \mu^2 \quad \text{for all} \quad x = (\sigma, \eta, z).$$

From (7.3.9) and the positivity of H it follows that

$$\dot{V}(x) \leq -\varepsilon \mid z \mid^2 -\mu_1 \eta^2 \quad \text{for all} \quad x \in \mathbf{R}^{n+2}.$$

Hence V is non-increasing along solutions of (7.2.1) and $\dot{V}(x(t)) \equiv 0$ for some solution $x(\cdot)$ implies that $z(t) \equiv 0, \eta(t) \equiv 0$ and, by (7.2.1), $\sigma(t) \equiv const$. Thus we can use Theorem 1.1.3 to conclude that system (7.2.1) is quasi-monostable. Since the set Ω contains the only equilibrium $(\sigma_1, 0, 0)$ the assertion of the theorem follows. ∎

In the next theorem we provide conditions which imply that every solution of (7.2.1) converges to an equilibrium.

Theorem 7.3.2 *Suppose that in* (7.2.1) $\mu_1 > 0$ *and the stationary set of* (7.2.1) *consists of isolated points. Let there exist a number* $\lambda \in (0, \mu_1)$ *such that the conditions (i) and (ii) of Theorem 7.3.1 are fulfilled and the associated second-order equation* (7.3.8) *is quasi-gradient-like. Then system* (7.2.1) *is gradient-like, i.e. any solution converges to an equilibrium.*

Proof Proceeding as in the proof of Theorem 7.3.1 we can use Theorem 1.1.5 which says that system (7.2.1) is monostable. It remains to show that every solution of (7.2.1) is bounded. Since conditions (i) and (ii) are satisfied we may apply Theorem 1.4.1 to conclude that there exists a positive definite matrix H and a number ε_1 such that

$$2z^* H[(A + \lambda I)z + B\xi] + 2z^* C\xi \leq -\varepsilon_1 \mid z \mid^2 \tag{7.3.11}$$

for all $(z, \xi) \in \mathbf{R}^n \times \mathbf{R}^m$. Let us consider now the function $U_\nu : \mathbf{R}^n \times \mathbf{R} \times \mathbf{R} \to \mathbf{R}$ defined by $U_\nu(x) = z^* H z + \eta^2 - \nu$ for $x = (\sigma, \eta, z)$, where ν is an arbitrary number satisfying

$$\nu > \frac{1}{4\lambda(\mu_1 - \lambda)} \max_{\sigma \in [0, \Delta]} \varphi(\sigma)^2. \tag{7.3.12}$$

Using this, (7.2.2) and (7.3.11) it follows that the derivative of U_ν with respect to (7.2.1) satisfies the inequality

$$\dot{U}_\nu(x) + 2\lambda U_\nu(x) \leq -2(\mu_1 - \lambda)\eta^2 - 2\varphi(\sigma)\eta - 2\lambda\nu \leq 0$$

for all $x = (\sigma, \eta, z)$. Then by Lemma 3.1.1 the level set

$$C^\nu = \{x \: : \: U_\nu(x) \leq 0\}$$

is positively invariant for (7.2.1) provided that ν satisfies (7.3.12).

It follows that any solution $x(\cdot) = (\sigma(\cdot), \eta(\cdot), z(\cdot))$ of (7.2.1) is bounded with respect to $z(\cdot)$ and $\eta(\cdot)$ on $\mathbf{R}_+$. In order to show that the component $\sigma(\cdot)$ is also bounded we want to apply Lemma 5.3.1 to our system. Putting

$$\varepsilon = \mu_1 - \lambda, \quad \Psi(\sigma) = \varphi(\sigma), \quad w(t) = \frac{1}{2}[z(t)^* H z(t) + \eta(t)^2]$$

and noticing that by (7.2.2) and (7.3.12)

$$\dot{w}(t) + 2\lambda w(t) + \varepsilon \dot{\sigma}(t)^2 + \varphi(\sigma(t))\dot{\sigma}(t) \leq 0$$

for all $t \geq 0$, we conclude by Lemma 5.3.1 that $\sigma(t)$ is bounded on $\mathbf{R}_+$. $\blacksquare$

We now turn to the problem of finding sufficient conditions for the existence of cycles of the first kind in system (7.2.1).

Theorem 7.3.3 *Suppose that there is a $\mu > 0$ satisfying*

$$| f(\sigma) |^2 \leq \mu, \qquad \sigma \in \mathbf{R}. \tag{7.3.13}$$

Suppose also that there exists a number $\lambda > \max\{0, \mu_2\}$ such that the following conditions are satisfied:

(i) $\det[\mu^{-1}(\mu_2 - \lambda)I + \operatorname{Re} K(i\omega - \lambda)] \neq 0$ *for all $\omega \in \mathbf{R}$;*

(ii) *matrix $+\lambda I$ is Hurwitzian;*

(iii) *the equation*
$$\ddot{\vartheta} + \lambda \dot{\vartheta} + \varphi(\vartheta) = 0 \tag{7.3.14}$$

has a circular solution $\vartheta(\cdot)$ with $\dot{\vartheta}(t) > 0$ for all $t > 0$ and satisfying the initial conditions $\vartheta(0) = \sigma_0, \dot{\vartheta}(0) = 0$.

Then for arbitrary $\delta > 0$ there exists a circular solution $(\sigma(\cdot), \eta(\cdot), z(\cdot))$ of system (7.2.1) which satisfies $\sigma(0) = \sigma_0, | \eta(0) | + | z(0) | < \delta$.

If in addition to the above assumptions $\mu_1 > 0$ and

$$\operatorname{Re} K(i\omega) > 0 \quad \text{for all} \quad \omega \in \mathbf{R}, \qquad \lim_{\omega \to \infty} \omega^2 \operatorname{Re} K(i\omega) > 0, \tag{7.3.15}$$

then system (7.2.1) has a circular cycle of the second kind.

Proof Since $\lim\limits_{\omega \to \infty} \operatorname{Re} K(i\omega - \lambda) = 0$ it follows from condition (i) that there exists a positive number $\nu_0 < \lambda - \mu_2$ such that

$$\mu^{-1}\nu_0 I + \operatorname{Re} K(i\omega - \lambda) < 0 \tag{7.3.16}$$

for all $\omega \in \mathbf{R}$. From this and condition (ii) it follows according to Theorem 1.4.4 and Lemma 1.2.1 that there exist an $n \times n$ matrix $H = H^* > 0$ and a number $\varepsilon > 0$ such that

$$2z^* H[(A + \lambda I)z + B\xi] + 2z^* C\xi - 2\mu^{-1}\nu_0 | \xi |^2 \leq -\varepsilon | z |^2 \tag{7.3.17}$$

for all $(z, \xi) \in \mathbf{R}^n \times \mathbf{R}^m$. Let $x(\cdot) = (\sigma(\cdot), \eta(\cdot), z(\cdot))$ be an arbitrary solution of (7.2.1). We make use of Lemma 6.2.1 and put

$$w(t) = \frac{1}{2}[z(t)^* H z(t) - \eta(t)^2], \qquad \nu = \frac{1}{2}, \qquad \Psi(\vartheta) = \varphi(\vartheta).$$

In order to define the required function F for Lemma 6.2.1 we consider the first-order equation

$$\frac{dF}{d\vartheta}F(\vartheta) + \lambda F(\vartheta) + \varphi(\vartheta) = 0, \qquad (7.3.18)$$

which corresponds to (7.3.14) and choose the solution F of (7.3.18) with $F(\sigma_0) = 0$. We also put $\vartheta_1 = \sigma_0$ and $\vartheta_2 = +\infty$. Let us show now that the required assumptions of Lemma 6.2.1 are satisfied. The conditions (i) to (iv) are obvious. Condition (v) is satisfied since $H > 0$ (note, that $t_1 = 0, t_2 = +\infty$). To verify condition (vi) we write

$$\dot{w}(t) + 2\lambda w(t) - \Psi(\sigma(t))\dot{\sigma}(t) \;=\; z(t)^* H[(A + \lambda I)z(t) + Bf(\sigma(t))\eta(t)] - \lambda\eta(t)^2 -$$
$$-\dot{\sigma}(t)[-g(\eta(t),\sigma(t)) + z(t)^* C f(\sigma(t)) - \varphi(\sigma(t))] - \varphi(\sigma(t))\dot{\sigma}(t),$$

$t \geq 0$. Thus by (7.2.2) and (7.3.17) we get

$$\begin{aligned}
\dot{w}(t) + 2\lambda w(t) - \Psi(\sigma(t))\dot{\sigma}(t) \;&\leq\; -\lambda\eta(t)^2 + g(\eta(t),\sigma(t))\eta(t) + \mu^{-1}\nu_0 \mid f(\sigma(t)) \mid^2 \eta(t)^2 \\
&\leq\; (-\lambda + \mu_2 + \nu_0)\eta(t)^2 \\
&<\; 0
\end{aligned}$$

for $t \geq 0$. So condition (vi) is satisfied. Note also that for any arbitrary small $\delta > 0$ there exist a vector z_0 and a number η_0 such that the inequalities

$$\eta_0 > 0, \mid \eta_0 \mid + \mid z_0 \mid < \delta \qquad \text{and} \qquad z_0^* H z_0 - \eta_0^2 < 0 \qquad (7.3.19)$$

are satisfied. Hence any solution $x(\cdot) = (\sigma(\cdot),\eta(\cdot),z(\cdot))$ of (7.2.1) starting for $t = 0$ in $x_0 = (\sigma_0,\eta_0,z_0)$ satisfies $\dot{\sigma}(t) \geq F(\sigma(t))$ for all $t \geq 0$. Since $F(\cdot)$ corresponds to a circular solution of (7.3.14) it follows from the latter inequality that the solution x of (7.2.1) with $x(0) = x_0$ is a circular one. Let us prove now the existence of a cycle of the second kind in system (7.2.1). From Lemma 6.2.1 it follows that any solution $x = (\sigma,\eta,z)$ of (7.2.1) with $\sigma(0) \geq \sigma_0, \eta(0) > 0$ and $z(0)^* H z(0) - \eta(0)^2 < -F(\sigma(0))^2$ satisfies

$$z(t)^* H z(t) - \eta(t)^2 + F(\sigma(t))^2 < 0, \quad t \geq 0, \qquad (7.3.20)$$

$$\dot{\sigma}(t) \geq F(\sigma(t))^2, \quad t \geq 0. \qquad (7.3.21)$$

Remember (cf. the proof of Theorem 6.2.2 that the set

$$\Omega = \left\{ x = (\sigma,\eta,z) \;:\; z^* H z - \eta^2 + F(\sigma)^2 < 0, \eta > 0, \sigma \geq \sigma_0 \right\}$$

and its closure $\overline{\Omega}$ are positively invariant for (7.2.1). By Theorem 1.4.1, there are a matrix $M = M^* > 0$ and a number $\varepsilon_1 > 0$ such that

$$2z^* M(Az + b\xi) + z^* C\xi \leq -\varepsilon_1 \mid z \mid^2, \qquad z \in \mathbf{R}^n, \xi \in \mathbf{R}^m. \qquad (7.3.22)$$

Since φ is bounded it follows from (7.2.2), (7.3.22) and the positivity of μ_1 that there exist numbers $\gamma > 0$ and $\varrho > 0$ such that the function $U_\varrho : \mathbf{R}^{n+2} \to \mathbf{R}$ defined by

$$U_\varrho(x) = z^* M z + \frac{1}{2}\eta^2 - \varrho$$

has a derivative with respect to (7.2.1) which satisfies

$$\dot{U}_\varrho(x) + \gamma U_\varrho(x) \leq 0$$

for all $x \in \mathbf{R}^{n+2}$. Thus by Lemma 3.1.1, the set $C^\varrho := \{x \; : \; U_\varrho(x) \leq 0\}$ is positively invariant. Clearly, that $C^\varrho \cap \overline{\Omega}$ is also invariant. Take now a $\sigma_1 > \sigma_0$ so near that $F(\sigma_1 + \Delta) > F(\sigma_1)$. This is possible since F is continuous on $[\sigma_0, +\infty)$ and $F(\sigma_0) = 0, F(\sigma) > 0$ for $\sigma > \sigma_0$ is valid. It follows from (7.3.21) that for any point $p \in \Omega_1 = \overline{\Omega} \cap C^\varrho \cap \{\sigma = \sigma_1\}$ there exists a time $t = t(p) > 0$ such that

$$x(t(p), p) \in \Omega_2 := \overline{\Omega} \cap C^\varrho \cap \{\sigma = \sigma_1 + \Delta\}$$

and

$$x(t, p) \in \Omega_2 \qquad \text{for all} \qquad 0 \leq t < t(p).$$

Thus we can define a transformation $T : \Omega_1 \to \Omega_2$ by $T_p = x(t(p), p)$. To show that Ω_1 is convex we can repeat the argument of Lemma 6.2.3. Introducing as above the transformation $Q : \Omega_2 \to \Omega_1$ by $Q(\sigma, \eta, z) = (\sigma - \Delta, \eta, z)$ (this is possible since $F(\sigma_1 + \Delta) > F(\sigma_1)$) we get the mapping $(Q \circ T)\Omega_1 \subset \Omega_1$ which is continuous by the facts that $\partial\Omega_2$ is without contact with the vector-field of (7.2.1) and the solution depends continuously on the initial conditions. By Brouwer's theorem there exists a $\overline{p} = (\overline{\sigma}, \overline{\eta}, \overline{z}) \in \Omega_1$ with $(Q \circ T)(\overline{p}) = \overline{p}$. Clearly, $\sigma(t(\overline{p}), \overline{p}) = \overline{\sigma} + \Delta$, $\eta(t(\overline{p}), \overline{p}) = \overline{\eta}$ and $z(t(\overline{p}), \overline{p}) = \overline{z}$. $\blacksquare$

Consider now the special nonlinearity

$$\varphi(\sigma) = \sin(\sigma + \Sigma_0) - \sin \Sigma_0, \tag{7.3.23}$$

where $\Sigma_0 \in (0, \frac{\pi}{2})$ is fixed. Using Theorem 4.3.3 and Theorem 7.3.1 we obtain immediately the

Corollary 7.3.1 Suppose $\mu_1 > 0$ and there exists a number $\lambda \in (0, \mu_1)$ such that conditions (i), (ii) of Theorem 7.3.1 are valid and $\sqrt{\lambda_1(\mu_1 - \lambda)} \geq \sin(\frac{1}{2}\Sigma_0)$. Then system (7.2.1), (7.3.23) is gradient-like.

From Theorem 4.3.3 and Theorem 7.3.3 we get the

Corollary 7.3.2 Suppose that (7.3.13) is true and there exists a number $\lambda > \max\{0, \mu_2\}$ satisfying $\lambda^2 \leq \sqrt{3\cos^3 \Sigma_0 + 1} - 2\cos \Sigma_0$. Suppose further that conditions (i) and (ii) of Theorem 7.3.3 are satisfied. Then for any small $\delta > 0$ there exists a circular solution $x(\cdot) = (\sigma(\cdot), \eta(\cdot), z(\cdot))$ of (7.2.1) which satisfies $|\eta(0)| + |z(0)| < \delta$. If in addition $\mu_1 > 0$ and the frequency-domain condition (7.3.15) is satisfied then system (7.2.1) with nonlinearity (7.3.23) has a circular cycle of the second kind.

The following examples will show that the conditions of Theorems 7.3.1 to 7.3.3 can be verified.

Example 7.3.1 Suppose that the transient behavior of a synchronous machine is described by the system [157]

$$\begin{aligned}
\dot{\vartheta} &= \eta, \\
\dot{\eta} &= m(\vartheta_0, I_0) - m(\vartheta + \vartheta_0, z + I_0) - [\alpha_1 + \alpha_2 \cos 2(\vartheta + \vartheta_0)]\eta, \\
\dot{z} &= -\alpha_4 z + \alpha_3 \eta \sin(\vartheta + \vartheta_0), \\
m(\sigma, z) &= \alpha_3 z \sin \sigma \quad (\sigma, z \in \mathbf{R}).
\end{aligned} \tag{7.3.24}$$

Here $\alpha_1, \alpha_2, \alpha_3 = I_0^{-1}, \alpha_4$ and ϑ_0 are non-negative parameters. It is clear that system (7.3.24) can be transformed into system (7.2.1) by introducing the new variable $\sigma = \vartheta + \vartheta_0$ and defining the functions

$$g(\eta, \sigma) = (\alpha_1 + \alpha_2 \cos 2\sigma)\eta, \quad \varphi(\sigma) = \sin \sigma - m(\vartheta_0, I_0), \quad f(\sigma) = \sin \sigma, \qquad \sigma, \eta \in \mathbf{R}.$$

If we also define $A = -\alpha_4$, $B = \alpha_3$ and $C = -\alpha_3$ we get $K(p) = \alpha_3^2(p+\alpha_4)^{-1}$, $\det(pI-A) = p+\alpha_4$. Clearly, that

$$(\alpha_1 - \alpha_2) \le \frac{g(\eta,\sigma)}{\eta} \le (\alpha_1 + \alpha_2), \quad \eta \ne 0, \sigma \in \mathbf{R}.$$

In order to satisfy conditions (i) and (ii) of Theorem 7.3.1 we have to take $\lambda \in (0, \alpha_4)$. By Corollary 7.3.1 it follows that system (7.3.24) is gradient-like if $\alpha_1 > \alpha_2$ and $\sqrt{\Gamma} > \sin \frac{1}{2}\vartheta_0$, where

$$\Gamma = \begin{cases} \frac{1}{4}(\alpha_1 - \alpha_2)^2 & \text{if } \alpha_1 - \alpha_2 < 2\alpha_4, \\ \alpha_4(\alpha_1 - \alpha_2 - \alpha_4) & \text{if } \alpha_1 - \alpha_2 \ge 2\alpha_4. \end{cases}$$

Because $\mu = 1$ the conditions (i) and (ii) of Theorem 7.3.3 are satisfied if for $\alpha_1 + \alpha_2 < \lambda < \alpha_4$ we have

$$\lambda - (\alpha_1 + \alpha_2) - \alpha_3^2(\alpha_4 - \lambda)[\omega^2 + (\alpha_4 - \lambda)^2]^{-1} > 0 \tag{7.3.25}$$

for all $\omega > 0$, which is possible if

$$\alpha_4 > \alpha_1 + \alpha_2 + 2\alpha_3. \tag{7.3.26}$$

By Corollary 7.3.2 system (7.3.24) has circular solutions provided that (7.3.26) is valid and the inequality

$$\lambda_0^2 \le \sqrt{3\cos^2 \vartheta_0 + 1} - 2\cos \vartheta_0$$

is satisfied, where

$$\lambda_0 = \frac{1}{2}(\alpha_1 + \alpha_2 + \alpha_4) - \frac{1}{2}\sqrt{(\alpha_1 + \alpha_2 - \alpha_4)^2 - 4\alpha_3^2}.$$

If in addition $\alpha_1 > \alpha_2$ then system (7.3.24) has a cycle of the second kind

Example 7.3.2 Let us examine the system from [71]

$$\begin{aligned} \dot{\sigma} &= \eta, \\ \dot{\eta} &= 4.5 \cdot 10^{-3}\eta - 0.815 \cdot 10^{-3} \sin \sigma - 0.12 \cdot 10^{-3} \sin 2\sigma - 0.4 \cdot 10^{-3} z \sin \sigma + MH^{-1}, \\ \dot{z} &= -2.4 \cdot 10^{-3}z + 2.57\eta \sin \sigma, \\ H &= 2180. \end{aligned}$$

$$\tag{7.3.27}$$

Choosing

$$\begin{aligned} \varphi(\sigma) &= -T + \sin \sigma + 0.148 \sin 2\sigma, \\ f(\sigma) &= \sin \sigma, \\ g(\eta,\sigma) &= \mu_1\eta, \quad T = M(0.815 \cdot 10^{-3}M)^{-1}, \\ \mu_1 &= \mu_2 = 0.158, \quad A = -0.084, \quad m = n = 1 \end{aligned}$$

we can write system (7.3.27) in the form (7.2.1). If we take now $\lambda = \frac{1}{2}\mu_1 < -A$ the conditions (i) and (ii) of Theorem 7.3.1 are fulfilled. To define the reduction-type equation of condition (iii) we put $a = 2\sqrt{(\mu_1 - \lambda)\lambda} = \mu_1 = 0.158$. Thus for any $a > a_{cr}$ Theorem 7.3.2 says, that every solution of (7.3.27) tends to an equilibrium for $t \to +\infty$. If $a < a_{cr}$ we may use an estimate of $\sigma_*(a)$ from below which is based on the following inequality for the representation $\widetilde{F}_0$ of the stable manifold $W^s(\sigma_2, 0)$ in the region $\eta > 0, \sigma \in [\sigma_1, \sigma_2]$ (see [20]):

$$\widetilde{F}_0(\sigma)^2 > 2 \int\limits_\sigma^{\sigma_2} \varphi(\vartheta)\, d\vartheta + 2a \int\limits_\sigma^{\sigma_2} (2 \int\limits_\vartheta^{\sigma_2} \varphi(\omega)d\omega)^{\frac{1}{2}}\, d\vartheta. \tag{7.3.28}$$

Let F be a solution of (7.3.2) such that $F(\sigma_0) = 0$ and $F(\sigma) > 0$ for $\sigma \in (\sigma_0, \sigma_2)$ where $\sigma_0 \in (\sigma_*, \sigma_1)$. It follows from (7.3.2) that

$$F(\sigma)^2 \le 2 \int\limits_\sigma^{\sigma_2} \varphi(\vartheta)\, d\vartheta \qquad \text{for} \qquad \sigma \in [\sigma_1, \sigma_0].$$

Thus if

$$\int\limits_{\sigma_0}^{\sigma_2} \varphi(\vartheta)\, d\vartheta + a \int\limits_{\sigma_1}^{\sigma_2} \left(2 \int\limits_{\vartheta}^{\sigma_2} \varphi(\omega)d\omega\right)^{\frac{1}{2}} d\vartheta \geq 0 \tag{7.3.29}$$

then $F(\sigma_1) < \widetilde{F}_0(\sigma_1)$. By the uniqueness of solutions it follows that $\sigma_* < \sigma_0$. Replacing the second term of (7.3.29) by an estimate from below we obtain the relation

$$\int\limits_{\sigma_0}^{\sigma_2} \varphi(\vartheta)\, d\vartheta + \left\{3 \max_{\vartheta \in [\sigma_1,\sigma_2]} \varphi(\vartheta)\right\}^{-1} a \left\{2 \int\limits_{\sigma_1}^{\sigma_2} \varphi(\vartheta)\, d\vartheta\right\}^{\frac{3}{2}} \geq 0. \tag{7.3.30}$$

Thus the value of σ_0 for system (7.3.28) can be defined by solving the equation

$$\Phi(\sigma_2) - \Phi(\sigma_0) + a \left\{3 \max_{\sigma \in [\sigma_1,\sigma_2]} [-T + \sin\sigma + 0.148 \sin 2\sigma]\right\}^{-1} \{2(\Phi(\sigma_2) - \Phi(\sigma_1))\}^{\frac{3}{2}} = 0, \tag{7.3.31}$$

where $\Phi(\sigma) = -T\sigma - \cos\sigma - 0.074 \cos 2\sigma$. Let us compare our result with a result obtained on the basis of the so-called area method ([157]). By this method one constructs a Lyapunov function for (7.2.1) of the type

$$V(\sigma, \eta, z) = z^* H z + \eta^2 + 2 \int\limits_{\sigma_1}^{\sigma} \varphi(\vartheta)\, d\vartheta,$$

where the $n \times n$ matrix $H = H^*$ positive definite. It is not difficult to show that the level sets $K^c := \{(\sigma, \eta, z) \; : \; V(\sigma, \eta, z) = c\}$ are closed surfaces with the point $(\sigma_1, 0, 0)$ inside provided that $0 \leq c \leq 2 \int\limits_{\sigma_1}^{\sigma_2} \varphi(\vartheta)\, d\vartheta$. It follows that the part of the surface

$$\left\{(\sigma, \eta, z) \; : \; z^* H z + \eta^2 = 2 \int\limits_{\sigma}^{\sigma_2} \varphi(\vartheta)\, d\vartheta, \quad \sigma \in [\sigma_0, \sigma_1]\right\}$$

with σ_0 such that

$$\int\limits_{\sigma_0}^{\sigma_1} \varphi(\vartheta)\, d\vartheta = - \int\limits_{\sigma_1}^{\sigma_2} \varphi(\vartheta)\, d\vartheta$$

is closed and includes the point $(\sigma_1, 0, 0)$. More than that, this surface bounds the domain of attraction of $(\sigma_1, 0, 0)$. Thus according to the area method a boundary of the attraction region can be obtained by the equation

$$\Phi(\sigma_2) - \Phi(\sigma_0) = 0. \tag{7.3.32}$$

It is evident that (7.3.31) gives a more precise result than (7.3.32). For example if $T = 0.5$ then formula (7.3.31) gives $\sigma_0 = -45°$ and formula (7.3.32) gives $\sigma_0 = -38°$. For $T = 0.4$ (7.3.31) gives $\sigma_0 = -61.5°$ and (7.3.32) $\sigma_0 = -53°$.

7.4 Synchronous Machines Equations with a Forcing Term

In this section we consider the system

$$\begin{aligned}
\dot\sigma &= \eta, \\
\dot\eta &= -g(\eta,\sigma) + z^* C f(\sigma) - \varphi(\sigma) - q(t), \\
\dot z &= Az + B f(\sigma)\eta,
\end{aligned} \tag{7.4.1}$$

where $q : \mathbf{R}_+ \to \mathbf{R}$ is C^1. We assume the assumptions concerning system (7.2.1). In particular we suppose that

$$\mu\eta^2 \leq g(\eta,\sigma)\eta, \qquad (\eta,\sigma) \in \mathbf{R} \times \mathbf{R}, \tag{7.4.2}$$

where $\mu > 0$ is a constant. Because (7.4.1) is non-autonomous a main question of global behavior is the boundedness of solutions on $\mathbf{R}_+$.

Theorem 7.4.1 *Suppose that there exist numbers γ_1, γ_2 and $\gamma \in (0,\mu)$ such that the following conditions are valid:*

(i) *$[\varphi(\sigma) + \gamma_j]^2 + \varphi'(\sigma)^2 \neq 0$ for all $\sigma \in \mathbf{R}$ and $j = 1, 2$;*

(ii) *$\operatorname{Re} K(i\omega - \lambda) > 0$ for all $\omega \in \mathbf{R}$; $\lim\limits_{\omega \to \infty} \omega^2 \operatorname{Re} K(i\omega - \lambda) > 0$;*

(iii) *$A + \lambda I$ is a Hurwitz matrix;*

(iv) *any solution of $\ddot{\vartheta} + 2\sqrt{\lambda(\mu - \lambda)}\dot{\vartheta} + \varphi(\vartheta) + \gamma_j = 0$ $(j = 1, 2)$ is bounded on $[0, +\infty)$;*

(v) *$\gamma_2 \leq q(t) + (2\lambda)^{-1}\dot{q}(t) \leq \gamma_1$ for $t \geq 0$;*

(vi) *$\gamma_2 \leq q(0) \leq \gamma_1$.*

Then any solution of (7.4.1) is bounded on $\mathbf{R}_+$.

Proof Because of conditions (ii) and (iii) there exists by Theorem 1.4.1 an $n \times n$ matrix $H = H^* > 0$ and a number $\varepsilon > 0$ such that

$$2z^*H[(A + \lambda I)z + B\xi] + z^*C\xi \leq -\varepsilon \mid z \mid^2, \quad (z,\xi) \in \mathbf{R}^{n+m}. \tag{7.4.3}$$

In order to use Lemma 5.5.1 we consider an arbitrary solution (σ, η, z) of (7.4.1) and introduce the functions

$$\Psi := \varphi, \quad w(t) = z(t)^*Hz(t) + \frac{1}{2}\eta(t)^2$$

and the number $\varepsilon = \mu - \lambda$. The conditions (iv), (v) and (vi) of Lemma 5.5.1 coincide with conditions (iv), (v) and (vi) of the present theorem. The condition (ii) of the lemma is true since $H > 0$ and the condition (iii) follows from (7.2.2) and (7.2.3). By Lemma 5.5.1 $\sigma(\cdot)$ is bounded on $\mathbf{R}_+$. To show that $\eta(\cdot)$ and $z(\cdot)$ are also bounded we consider the function $U : \mathbf{R}^n \times \mathbf{R} \to \mathbf{R}$ defined by $U(z,\eta) = z^*Hz + \eta^2 - \nu$, where ν is a number which satisfies the inequality

$$\nu > \mid 4\lambda(\mu - \lambda) \mid^{-1} \left(\max_{\sigma \in [0,\Delta]} \varphi(\sigma) + m \right)^2 \tag{7.4.4}$$

and $\mid q(t) \mid \leq m$ on $\mathbf{R}_+$. It is not difficult to see that along the solution (σ, η, z)

$$\frac{d}{dt}U(z(t),\eta(t)) + 2\lambda U(z(t),\eta(t)) \leq -2(\mu - \lambda)\eta^2 - 2\eta\varphi(\sigma(t)) + q(t) - 2\lambda\nu$$

for $t \geq 0$. Because of (7.4.4) we have

$$\frac{d}{dt}U(z(t),\eta(t)) + 2\lambda U(z(t),\eta(t)) \leq 0. \tag{7.4.5}$$

By Lemma 3.1.1 it follows that the set

$$C^\nu := \{(\eta,z) \,:\, U(z,\eta) \leq 0\}$$

is invariant for the components $\eta(\cdot)$ and $z(\cdot)$, provided that ν satisfies (7.4.4). Since C^ν is bounded the assertion of the theorem follows. ∎

7.5 The Equation of a Synchronous Machine with a Speed Governor

In this section the equations of a synchronous machine with a speed governor [56, 125] are considered. For this purpose we consider a concrete system consisting of three equations taken from [8, 55].

$$
\begin{aligned}
\dot\sigma &= \omega_0\eta, \\
T\dot\eta + D\eta &= P_{mec} - \left[\frac{zU\sin\sigma}{x_d'} + \left(\frac{1}{x_q} - \frac{1}{x_d'}\right)U^2\sin\sigma\cos\sigma\right], \\
T_{do}'\dot z &= e_f - \frac{x_d}{x_d'}z + \frac{x_d - x_d'}{x_d'}U\cos\sigma,
\end{aligned}
\tag{7.5.1}
$$

where ω_0, T, D, P_{mec}, U, x_q, x_d, x_d', e_f, T_{do}' are constants. Let us introduce with respect to (7.5.1) the 2π-periodic nonlinearity

$$
\varphi_0(\sigma) = \frac{e_f U}{x_d}\sin\sigma + \left(\frac{1}{x_d} - \frac{1}{x_d'}\right)U^2\sin\sigma\cos\sigma - P_{mec}
$$

and the real-valued C^1-function

$$
V_0(\sigma,\eta,z) = \frac{1}{2}\omega_0\eta^2 + \frac{1}{T}\left\{\frac{1}{2}\frac{x_d}{x_d'(x_d - x_d')}\left[z - \frac{x_d'}{x_d}e_f - \frac{x_d - x_d'}{x_d}U\cos\sigma\right]^2 + \int\limits_{\bar\sigma}^{\sigma}\varphi_0(\vartheta)\,d\vartheta\right\}.
$$

A direct checking shows that system (7.5.1) can be written as follows

$$
\begin{aligned}
\dot\sigma &= \frac{\partial V_0}{\partial\eta}, \\
\dot\eta &= -\frac{\partial V_0}{\partial\sigma} - \frac{D}{T}\eta, \\
\dot z &= -\frac{T}{T_{do}'}(x_d - x_d')\frac{\partial V_0}{\partial z}.
\end{aligned}
\tag{7.5.2}
$$

Thus (7.5.2) may be considered as a Hamiltonian system with a dissipation term coupled with a gradient system. Taking this accont we can study any system written in the form

$$
\dot\sigma = \frac{\partial V}{\partial\eta}, \quad \dot\eta = -\frac{\partial V}{\partial\sigma} - \alpha\eta, \quad \dot z = -B_0\left(\frac{\partial V^*}{\partial z}\right),
\tag{7.5.3}
$$

where $\partial V/\partial x$ is a row vector, B_0 is a positive definite matrix, α is a number and $V : \mathbf{R}\times\mathbf{R}\times\mathbf{R}^m \to \mathbf{R}$ is given by

$$
V(\sigma,\eta,z) = \frac{1}{2}[\omega_0\eta^2 + z^*Hz + z^*f(\sigma) + f(\sigma)^*z + \Phi(\sigma)].
$$

In this function H is a positive definite $m\times m$ matrix, $f : \mathbf{R} \to \mathbf{R}^m$, $\Phi : \mathbf{R} \to \mathbf{R}$ are C^1 functions. If in the equation of a synchronous machine the prime mover and the speed governor are taken into account we get a more general system

$$
\begin{aligned}
\dot\sigma &= \frac{\partial V}{\partial\eta}, \\
\dot\eta &= -\frac{\partial V}{\partial\sigma} - (\alpha + \beta)\eta + q^*y, \\
\dot z &= -B_0\frac{\partial V^*}{\partial z}, \\
\dot y &= A_0 y - b\eta,
\end{aligned}
\tag{7.5.4}
$$

where in addition to the above A_0 is a certain matrix, b and q are constant vectors and β is a number. In order to derive sufficient conditions for the gradient-like behavior of system (7.3.5) we apply the reduction-type Lemma 5.3.1 and the monostability Theorem 1.1.4. Define the functions

$$\Psi(\sigma) \;=\; \omega_0[\Psi'(\sigma) - f(\sigma)^* H^{-1} f'(\sigma)] \qquad \text{and}$$
$$W(\sigma,\eta,z,y) \;=\; \tfrac{1}{2}\omega_0^2\eta^2 + \tfrac{\omega_0}{2}\left\{[z + H^{-1}f(\sigma)]^* H[z + H^{-1}f(\sigma)] + \omega_0 y^* G y\right\}, \tag{7.5.5}$$

where $G = G^*$ is a poisitive definite matrix. Let (σ,η,z,y) be an arbitrary solution of (7.5.5) and introduce the function $w : \mathbf{R}_+ \to \mathbf{R}$ by $w(t) = W(\sigma(t),\eta(t),z(t),y(t))$. Consider also a reduction-type equation

$$\ddot{\vartheta} + a\dot{\vartheta} + \Psi(\vartheta) = 0 \tag{7.5.6}$$

with parameter $a > 0$ and denote by $a_{cr} = a_{cr}(\Psi)$ the bifurcation value defined in Section 2.2. It is convenient for us to have the equation

$$4\lambda^2 - 4(\alpha + \beta)\lambda + a_{cr}^2 = 0. \tag{7.5.7}$$

Note that if $\alpha + \beta > a_{cr}$ equation (7.5.7) has the two real roots $\lambda_1 < \lambda_2$. Finally we denote by $\lambda_{min}(HB_0)$ the smallest eigenvalue of HB_0 and introduce the matrix

$$\widetilde{A} := \begin{bmatrix} -(\alpha + \beta) & q^* \\ b & A_0 \end{bmatrix}.$$

The following result is obtained in [56].

Theorem 7.5.1 *Suppose that $B_0 > 0$, $\widetilde{A}$ is Hurwitzian, $\alpha + \beta > a_{cr}$, $\lambda_{min}(HB_0) > \lambda_1$ and the following conditions are satisfied:*

(i) *the function Ψ defined by (7.5.5) has exactly two zeros on the period and $\Psi(\sigma)^2 + [\Psi'(\sigma)]^2 \neq 0$ for all $\sigma \in \mathbf{R}$;*

(ii) *there exist real numbers λ and ε such that $\lambda_1 < \lambda < \min\{\lambda_2, \lambda_{min}(HB_0)\}$, $\frac{a_{cr}^2}{4\lambda} < \varepsilon < \alpha + \beta - \lambda$ and the following hypotheses are valid:*

    a) $A_0 + \lambda I$ *is Hurwitzian;*

    b) *the pair* $(A_0 + \lambda I, b)$ *is controllable;*

    c) $\alpha + \beta - \varepsilon - \lambda + \operatorname{Re} q^*(-\lambda I + i\omega I - A_0)^{-1} b > 0$ *for all* $\omega \geq 0$.

Then system (7.5.4) is gradient-like.

Proof Let us verify the hypotheses of Lemma 5.3.1. Since H and G are positive definite condition (ii) of the lemma is fulfilled. In order to verify condition (iv) one has to compute for an arbitrary solution (σ,η,z,y) the term

$$\dot{w} + 2\lambda w + \varepsilon\dot{\sigma}^2 + \Psi(\sigma(t))\dot{\sigma} \;=\; -\omega_0 \frac{\partial V}{\partial z}(B_0 - \lambda H^{-1})(\frac{\partial V}{\partial z})^* + \omega_0^2 \eta q^* y - \omega_0^2(\alpha + \beta - \varepsilon - \lambda)^2 +$$
$$+ \omega_0 y^* G(A_0 y - b\eta) + \omega_0(A_0 y - b\eta)^* G y + 2\lambda\omega_0 y^* G y.$$

Clearly, that $B_0 - \lambda H^{-1} > 0$. Condition (ii) guarantees (Theorem 1.4.1) that there exists a matrix $G = G^*$ such that

$$2y^* G[(A_0 + \lambda I)y - b\eta] + \omega_0(\alpha + \beta - \varepsilon - \lambda)\eta^2 + \omega_0\eta q^* y > 0 \tag{7.5.8}$$

for all $y \neq 0, \eta \neq 0$. The choice of λ and ε guarantees also that condition (i) of Lemma 5.3.1 is satisfied. From this lemma it follows that the solution component $\sigma(\cdot)$ is bounded on $\mathbf{R}_+$.

168

Because $\widetilde{A}$ is Hurwitzian the functions $\eta(\cdot)$ and $z(\cdot)$ are also bounded on $\mathbf{R}_+$. Let us use now Theorem 1.1.4. Introducing the function V_1 by $V_1(\sigma,\eta,z,y) = V(\sigma,\eta,z) + y^*Gy$ we get for the derivative of V_1 with respect to system (7.5.4)

$$\dot{V}_1(\sigma,\eta,z,y) = -\frac{\partial V}{\partial z}B_0\left(\frac{\partial V}{\partial z}\right)^* - (\alpha+\beta)\omega_0\eta^2 + \omega_0\eta q^*y + 2y^*G(A_0y - b\eta). \tag{7.5.9}$$

Using now inequality (7.5.8) we receive

$$\dot{V}_1 \leq -\frac{\partial V}{\partial z}B_0\left(\frac{\partial V}{\partial z}\right)^* - \lambda\omega_0\eta^2 - 2y^*Gy \leq 0$$

for all (σ,η,z,y). Suppose that $V_1(\sigma(t),\eta(t),z(t),y(t)) \equiv const$ on $\mathbf{R}$ for some solution. It follows immediately that this solution is constant. Therefore system (7.5.4) is quasi-gradient-like. The stationary set has the form $\{(\sigma,\eta,z,y) : \eta = 0,\ y = 0,\ z + H^{-1}f(\sigma) = 0\}$. Hence the limit

$$\lim_{t\to\infty}[\Phi'(\sigma(t)) - f(\sigma(t))^*H^{-1}f'(\sigma(t))]$$

exists. From here and the fact that the singular points of Ψ are isolated we conclude that $\sigma(t)$ must have a limit. $\blacksquare$

7.6 The Dynamics of Two Coupled Synchronous Machines

In this section we investigate a system considered in [157], [91]

$$\begin{aligned}
\dot{\sigma} &= \eta, \\
\dot{\eta} &= -\alpha_1\eta + \tfrac{\alpha_4}{\alpha_9}[\alpha_5\alpha_6\sin\sigma_0 - (\alpha_5 + z_1)(\alpha_6 + z_2)\sin(\sigma + \sigma_0)], \\
\alpha_2\dot{z}_1 + \alpha_4\cos(\sigma + \sigma_0)\dot{z}_2 &= \alpha_4(\alpha_6 + z_2)\eta\sin(\sigma + \sigma_0) - \alpha_7 z_1, \\
\alpha_3\dot{z}_2 + \alpha_4\cos(\sigma + \sigma_0)\dot{z}_1 &= \alpha_4(\alpha_5 + z_1)\eta\sin(\sigma + \sigma_0) - \alpha_8 z_2,
\end{aligned} \tag{7.6.1}$$

where $\alpha_j,\ j = 1,2,\ldots,9$, are positive constants and $\sigma_0 \in [0,\frac{\pi}{2}]$ is fixed. System (7.6.1) describes the work of two synchronous machines in the case when the active resistances of stators windings are to zero, the four-terminal network which links the machines is purely reactive and the rotor damping moment of each machine is proportional to the sliding and the inertia moment of the rotor. We suppose that $\alpha_2\alpha_3 > \alpha_4^2$. This inequality is always fulfilled for systems with unramified gear [157]. In order to applicate Lemma 5.3.1 to our situation, we introduce the function $\Psi : \mathbf{R} \to \mathbf{R}$ by

$$\Psi(\sigma) = \alpha_4\alpha_5\alpha_6[\sin(\sigma + \sigma_0) - \sin\sigma_0]$$

and, with respect to an arbitrary solution $\sigma(\cdot),\eta(\cdot),z_1(\cdot),z_2(\cdot)$ of (7.6.1) the function

$$w(t) = \frac{1}{2}\alpha_9\eta(t)^2 + \frac{1}{2}\alpha_2 z_1(t)^2 + \frac{1}{2}\alpha_3 z_2(t)^2 + \alpha_4\cos(\sigma(t) + \sigma_0)z_1(t)z_2(t).$$

$t \geq 0$. Because of $\alpha_2\alpha_3 > \alpha_4^2$ condition (ii) of Lemma 5.3.1 is fulfilled. A direct computation shows that

$$\dot{w}(t) = -\alpha_7 z_1(t)^2 - \alpha_8 z_2(t)^2 - \alpha_1\alpha_9\eta(t)^2 - \Psi(\sigma(t))\eta(t).$$

Thus if the positive numbers λ and ε satisfy the inequalities

$$\alpha_7 - \lambda\alpha_2 > 0, \quad \alpha_8 - \lambda\alpha_3 > 0, \quad \alpha_9(\alpha_1 - \lambda) \geq \varepsilon, \quad (\alpha_7 - \lambda\alpha_2)(\alpha_8 - \lambda\alpha_3) \geq \lambda^2\alpha_4 \tag{7.6.2}$$

then the inequality
$$\dot{w}(t) + 2\lambda w(t) + \Psi(\sigma(t))\eta(t) + \varepsilon\eta(t)^2 < 0$$
is valid for $t > 0$. So condition (iii) of the lemma is fulfilled. It follows that if there exist positive numbers λ, ε such that (7.6.2) is satisfied and any solution of the reduction-type equation

$$\ddot{\vartheta} + 2\sqrt{\lambda\varepsilon}\,\dot{\vartheta} + \alpha_4\alpha_5\alpha_6[\sin(\vartheta + \sigma_0) - \sin\sigma_0] = 0 \tag{7.6.3}$$

is bounded on $\mathbf{R}_+$ the considered solution component $\sigma(t)$ of (7.6.1) is by Lemma 5.3.1 bounded on $\mathbf{R}_+$. To answer the question, for which parameters equation (7.6.3) is Lagrange stable, we can use the Böhm-Hayes theorem (Theorem 5.3.3 together with Remark 5.3.1). Introduce the number

$$\lambda_0 = \frac{\alpha_2\alpha_8 + \alpha_3\alpha_7 - \sqrt{(\alpha_2\alpha_8 - \alpha_3\alpha_7)^2 + 4\alpha_7\alpha_8\alpha_4^2}}{2(\alpha_2\alpha_3 - \alpha_4^2)}$$

and define

$$\lambda_0 = \begin{cases} \lambda_0 & \text{if } \lambda < \frac{\alpha_1}{2}, \\ \frac{\alpha_1}{2} & \text{if } \lambda \geq \frac{\alpha_1}{2}, \end{cases} \qquad \varepsilon = \alpha_9(\alpha_1 - \lambda).$$

Theorem 5.3.3 says, that if

$$\left(\sin\frac{\sigma_0}{2}\right)^2 \leq \frac{\alpha_9\Gamma}{\alpha_4\alpha_5\alpha_6}, \tag{7.6.4}$$

where

$$\Gamma = \begin{cases} \lambda_0(\lambda_1 - \lambda_0) & \text{if } \quad 2\lambda_0 < \alpha_1, \\ 0.25\alpha_1^2 & \text{if } \quad 2\lambda_0 \geq \alpha_1, \end{cases}$$

then any solution $\sigma(\cdot)$ of (7.6.3) is bounded on $\mathbf{R}_+$, i.e. $|\sigma(t)| < m$ for $t > 0$. Let us demonstrate now that the solution components $\eta(\cdot), z_1(\cdot), z_2(\cdot)$ of (7.6.1) are also bounded on $\mathbf{R}_+$. To do this we introduce the function $U : \mathbf{R}^4 \to \mathbf{R}$ defined by

$$U(x) = \frac{1}{2}\alpha_9\eta^2 + \frac{1}{2}\alpha_2 z_1^2 + \frac{1}{2}\alpha_3 z_2^2 + \alpha_4 z_1 z_2 \cos(\sigma + \sigma_0) + \alpha_4\alpha_5\alpha_6 \int_0^\sigma [\sin(\vartheta + \sigma_0) - \sin\sigma_0]\, d\vartheta - \nu,$$

$x = (\sigma, \eta, z_1, z_2)$, where the parameter ν satisfies

$$\nu > \alpha_4\alpha_5\alpha_6 m. \tag{7.6.5}$$

Note that for the derivative of U with respect to system (7.6.1) we have

$$\dot{U}(x) + 2\lambda U(x) = -\left\{\alpha_9(\alpha_1 - \lambda)\eta^2 + (\alpha_7 - \lambda\alpha_2)z_1^2 + (\alpha_8 - \lambda\alpha_3)z_2^2 + 2\lambda\alpha_4 z_1 z_2 \cos(\sigma + \sigma_0)\right\}$$

$$+ 2\lambda\alpha_4\alpha_5\alpha_6 \int_0^\sigma [\sin(\vartheta + \sigma_0) - \sin\sigma_0]\, d\vartheta - 2\lambda\nu, \quad x = (\sigma, \eta, z_1, z_2) \in \mathbf{R}^4.$$

The inequalities (7.6.2) and (7.6.5) imply that for an arbitrary solution $x(\cdot)$ of (7.6.1)

$$\dot{U}(x(t)) + 2\lambda U(x(t)) \leq 0 \quad \text{for all} \quad t \geq 0.$$

By Lemma 3.1.1 we conclude that the set $\{x : U(x) \leq 0\}$ is positively invariant for the solutions of (7.6.1) proposed that (7.6.5) is satisfied. Thus for $\alpha_3\alpha_2 > \alpha_4^2$ the components η, z_1, z_2 are also bounded on $\mathbf{R}_+$.

Let us now show that system (7.6.1) is in fact, under the stated conditions, quasi-monostable. Note that the derivative of $V := U + \nu$ with respect to system (7.6.1) is given by

$$\dot{V}(x) = -\alpha_9\alpha_1\eta^2 - \alpha_7 z_1^2 - \alpha_8 z_2^2, \quad x = (\sigma, \eta, z_1, z_2) \in \mathbf{R}^4,$$

which is non-positive. Because of $\alpha_j > 0$, $j = 1, \ldots, 9$, it follows from $\dot{V}(x(t)) \equiv 0$ for a solution of (7.6.1) that $\eta(\cdot) = z_1(\cdot) = z_2(\cdot) = 0$. By Theorem 1.1.3 we see that (7.6.1) is quasi-monostable.

Remark 7.6.1 The material of this Chapter goes back to [83, 81, 82, 36, 56].

170

Chapter 8

Integro-Differential Equations

The aim of this chapter is to extend some results of Chapters 1 – 7 concerning bouncedness, convergence and quasiconvergence to a class of integro-differential equations with retarded argument which arises from phase synchronization problems. Our aim is to apply ordinary differential equation methods such as the Bakaev-Guzh technique and non-local reduction for the global behavior investigation of functional-differential equations. As in the the ordinary differential equation case we will use auxiliary Lyapunov functionals of the Popov type. The material of this chapter is due to [100, 102, 101, 103].

8.1   General Setting

Let us consider the Volterra integro-differential equation

$$\dot{\sigma}(t) = \alpha(t) + \varrho\varphi(\sigma(t-h)) - \int_0^t \gamma(t-\tau)\varphi(\sigma(\tau))\,d\tau \tag{8.1.1}$$

with $\alpha : [0, +\infty) \to \mathbf{R}$ continuous and bounded, $\gamma : [0, +\infty) \to \mathbf{R}$ is $L^1(0, +\infty)$, $\varphi : \mathbf{R} \to \mathbf{R}$ is Lipschitz continuous and Δ-periodic; ϱ and $h \geq 0$ are constant. Note that pendulum-like systems in the second canonical form obtained in Section 2.1 can be written in the form (8.1.1).

The Cauchy problem for equation (8.1.1) is formulated by means of the initial condition

$$\sigma(t) = \sigma_0(t) \qquad (t \in [-h, 0]), \tag{8.1.2}$$

where $\sigma_0 : [-h, 0] \to \mathbf{R}$ is an arbitrary given continuous function. The function $\sigma(\cdot) = \sigma(\cdot, \sigma_0)$ is said to be a *solution* of (8.1.1), (8.1.2) on $[0, T)$ if $T > 0$, $\sigma(\cdot, \sigma_0) : [-h, T) \to \mathbf{R}$ is continuous, $\sigma(t, \sigma_0) = \sigma_0(t)$ on $[-h, 0]$, $\sigma(\cdot, \sigma_0)$ is continuously differentiable for $t \in (0, T)$ and $\sigma(\cdot, \sigma_0)$ satisfies (8.1.1) on $[0, T)$. A solution $\sigma(\cdot, \sigma_0)$ with $\sigma(t, \sigma_0) = const$ for $t \in [-h, +\infty)$ is called a *stationary solution* of (8.1.1).

The following result shows that equation (8.1.1) has global solutions.

Proposition 8.1.1 *Under the above conditions there exists a unique solution of the Cauchy problem* (8.1.1), (8.1.2) *on* $[0, +\infty)$.

Proof On the interval $[0, h]$ the problem (8.1.1)-(8.1.2) can be rewritten in the form

$$\dot{\sigma} = \alpha(t) + \varrho\varphi(\sigma_0(t-h)) - \int_0^t \gamma(t-\tau)\varphi(\sigma(\tau))\,d\tau, \qquad \sigma(0) = \sigma_0(0), \tag{8.1.3}$$

or as an equivalent integral equation

$$\sigma(t) = \sigma_0(0) + \int\limits_0^t \left\{ \beta(s) - \int\limits_0^s \gamma(s-\tau)\varphi(\sigma(\tau))\,d\tau \right\} ds, \tag{8.1.4}$$

where we used the notation

$$\beta(s) = \alpha(s) + \varrho\varphi(\sigma_0(s-h)).$$

For arbitrary $\delta \in (0, h)$ we consider the Banach space $C([0,\delta], \mathbf{R})$ of continuous functions with the maximum norm

$$\|\xi\| = \max_{t\in[0,\delta]} |\xi(t)|.$$

Note that under our assumptions

$$\sup_{\substack{\delta > 0 \\ s \in [0,\delta] \\ \xi \in C([0,\delta], \mathbf{R})}} \left| \beta(s) - \int\limits_0^s \gamma(s-\tau)\varphi(\xi(\tau))\,d\tau \right| =: M_0 < \infty. \tag{8.1.5}$$

Let us define the operator

$$A : C([0,\delta], \mathbf{R}) \to C([0,\delta], \mathbf{R})$$

by

$$(A\xi)(t) = \sigma_0(0) + \int\limits_0^t \left\{ \beta(s) - \int\limits_0^s \gamma(s-\tau)\varphi(\xi(\tau))\,d\tau \right\} ds.$$

We show that for δ sufficiently small A is a contractive self-mapping of the closed ball $S = \{\xi \in C([0,\delta], \mathbf{R}) : \|\xi - \xi_0\| \le 1\}$ with center in the constant function $\xi_0 \equiv \sigma_0(0)$. To see that $A(S) \subset S$, note that $A\xi$ is a continuous function and that for $t \in [0,\delta]$ we have

$$|(A\xi)(t) - \sigma_0(0)| \le \delta M_0.$$

It follows that we have to require $\delta M_0 < 1$. To see that A is contractive, note that

$$\|A\xi_1 - A\xi_2\| \le \sup_{0\le t\le\delta} \left| \int\limits_0^t \gamma(s-\tau)\left(\varphi(\xi_1(\tau)) - \varphi(\xi_2(\tau))\right) ds \right| \le M_1\delta\|\xi_1 - \xi_2\|,$$

where the constant M_1 depends only on the Lipschitz constant of φ and the number $\int\limits_0^{+\infty} |\gamma(s)|\,ds$.

Thus, if $\delta < \min\left\{ \frac{1}{M_0}, \frac{1}{M_1} \right\}$, then A is a contraction on S with a unique fixed point σ_δ which is the unique solution of (8.1.1), (8.1.2) on $[0,\delta)$.

Now, considering (8.1.1) with the new initial condition $\sigma(t) = \sigma_\delta(t)$ $(t \in [-h,\delta])$, by the same argument as above we obtain the unique solution of (8.1.1), (8.1.2) on $[-h, 2\delta)$ (inequality (8.1.5) is of crucial importance here) which coincides with σ_δ on $[-h,\delta]$. Repeating this argument, we obtain the unique solution of (8.1.1), (8.1.2) on any finite interval $[0,T)$ and, consequently, on $[0,+\infty)$. ∎

The conditions on φ guarantee that for a solution $\sigma = \sigma(t,\sigma_0)$ the function $\varphi \circ \sigma$ is uniformly continuous on $[0,+\infty)$ and that $\dot\sigma$ is bounded on $[0,+\infty)$. In the subsequent part of this section we make additional assumptions on α and γ:

(A1) $\alpha(t) \to 0$ as $t \to \infty$ and there is a number $\kappa_1 > 0$ such that $t \mapsto \alpha(t)e^{\kappa_1 t}$ is $L^2(0, +\infty)$.

(A2) There is a number $\kappa_2 > 0$ such that the function $t \mapsto \gamma(t)e^{\kappa_2 t}$ is $L^2(0, +\infty)$.

From (A1) it follows that for a solution σ of (8.1.1) $\dot{\sigma}$ is uniformly continuous. Using this uniform continuity and the Barbalat lemma (Theorem 2.1.3, p. 16) we establish the following assertions, in which σ denotes a solution of (8.1.1), (8.1.2) on $[0, +\infty)$.

Proposition 8.1.2 *If $\dot{\sigma} \in L^2(0, +\infty)$ then $\dot{\sigma}(t) \to 0$ as $t \to +\infty$.*

Proposition 8.1.3 *If $\varphi \circ \sigma \in L^2(0, +\infty)$ then $\varphi(\sigma(t)) \to 0$ as $t \to +\infty$.*

Taking in equation (8.1.1) (with $\alpha(t) \equiv 0$) the formal Laplace transforms $\widetilde{\sigma}$, $\widetilde{\xi}$ of σ and $\varphi(\sigma(\cdot))$, respectively, we can define the transfer function χ by

$$\widetilde{\sigma}(s) = -\chi(s)\widetilde{\xi}(s)$$

with

$$\chi(s) = \frac{1}{s}K(s)$$

and

$$K(s) = -\varrho e^{-sh} + \int\limits_0^{+\infty} \gamma(t)e^{-st}\, dt.$$

8.2　A priori Integral Estimates

In the case of ordinary differential equations the frequency-domain condition guarantees that certain Lyapunov function along the solutions of the system is bounded from above. In an analogous way frequency-domain conditions are used to show the boundedness from above of certain Lyapunov functionals defined on the solutions of (8.1.1).

In preparing theorems on convergence we have to prove a number of lemmas.

Lemma 8.2.1 *Suppose there exist numbers $\varepsilon \geq 0$ and $\delta \geq 0$ such that*

$$\operatorname{Re} K(i\omega) - \varepsilon|K(i\omega)|^2 \geq \delta \qquad (\omega \in \mathbf{R}). \tag{8.2.1}$$

Then for an arbitrary solution σ of (8.1.1), (8.1.2) the integral

$$I_{1T} := \int\limits_0^T \left\{ \dot{\sigma}(t)\varphi(\sigma(t)) + \varepsilon\dot{\sigma}(t)^2 + \delta\varphi(\sigma(t))^2 \right\}\, dt$$

is bounded from above by a constant independent of T.

Proof Let us denote $\varphi \circ \sigma$ by η and define for arbitrary $T > 0$ the functions

$$\eta_T(t) = \begin{cases} 0, & t \notin [0, T] \\ \eta(t), & t \in [0, T] \end{cases}$$

and

$$\zeta_T(t) = \begin{cases} 0, & t < 0 \\ \varrho\eta_T(t - h) - \int\limits_0^t \gamma(t - \tau)\eta_T(\tau)\, d\tau, & t \geq 0. \end{cases}$$

Notice that for $t \in [0, T]$ we have

$$\dot{\sigma}(t) = \sigma^0(t) + \zeta_T(t), \tag{8.2.2}$$

where

$$\sigma^0(t) = \begin{cases} \alpha(t) + \varrho\varphi(\sigma(t-h)), & t \in [0,h] \\ \alpha(t), & t > h. \end{cases}$$

It is obvious that for every $T > 0$ the function η_T is in $L^2(\mathbf{R})$. Because of assumption (A2) the function ζ_T belongs also to $L^2(\mathbf{R})$. Thus, there exist the Fourier transforms $\widetilde{\eta}_T$ and $\widetilde{\zeta}_T$. By the Parseval equality we get

$$\begin{aligned} A_T &:= \int_0^{+\infty} \left\{ \zeta_T \eta_T + \varepsilon \zeta_T^2 + \delta \eta_T^2 \right\} dt \\ &= \frac{1}{2\pi} \int_{-\infty}^{+\infty} \left\{ \widetilde{\zeta}_T(i\omega)^* \widetilde{\eta}_T(i\omega) + \varepsilon |\widetilde{\zeta}_T(i\omega)|^2 + \delta |\widetilde{\eta}_T(i\omega)|^2 \right\} d\omega. \end{aligned} \tag{8.2.3}$$

Using the equation

$$\widetilde{\zeta}_T(i\omega) = -K(i\omega)\widetilde{\eta}_T(i\omega), \qquad (\omega \in \mathbf{R})$$

we get from (8.2.3)

$$A_T = -\frac{1}{2\pi} \int_{-\infty}^{+\infty} \left\{ \operatorname{Re} K(i\omega) - \varepsilon |K(i\omega)|^2 - \delta \right\} |\widetilde{\eta}_T(i\omega)|^2 d\omega.$$

By the frequency-domain inequality (8.2.1) we therefore find that

$$A_T \leq 0. \tag{8.2.4}$$

Now use (8.2.2) to write

$$A_T = \int_0^T \left\{ (\dot{\sigma} - \sigma^0)\eta + \delta\eta^2 + \varepsilon(\dot{\sigma} - \sigma^0)^2 \right\} dt + \varepsilon \int_T^{+\infty} \zeta_T^2 \, dt.$$

Hence

$$A_T \geq I_{1T} + B_T, \tag{8.2.5}$$

where

$$B_T = \int_0^T \left\{ \sigma^0 \eta - \varepsilon(\sigma^0)^2 - 2\varepsilon\sigma^0\dot{\sigma} \right\} dt.$$

It follows from assumption (A1) and the boundedness of $\dot{\sigma}$ that $|B_T| \leq c_1$, independent of T.

Then it follows from (8.2.4) and (8.2.5) that $I_{1T} \leq c_1$, independent of T. ∎

Sometimes we suppose, in addition to the previous assumption upon φ, that this function is piece-wise C^1. One can expect to obtain more precise information on convergence by using in the Lyapunov functionals additional terms containing φ'.

Convention 8.2.1 For simplicity we denote ess sup and ess inf by sup and inf, respectively.

Lemma 8.2.2 *Suppose φ is piece-wise C^1 and there exist numbers $\varepsilon \geq 0$, $\delta \geq 0$, $\tau \geq 0$, $\mu_1 \leq \inf_{\sigma \in [0,\Delta]} \varphi'(\sigma)$ and $\mu_2 \geq \sup_{\sigma \in [0,\Delta]} \varphi'(\sigma)$ such that for all $\omega \in \mathbf{R}$*

$$\operatorname{Re} \left\{ K(i\omega) - \tau[K(i\omega) + \mu_1^{-1}(i\omega)]^*[K(i\omega) + \mu_2^{-1}(i\omega)] \right\} - \varepsilon|K(i\omega)|^2 \geq \delta. \tag{8.2.6}$$

Then, for an arbitrary solution σ of (1.1), the integral I_{1T} defined in the statement of Lemma 8.2.1 is bounded from above by a constant independent of T.

174

Remark 8.2.1 In the assumptions of Lemma 8.2.2 it is possible to take $\mu_1 = -\infty$ or $\mu_2 = +\infty$ (but not $\mu_1 = -\infty$ and $\mu_2 = +\infty$, simultaneously.) In such a case one has to choose $\mu_1^{-1} = 0$ and $\mu_2^{-1} = 0$, respectively.

Proof of Lemma 8.2.2 Let us define

$$
\eta(t) = \begin{cases} 0, & t < 0 \\ \varphi(\sigma(t)), & t \geq 0 \end{cases}
$$

and for $T > 1$ the cut-off function

$$
\eta_T(t) = \begin{cases} \mu(t)\eta(t), & t \in [0, T] \\ \mu(t)\eta(t)e^{c(T-t)}, & t \notin [0, T], \end{cases}
$$

where $c > 0$ is a constant and

$$
\mu(t) = \begin{cases} 0, & t < 0 \\ t, & 0 \leq t \leq h \\ 1, & t > h. \end{cases}
$$

Furthermore, introduce the functions

$$
\eta_\mu(t) := \mu(t)\eta(t),
$$

$$
\zeta(t) := \varrho\eta_\mu(t - h) - \int_0^t \gamma(t - \tau)\eta_\mu(\tau)\,d\tau
$$

and

$$
\sigma^0(t) := \alpha(t) - \varrho[\mu(t - h) - 1]\eta(t - h) - \int_0^t \gamma(t - \tau)(1 - \mu(\tau))\eta(\tau)\,d\tau
$$

so that

$$
\dot\sigma(t) = \sigma^0(t) + \zeta(t). \tag{8.2.7}
$$

Let us also introduce the function

$$
\zeta_T(t) = \begin{cases} 0, & t < 0 \\ \varrho\eta_T(t - h) - \int_0^t \gamma(t - \tau)\eta_T(\tau)\,d\tau & t \geq 0, \end{cases}
$$

which coincides with $\zeta(t)$ on $[0, T]$. It is obvious that for every $T > 1$ the functions η_T and $\dot\eta_T$ belong to $L^2(\mathbf{R})$ and it is easy to show that $\zeta_T \in L^2(\mathbf{R})$.

Let us consider the functional

$$
A_T = \int_0^\infty \left\{ \zeta_T\eta_T + \delta\eta_T^2 + \varepsilon\zeta_T^2 + \tau[\zeta_T - \mu_1^{-1}\dot\eta_T][\zeta_T - \mu_2^{-1}\dot\eta_T] \right\}\,dt.
$$

By means of the Parsival equality for the Fourier transforms $\tilde\zeta_T$, $\tilde{\dot\eta}_T$, and $\tilde\eta_T$ satisfying

$$
\tilde\zeta_T(i\omega) = -K(i\omega)\tilde\eta_T(i\omega), \quad \tilde{\dot\eta}_T(i\omega) = i\omega\tilde\eta_T(i\omega) \qquad (\omega \in \mathbf{R})
$$

we get

$$
\begin{aligned}
A_T = \ & \frac{1}{2\pi} \int_{-\infty}^{+\infty} \left\{ -\operatorname{Re} K(i\omega) + \delta + \varepsilon|K(i\omega)|^2 + \right. \\
& \left. + \tau\operatorname{Re}[K(i\omega) + \mu_1^{-1}(i\omega)]^*[K(i\omega) + \mu_2^{-1}(i\omega)] \right\} |\tilde\eta_T(i\omega)|^2\,d\omega.
\end{aligned}
$$

From (8.2.1) it follows that

$$A_T \leq 0 \qquad \text{for all} \qquad T > 1. \tag{8.2.8}$$

Let us split the integral A_T in the following way

$$A_T = \sum_{i=1}^{6} A_{Ti},$$

where

$$
\begin{aligned}
A_{T1} &= \int\limits_{T}^{+\infty} \left\{ \zeta_T \eta_T + \delta \dot{\eta}_T^2 + \tau \eta_T^2 k_1^{-1} k_2^{-1} - (\mu_1^{-1} + \mu_2^{-1}) \zeta_T \dot{\eta}_T \right\} dt, \\
A_{T2} &= \tau \int\limits_{0}^{T} (\dot{\sigma} - \mu_1^{-1} \dot{\eta})(\dot{\sigma} - \mu_2^{-1} \dot{\eta}) \, dt, \\
A_{T3} &= \int\limits_{0}^{1} \left\{ (\mu - 1)\dot{\sigma}\eta + \delta(\mu^2 - 1)\eta^2 + \tau \mu_1^{-1} \mu_2^{-1} (\dot{\eta}_\mu^2 - \dot{\eta}^2) - \tau(\mu_1^{-1} + \mu_2^{-1})(\dot{\eta} - \dot{\eta}_\mu) \right\} dt, \\
A_{T4} &= \int\limits_{0}^{T} \left\{ \sigma^0 \eta_\mu + (\varepsilon + \tau)(\sigma^0)^2 - 2\varepsilon \sigma^0 \dot{\sigma} + \tau(\mu_1^{-1} + \mu_2^{-1})\dot{\eta}_\mu \sigma^0 \right\} dt, \\
A_{T5} &= \int\limits_{0}^{T} \dot{\sigma}\eta + \varepsilon \dot{\sigma}^2 + \delta \eta^2 \, dt, \\
A_{T6} &= \int\limits_{T}^{+\infty} (\varepsilon + \tau)\zeta_T^2 \, dt.
\end{aligned}
$$

It is obvious that $|A_{T1}| \leq c_1$, where c_1 is independent of T. Note that $A_{T2} \geq 0$ since $\mu_1 \leq \varphi'(\sigma) \leq \mu_2$ a.e. on $\mathbf{R}$. It is evident that $|A_{T3}| \leq c_3$, where c_3 does not depend on T. The same property can be established with respect to A_{T4} if one takes into consideration that $t \mapsto \sigma^0(t)e^{\kappa_0 t}$ belongs to $L^2(0, +\infty)$. It follows now from (8.2.8) that $A_{T5} + A_{T2} \leq c_4$ and, consequently, $A_{T5} \leq c_4$, where c_4 does not depend on T. ∎

The inequality $A_{T2} + A_{T5} \leq const$ just established in the proof immediately yields the following

Corollary 8.2.1 *Suppose the conditions of Lemma 8.2.2 are fulfilled and σ is a solution of (8.1.1). Then the integral*

$$
\begin{aligned}
I_T \; := \; \int\limits_{0}^{T} & \left\{ \dot{\sigma}(t)\varphi(\sigma(t)) + \varepsilon \dot{\sigma}(t)^2 + \delta \varphi(\sigma(t))^2 + \right. \\
& \left. + \tau \big(\mu_1^{-1}\varphi(\sigma(t))^{\cdot} - \dot{\sigma}(t) \big)\big(\mu_2^{-1}\varphi(\sigma(t))^{\cdot} - \dot{\sigma}(t) \big) \right\} dt
\end{aligned}
$$

is bounded from above by a constant, independent of T.

In the next part a number of theorems concerning the convergence of solutions of (8.1.1) will be proved on the basis of the Lemmata 8.2.1 and 8.2.2. In addition to the conditions (A1) – (A2) we fix the following assumption upon φ.

(A3) The set $\mathcal{E}$ of zeros of φ on $[0, \Delta]$ is closed and for any $x_1, x_2 \in \mathcal{E}$ satisfying $x_1 < x_2$ there exists an $y \in \mathbf{R} \backslash \mathcal{E}$ such that $x_1 < y < x_2$.

Theorem 8.2.1 *Suppose that φ is piece-wise C^1 and there exist numbers $\delta > 0$, $\varepsilon > 0$ and $\mu_1 \leq \inf\limits_{\sigma \in [0,\Delta]} \varphi'(\sigma)$, $\mu_2 \geq \sup\limits_{\sigma \in [0,\Delta]} \varphi'(\sigma)$ such that (8.2.1) is true. Suppose the solution σ of (8.1.1) is bounded on $[0, +\infty)$. Then the following relations take place:*

$$\dot{\sigma} \in L^2(0, +\infty), \tag{8.2.9}$$

176

$$\dot{\sigma}(t) \to 0 \qquad as \qquad t \to +\infty, \tag{8.2.10}$$

$$\sigma(t) \to \sigma_1 \qquad as \qquad t \to +\infty, \tag{8.2.11}$$

where σ_1 is a zero of φ. If $(8.2.1)$ is fulfilled with $\delta = 0$ and $\varepsilon > 0$ (μ_1, μ_2 unchanged) then $(8.2.9)$ and $(8.2.10)$ are true.
If $(8.2.1)$ is fulfilled with $\delta > 0$ and $\varepsilon = 0$ (μ_1, μ_2 unchanged) then relation $(8.2.11)$ is true.

Proof According to Lemma 8.2.2 the integral I_{1T} from Lemma 8.2.1 is bounded, independently of T. Let us use the representation

$$I_{1T} = \varepsilon \int_0^T \dot{\sigma}(t)^2\, dt + \delta \int_0^T \varphi(\sigma(t))^2\, dt + \int_{\sigma(0)}^{\sigma(T)} \varphi(\sigma)\, d\sigma.$$

Since the functions σ and φ are bounded, the last term in this representation is also bounded. It follows that

$$\varepsilon \int_0^T \dot{\sigma}(t)^2\, dt + \delta \int_0^T \varphi(\sigma(t))^2\, dt \le c_1,$$

where c_1 is independent of T. Consequently, we find that $\varepsilon \ne 0$ implies $\dot{\sigma} \in \mathrm{L}^2(0, +\infty)$ and $\delta \ne 0$ implies $\varphi \circ \sigma \in \mathrm{L}^2(0, +\infty)$. Using the Propositions 8.1.2 and 8.1.3 we get, in the case $\varepsilon \ne 0$, that $\dot{\sigma}(t) \to 0$ as $t \to +\infty$ and, in the case $\delta \ne 0$, that $\varphi(\sigma(t)) \to 0$ as $t \to +\infty$. It follows that $(8.2.11)$ is true. ∎

We now prove a supplementary proposition which, for a solution σ of $(8.1.1)$, connects the behavior of $\varphi(\sigma(t))$ and $\dot{\sigma}(t)$ as $t \to +\infty$.

Lemma 8.2.3 *Suppose that φ is C^1 and $K(0) = -\varrho + \int_0^{+\infty} \gamma(t)\, dt \ne 0$. Then for an arbitrary solution σ of $(8.1.1)$ the inclusion $\dot{\sigma} \in \mathrm{L}^2(0, +\infty)$ implies that $\varphi(\sigma(t)) \to 0$ as $t \to +\infty$.*

Proof Define the function $\eta = \varphi \circ \sigma$. From $(8.1.1)$ it follows that

$$
\begin{aligned}
\dot{\sigma} - \alpha(t) &= -\varrho\eta(t - h) + \int_0^t \gamma(t - \tau)\eta(\tau)\, d\tau \\
&= -\varrho\eta(t - h) + \eta(t) \int_0^{+\infty} \gamma(\tau)\, d\tau - \\
&\quad -\eta(0) \int_t^{+\infty} \gamma(\tau)\, d\tau - \int_0^t \left\{ \dot{\eta}(\tau) \int_{t-\tau}^{+\infty} \gamma(s)ds \right\}\, d\tau.
\end{aligned}
\tag{8.2.12}
$$

The left-hand side of $(8.2.12)$ tends to zero as $t \to +\infty$ since, because of assumption (A1), $\alpha(t)$ and $\dot{\sigma}(t)$ tend to 0 as $t \to +\infty$. Notice that $\dot{\sigma} \in \mathrm{L}^2(0, +\infty)$ implies $\dot{\eta} \in \mathrm{L}^2$. From assumption (A2) it follows that $t \mapsto \int_t^{+\infty} \gamma(\tau)\, d\tau$ is L^2. According to the well-known fact that the convolution of two L^2-functions tends to zero we get

$$\int_t^{+\infty} \left\{ \dot{\eta}(\tau) \int_{t-\tau}^{+\infty} \gamma(s)ds \right\}\, d\tau \to 0 \qquad as \qquad t \to +\infty.$$

The assumption (A2) guarantees that $\int_t^{+\infty} \gamma(\tau)\, d\tau \to 0$ as $t \to +\infty$. Define $\beta := \int_0^{+\infty} \gamma(\tau)\, d\tau$.

We want to show that from $\lim\limits_{t\to+\infty}(-\varrho\eta(t-h)+\beta\eta(t))=0$ follows $\lim\limits_{t\to+\infty}\eta(t)=0$. Suppose that there is a sequence $t_j\to+\infty$ such that $\eta(t_j)\not\to 0$. Consider

$$-\varrho\eta(t_j-h)+\beta\eta(t_j)=(-\varrho+\beta)\eta(t_j)-\varrho\varphi'(\sigma(\bar{t}_j))\dot\sigma(\bar{t}_j) \qquad (8.2.13)$$

where we have used the mean value theorem ($\bar{t}_j\in(t_j-h,t_j)$). From (8.2.13) it follows that $\dot\sigma(\bar{t}_j)\not\to 0$ as $j\to+\infty$. On the other hand, we know that $\dot\sigma$ is bounded, uniformly continuous and L^2 which implies $\lim\limits_{t\to+\infty}\dot\sigma(t)=0$ (Proposition 8.1.2), a contradiction. $\blacksquare$

Theorem 8.2.2 *Suppose φ is C^1 and there exist numbers $\varepsilon>0$, $\tau\geq 0$, $\mu_1\leq\inf\limits_{\sigma\in[0,\Delta]}\varphi'(\sigma)$ and $\mu_2\geq\sup\limits_{\sigma\in[0,\Delta]}\varphi'(\sigma)$ such that condition (8.2.1) is satisfied with $\delta=0$. Also suppose $K(0)\neq 0$. Then every solution of (8.1.1) which is bounded on $[0,+\infty)$ satisfies (8.2.9) – (8.2.11).*

8.3 Bakaev-Guzh Technique

The results obtained in this section may be regarded as a certain extension of Theorem 4.3.2, p. 70, to integro-differential equations. We use here the same method as in Chapter 4 for constructing the Lyapunov functionals.

Theorem 8.3.1 *Consider equation (8.1.1) and suppose that φ is piece-wise C^1 and there exist numbers $\varepsilon>0$, $\delta>0$, $\tau\geq 0$, $\mu_1\leq\inf\limits_{\sigma\in[0,\Delta]}\varphi'(\sigma)$ and $\mu_2\geq\sup\limits_{\sigma\in[0,\Delta]}\varphi'(\sigma)$ such that for the transfer function K of (8.1.1) the inequality*

$$\mathrm{Re}\left\{K(i\omega)-\tau[K(i\omega)+\mu_1^{-1}i\omega]^*[K(i\omega)+\mu_2^{-1}i\omega]\right\}-\varepsilon|K(i\omega)|^2\geq\delta \qquad (\omega\in\mathbf{R}) \qquad (8.3.1)$$

is true and at least one of the following two conditions holds:

(i) $4\tau\delta>\nu^2$, *where*

$$\nu=\int\limits_0^\Delta\varphi(\sigma)\,d\sigma\left/\int\limits_0^\Delta\sqrt{(1-\mu_1^{-1}\varphi'(\sigma))(1-\mu_2^{-1}\varphi'(\sigma))}|\varphi(\sigma)|\,d\sigma\right.;$$

(ii) $4\varepsilon\delta>\nu_0^2$, *where*

$$\nu_0=\int\limits_0^\Delta\varphi(\sigma)\,d\sigma\left/\int\limits_0^\Delta|\varphi(\sigma)|\,d\sigma\right. .$$

Then for any solution σ of (8.1.1) the following relations are true:

$$\dot\sigma(t)\to 0 \qquad as \qquad t\to+\infty, \qquad (8.3.2)$$

$$\sigma(t)\to\sigma_1 \qquad as \qquad t\to+\infty, \qquad (8.3.3)$$

where σ_1 is a zero of φ.

178

Proof Let us consider the case when condition (i) is fulfilled. According to Corollary 8.2.1 the inequality (8.3.1) implies that for an arbitrary solution σ of (8.1.1) we have

$$
\begin{aligned}
I_T \ :=\ & \int_0^T \{\dot{\sigma}(t)\varphi(\sigma(t)) + \varepsilon\dot{\sigma}(t)^2 + \delta\varphi(\sigma(t))^2 + \\
& + \tau[\dot{\sigma}(t) - \mu_1^{-1}\varphi(\sigma(t))^\cdot][\dot{\sigma}(t) - \mu_2^{-1}\varphi(\sigma(t))^\cdot]\} \, dt \\
\leq\ & c_1,
\end{aligned}
\tag{8.3.4}
$$

where c_1 is independent of T.

Let us define the auxiliary functions φ_1 and F_1 from Lemma 4.2.2, p. 67, by

$$
\varphi_1(\sigma) := \sqrt{(1 - \mu_1^{-1}\varphi'(\sigma))(1 - \mu_2^{-1}\varphi'(\sigma))}
$$

and

$$
F_1(\sigma) := \varphi(\sigma) - \nu|\varphi_1(\sigma)\varphi(\sigma)| \qquad (\sigma \in \mathbf{R}).
$$

Using the abbreviations $\eta = \varphi \circ \sigma$ and $\eta_1 = \varphi_1 \circ \sigma$ we get

$$
\int_0^T \left\{\varepsilon\dot{\sigma}(t)^2 + \delta\eta^2 + \tau\eta_1^2\dot{\sigma} + \nu|\eta_1\eta|\dot{\sigma}\right\} dt + \int_{\sigma(0)}^{\sigma(T)} F_1(\sigma) \, d\sigma.
$$

Employing the fact that $\int_0^\Delta F_1(\sigma) \, d\sigma = 0$ and that (8.2.4) is true, we get the inequality

$$
\int_0^T \left\{\varepsilon\dot{\sigma}(t)^2 + \delta\eta^2 + \tau\eta_1^2\dot{\sigma} + \nu|\eta_1\eta|\dot{\sigma}\right\} dt \leq c_2,
\tag{8.3.5}
$$

where c_2 is independent of T. Condition (i) guarantees that the quadratic form $\delta x^2 + \nu xy + \tau y^2$ $(x, y \in \mathbf{R})$ is positive definite. Therefore in this case we have from (8.3.5) that

$$
\dot{\sigma} \in \mathrm{L}^2(0, +\infty), \qquad \varphi \circ \sigma \in \mathrm{L}^2(0, +\infty).
\tag{8.3.6}
$$

Assume now that (ii) is satisfied. We introduce the function

$$
F_0(\sigma) := \varphi(\sigma) - \nu_0|\varphi(\sigma)| \qquad (\sigma \in \mathbf{R})
$$

and use for I_{1T} the representation

$$
I_{1T} = \int_0^T \left\{\varepsilon\dot{\sigma}(t)^2 + \delta\eta^2 + \nu_0|\eta|\dot{\sigma}\right\} dt + \int_{\sigma(0)}^{\sigma(T)} F_0(\sigma) \, d\sigma.
$$

Because of $\int_0^\Delta F_0(\sigma) \, d\sigma = 0$ we have

$$
\int_0^T \left\{\varepsilon\dot{\sigma}^2 + \delta\eta^2 + \nu_0|\eta|\dot{\sigma}\right\} dt \leq c_3,
\tag{8.3.7}
$$

where c_3 is independent of T. From (ii) we see that the quadratic form $\varepsilon x^2 + \nu_0 xy + \delta y^2$ $(x, y \in \mathbf{R})$ is positive definite. It follows then from (8.3.7) that, again, the relations (8.3.6) are true. Now we apply Propositions 8.1.2 and 8.1.3 to see that (8.3.6) implies (8.3.2) and (8.3.3). $\blacksquare$

8.4 Non-Local Reduction Principle

In this section for the L^2-stability analysis of equation (8.1.1) we shall construct Lyapunov functionals which include solutions of some second order ordinary differential equations of the same type as in the nonlocal reduction principle considered in the preceeding sections. Thus we are able to combine the integral a-priori estimates method of Sect. 8.2 with the nonlocal reduction principle to receive sufficient conditions for boundedness and convergence of solutions of (8.1.1).

Let us consider the ordinary differential system

$$\begin{aligned}
\dot{\vartheta} &= y, \\
\dot{y} &= -ay - \varphi(\vartheta),
\end{aligned} \qquad (8.4.1)$$

where $a > 0$ is a parameter and φ is the nonlinearity from (8.1.1). System (8.4.1) is equivalent to the second order equation

$$\ddot{\vartheta} + a\dot{\vartheta} + \varphi(\vartheta) = 0. \qquad (8.4.2)$$

We shall use solutions of the first order equation

$$y\frac{dy}{d\vartheta} + ay + \varphi(\vartheta) = 0, \qquad (8.4.3)$$

to design Lyapunov functionals for equation (8.1.1). Suppose that (8.4.1) has a saddle point in $(\vartheta^*, 0)$ with $\vartheta^* \in [0, \Delta)$, and consequently there exists a separatrix with the corresponding solution $y = \Phi(\vartheta, a)$.

Let us recall that for $a > a_{cr}$, i.e. in the case when (8.4.2) is Lagrange stable, there exists a solution $\Phi(\cdot, a)$ of (8.4.3) with the following properties:

(a) $\Phi(\vartheta^*, a) = 0$, $\Phi(\vartheta, a) \neq 0$ for $\vartheta \neq \vartheta^*$;

(b) $\Phi(\vartheta, a) \to +\infty$ as $\vartheta \to -\infty$,
    $\Phi(\vartheta, a) \to -\infty$ as $\vartheta \to +\infty$,

(c) If $\varphi'(\vartheta^*) \neq 0$ then $\Phi'(\vartheta^*, a) \neq 0$, where $\Phi'(\vartheta, a) := D_1\Phi(\vartheta, a)$.

Theorem 8.4.1 *Suppose that there exist positive numbers ε and $\lambda < \kappa_0 = \min\{\kappa_1, \kappa_2\}$ such that the following conditions are fulfilled:*

(i) *any solution of equation (8.4.2) with $a = 2\sqrt{\varepsilon\lambda}$ is bounded on $[0, +\infty)$;*

(ii)

$$\operatorname{Re} K(i\omega - \lambda) - \varepsilon|K(i\omega - \lambda)|^2 \geq 0 \quad (\omega \in \mathbf{R}). \qquad (8.4.4)$$

Then any solution $\sigma(\cdot, \sigma_0)$ of (8.1.1) is bounded on $[0, +\infty)$.

Proof We set $\eta := \varphi \circ \sigma$ and define for every $T > 0$ the functions

$$\eta_T(t) = \begin{cases} 0, & t \notin [0, T], \\ e^{\lambda t}\eta(t), & t \in [0, T], \end{cases}$$

and

$$\zeta_T(t) = \begin{cases} 0, & t < 0, \\ \varrho\eta_T(t - h) - \int\limits_0^t \gamma(t - \tau)\eta_T(\tau)\, d\tau, & t \geq 0. \end{cases}$$

180

Note that we can look at η_T and ζ_T as weighed functions η_T and ζ_T of Lemma 8.2.1. It is obvious that for every $T > 0$ the function η_T is L^2. Because of assumption (A2) and $\lambda < \mu_0$ we have also $\zeta_T \in L^2$. Applying the Parseval equality we get

$$A_T := \int_0^{+\infty} (\eta_T \zeta_T + \varepsilon \zeta_T^2)\, dt = -\frac{1}{2\pi} \int_{-\infty}^{+\infty} [\operatorname{Re} K(i\omega - \lambda) - \varepsilon |K(i\omega - \lambda)|^2]|\widetilde{\eta}_T(i\omega)|^2 d\omega,$$

where we used

$$\widetilde{\zeta}_T(i\omega) = -K(i\omega - \lambda)\widetilde{\eta}_T(i\omega) \quad (\omega \in \mathbf{R}).$$

Since (8.4.4) is fulfilled we have

$$A_T \leq 0 \quad \text{for all} \quad T > 0. \tag{8.4.5}$$

Taking into account that $\dot{\sigma}(t) = \sigma^0(t) + e^{-\lambda t}\zeta_T(t)$, where

$$\sigma^0(t) = \begin{cases} \alpha(t) + \varrho\eta(t - h), & t \in [0, h], \\ \alpha(t), & t > h, \end{cases}$$

we get from (8.4.5) the inequality

$$\int_0^T [\eta\dot{\sigma} + \varepsilon\dot{\sigma}^2]e^{2\lambda t}\, dt \leq \int_0^T [\eta\sigma^0 + 2\varepsilon\dot{\sigma}\sigma^0]e^{2\lambda t}\, dt. \tag{8.4.6}$$

By assumptions (A1) and (A2) we have

$$\sigma^0(t)e^{\kappa_0 t} \in L^2. \tag{8.4.7}$$

First consider the case $\lambda < \frac{\kappa_0}{2}$. Then from (8.4.6) and (8.4.7) it follows that

$$\int_0^T [\eta\dot{\sigma} + \varepsilon\dot{\sigma}]e^{2\lambda t}\, dt \leq c_1, \tag{8.4.8}$$

where c_1 is independent of T. Condition (i) of Theorem 8.4.1 guarantees that for $a = 2\sqrt{\lambda\varepsilon}$ the solution $\Phi(\cdot, 2\sqrt{\lambda\varepsilon})$ of (8.4.2) has the properties (a) – (c). Further we use the abbreviation $\Phi_k(\vartheta) := \Phi(\vartheta + k\Delta, 2\sqrt{\lambda\varepsilon})$. Let us consider the expression

$$I_{T,N} = \int_0^T [\eta\dot{\sigma} + \varepsilon\dot{\sigma}^2]e^{2\lambda t}\, dt + \frac{1}{2}\Phi_N^2(\sigma(T))e^{2\lambda T} - \frac{1}{2}\Phi_N(\sigma(0)),$$

where the natural number N will be chosen later on. We can represent $I_{T,N}$ in the form

$$I_{T,N} = \int_0^T \left\{ \eta\dot{\sigma} + \varepsilon\dot{\sigma}^2 + \Phi_N(\sigma(t))\Phi_N'(\sigma(t))\dot{\sigma} + \lambda\Phi_N^2(\sigma(t)) \right\} e^{2\lambda t}\, dt.$$

Denote the integrand of $I_{T,N}$ by $\Omega_N(\cdot)$ and consider it as a quadratic form of the three variables $\eta, \dot{\sigma}$ and $\Phi_N\Phi_N'$. We obtain that

$$\eta\dot{\sigma} + \varepsilon\dot{\sigma}^2 + \Phi_N\Phi_N'\dot{\sigma} = \left[\sqrt{\varepsilon}\dot{\sigma} + \frac{1}{2\sqrt{\varepsilon}}\eta + \frac{1}{2\sqrt{\varepsilon}}\Phi_N\Phi_N' \right]^2 - \frac{1}{4}[\eta + \Phi_N\Phi_N']^2,$$

so that for any natural N and $t \geq 0$ we have

$$\Omega_N(t) \geq \left[\lambda \Phi_N^2 - \frac{1}{4\varepsilon}(\eta + \Phi_N \Phi_N')^2 \right] e^{2\lambda t}$$

$$= \left[\sqrt{\lambda}\Phi_N - \frac{1}{2\sqrt{\varepsilon}}(\eta + \Phi_N \Phi_N') \right] \left[\sqrt{\lambda}\Phi_N + \frac{1}{2\sqrt{\varepsilon}}(\eta + \Phi_N \Phi_N') \right] e^{2\lambda t}.$$

As Φ_N is a solution of (8.4.2) for $a = 2\sqrt{\lambda\varepsilon}$, we have

$$\Omega_N(t) \geq 0 \qquad (t \geq 0),$$

and, consequently, for all $T \geq 0$

$$I_{T,N} \geq 0. \tag{8.4.9}$$

It follows from (8.4.8) and (8.4.9) that

$$e^{2\lambda T}\Phi_N^2(\sigma(t)) \geq \Phi_N^2(\sigma(0)) - c_1. \tag{8.4.10}$$

Because of property (b) of Φ we can choose N so large that

$$\vartheta^* - N\Delta < \sigma(0) < \vartheta^* + N\Delta$$

and

$$\vartheta_{\pm N}^2(\sigma(0)) > c_1.$$

Thus it follows from (8.4.10) that $\Phi_N(\sigma(T))$ and $\Phi_{-N}(\sigma(T))$ cannot vanish for any $T \geq 0$. It means that for all $T \geq 0$

$$\vartheta^* - N\Delta \leq \sigma(T) \leq \vartheta^* + N\Delta. \tag{8.4.11}$$

Consider now the case $\kappa_0/2 < \lambda < \kappa_0$. Let us choose $\overline{\varepsilon} < \varepsilon$ and $\overline{\lambda} < \lambda$ with $a_{cr} < 2\sqrt{\lambda\varepsilon}$. Define $\Phi_k(\cdot) = \Phi(\cdot, 2\sqrt{\lambda\overline{\varepsilon}}) + k\Delta$ and let $\overline{\vartheta}$ be such that $\varphi(\overline{\vartheta}) = 0$ and $\Phi_0(\overline{\vartheta}) = 0$. Let $\varepsilon_1 := \varepsilon - \overline{\varepsilon}$ and $\lambda_1 := \lambda - \overline{\lambda}$. It is not difficult to show that there exists a number ε_2 such that

$$\lambda_1 \Phi_0^2(\vartheta) - \varepsilon_2 \varphi^2(\vartheta) \geq 0 \quad \text{for all} \quad \vartheta \in \mathbf{R}. \tag{8.4.12}$$

Indeed, let D be a neighborhood of $\overline{\vartheta}$, such that $\Phi_0'(\vartheta) \neq 0$ for $\vartheta \in \overline{D}$, the closure of D. Let

$$A := \min_{\vartheta \in \overline{D}} \Phi_0'(\vartheta) \qquad \text{and} \qquad B := \min_{\vartheta \in \overline{D}} \Phi_0(\vartheta).$$

For $\vartheta \notin D$ it is sufficient for (8.4.12) to choose ε_2 as

$$\varepsilon_2 < \frac{\lambda_1 B^2}{\max_{\sigma} |\varphi(\sigma)|^2}.$$

For $\vartheta \in \overline{D}$ one may use the inequalities

$$|\Phi_0(\vartheta)| \geq A|\vartheta - \overline{\vartheta}| \qquad \text{and} \qquad |\varphi(\vartheta)| \leq L|\vartheta - \overline{\vartheta}|,$$

where L is a Lipschitz constant for φ. It follows that for $\vartheta \in \overline{D}$ it is, for (8.4.12), sufficient to choose $\varepsilon_2 < \frac{\lambda_1 A^2}{L^2}$. We see that (8.4.12) is true if ε_2 satisfies the last two inequalities. Using the inequality

$$ab \leq \frac{a^2}{4\beta} + \beta b^2$$

for $\beta > 0$ and $a, b \in \mathbf{R}$, we get from (8.4.6)

$$\int\limits_0^T [\eta\dot{\sigma} + \varepsilon\dot{\sigma}^2]e^{2\lambda t}\, dt \leq \int\limits_0^T [\varepsilon_2\eta^2 + \varepsilon_1\dot{\sigma}^2 + \varepsilon_3(\sigma^0)^2]e^{2\lambda t}\, dt,$$

where ε_3 depends on ε, ε_1 and ε_2. By (8.4.7) we conclude that

$$\int\limits_0^T [\eta\dot{\sigma} + \varepsilon\dot{\sigma}^2]e^{2\lambda t}\, dt - \int\limits_0^T \varepsilon_2\eta^2 e^{2\lambda t}\, dt \leq c_2, \qquad (8.4.13)$$

where c_2 is independent of T. Let us introduce the term

$$\overline{I}_{TN} = \int\limits_0^T [\eta\dot{\sigma} + \overline{\varepsilon}\dot{\sigma} - \varepsilon_2\eta^2]e^{2\lambda t}\, dt + \frac{1}{2}e^{2\lambda t}\Phi_N^2(\sigma(T)) - \frac{1}{2}\Phi_N^2(\sigma(0)).$$

We can write $\overline{I}_{TN}$ in the form

$$\overline{I}_{TN} = \int\limits_0^T \overline{\Omega}_N(t)\, dt + \int\limits_0^T [\lambda_1\Phi_N^2(\sigma(t)) - \varepsilon_2\eta^2(t)]e^{2\lambda t}\, dt,$$

where

$$\overline{\Omega}_N(t) = [\eta(t)\dot{\sigma}(t) + \overline{\varepsilon}\dot{\sigma}^2(t) + \Phi_N(\sigma(t))\Phi_N'(\sigma(t))\dot{\sigma}(t) + \overline{\lambda}\Phi_N^2(\sigma(t))]e^{2\lambda t}.$$

Arguing as above we can show that $\overline{\Omega}_N(t) \geq 0$ for $t \geq 0$. Because of the periodicity of φ it follows from (8.4.12) that for any $N = 1, 2, \cdots$ the inequality

$$\lambda_1\Phi_{\pm N}^2(\sigma) - \varepsilon_2\varphi^2(\sigma) \geq 0 \qquad (\sigma \in \mathbf{R})$$

holds.

Thus, $\overline{I}_{T,N} \geq 0$. Repeating now the argument used in the proof for the case $\lambda < \frac{\kappa_0}{2}$ and using the estimate (8.4.13) we get the boundedness of σ on $[0, +\infty)$. ∎

In the next theorem we suppose that φ is piecewise C^1.

Theorem 8.4.2 *Suppose there exist numbers $\lambda \in (0, \kappa_0)$, $\tau \geq 0$, $\varepsilon > 0$, $\mu_1 \leq \inf\limits_{\sigma \in [0,\Delta]} \varphi'(\sigma)$ and $\mu_2 \geq \sup\limits_{\sigma \in [0,\Delta]} \varphi'(\sigma)$ such that the following conditions hold:*

(i) *any solution of (8.4.2) with $a = 2\sqrt{\lambda\varepsilon}$ is bounded on $[0, +\infty)$;*

(ii)

$$\mathrm{Re}\left\{ K(i\omega - \lambda) - \varepsilon|K(i\omega - \lambda)|^2 - \right.$$
$$\left. - \tau[K(i\omega - \lambda) + \mu_1^{-1}(i\omega - \lambda)]^*[K(i\omega - \lambda) + \mu_2^{-1}(i\omega - \lambda)]\right\} \geq 0. \qquad (8.4.14)$$

Then any solution $\sigma = \sigma(\cdot, \sigma_0)$ of (8.1.1) is bounded on $[0, +\infty)$.

Proof Suppose $(\vartheta^*, 0)$ is a saddle of (8.4.1). Let us introduce the set $M = \{\vartheta^* + k\Delta, k \in \mathbf{Z}\}$. For the given solution σ we define the set $Z = \{t : t \geq 1, \sigma(t) \in M\}$. Suppose at first that Z is bounded and $T_1 = \sup Z$. Then for all $t \geq 0$ we have

$$|\sigma(t)| \leq \max\limits_{t \in [0,T_1]} |\sigma(t)| + \Delta$$

and the conclusion of the theorem follows. Let us now consider the case when Z is not bounded.

We use the functions μ, η_μ, σ^0 and ζ introduced in the proof of Lemma 8.4.1. Note also that the inequality (8.2.7) is satisfied. Suppose that $T \in Z$. Let us construct a cut-off function in the following way:

$$\eta_T(t) = \begin{cases} \eta_\mu(t), & t \in [0, T], \\ 0, & t \notin [0, T]. \end{cases}$$

Because of $\varphi(\sigma(T)) = 0$ for $T \in Z$ the function η_T is continuous and a.e. differentiable.

Let us define the function

$$\zeta_T(t) = \begin{cases} 0, & t < 0, \\ \varrho\eta_T(t - h) - \int_0^t \gamma(t - \tau)\eta_T(\tau)d\tau, & t \geq 0, \end{cases}$$

which coincides with ζ on $[0, T]$.

Denote the functions $\eta_T(t)e^{\lambda t}, \zeta_T(t)e^{\lambda t}$ and $\dot{\eta}_T(t)e^{\lambda t}$ by $\eta_{T\lambda}(\cdot), \zeta_{T\lambda}(\cdot)$ and $\dot{\eta}_{T,\lambda}$. Using the fact that $\eta_{T\lambda}, \zeta_{T\lambda}$ and $\dot{\eta}_{T\lambda}$ are L^2 we can define the integral

$$A_T := \int_0^{+\infty} \{\eta_{T\lambda}\zeta_{T\lambda} + \varepsilon\zeta_{T\lambda}^2 + \tau[\zeta_{T\lambda} - \mu_1^{-1}\dot{\eta}_{T,\lambda}][\zeta_{T\lambda} - \mu_2^{-1}\dot{\eta}_{T,\lambda}]\}dt.$$

Applying the Parseval equality we get for the Fourier transforms $\tilde{\eta}_{T\lambda}, \tilde{\zeta}_{T\lambda}$ and $\tilde{\dot{\eta}}_{T,\lambda}$ of $\eta_{T\lambda}, \zeta_{T\lambda}$ and $\dot{\eta}_{T,\lambda}$ the relation

$$A_T = -\frac{1}{2\pi}\int_{-\infty}^{+\infty} \mathrm{Re}\,\{K(i\omega) \;-\; \varepsilon|K(i\omega)|^2$$
$$-\; \tau[K(i\omega - \lambda) + \mu_1^{-1}(i\omega - \lambda)]^*[K(i\omega - \lambda) + \mu_2^{-1}(i\omega - \lambda)]\}|\tilde{\eta}_{T\lambda}(i\omega)|$$

where we have exployed the fact that

$$\tilde{\dot{\eta}}_{T,\lambda}(i\omega) = (i\omega - \lambda)\tilde{\eta}_{T\lambda}(i\omega)$$

and

$$\tilde{\zeta}_{T\lambda}(i\omega) = -K(i\omega - \lambda)\tilde{\eta}_{T\lambda}(i\omega) \quad \text{for} \quad \omega \in \mathbf{R}.$$

It follows from (8.4.14) that for all $T \geq 0$

$$A_T \leq 0 \tag{8.4.15}$$

Using the equality (8.2.7) we can split the integral A_T as follows:

$$A_T = \sum_{i=1}^5 A_{Ti},$$

where

$$A_{T1} = \int_0^T [\eta\dot{\sigma} + \varepsilon\dot{\sigma}^2]e^{2\lambda t}\,dt,$$

$$A_{T2} = \int_T^{+\infty} (\varepsilon + \tau)\zeta_{T\lambda}^2\,dt,$$

$$A_{T3} = \tau\int_0^T (\dot{\sigma} - \mu_1^{-1}\dot{\eta})(\dot{\sigma} - \mu_2^{-1}\dot{\eta})e^{2\lambda t}\,dt,$$

$$A_{T4} = \int_0^T \{-\varepsilon\sigma^0 - 2(\varepsilon + \tau)\sigma^0\dot{\sigma} + (\eta + \tau)(\sigma^0)^2 - \tau(\mu_1^{-1} + \mu_2^{-1})\dot{\eta}\sigma^0\}\,e^{2\lambda t}\,dt,$$

$$A_{T5} = \int_0^1 \{-\eta\zeta + \eta_\mu\zeta - \tau\mu_1^{-1}\mu_2^{-1}(\dot{\eta}_\mu^2 - \dot{\eta}^2) + \tau(\mu_1^{-1} + \mu_2^{-1})(\zeta\eta_\mu - \zeta\eta)\}\,e^{2\lambda t}\,dt.$$

184

As $\mu_1 \leq \varphi'(\sigma)/\sigma \leq \mu_2$ a.e. on $\mathbf{R}$ it is obvious that $A_{T3} \geq 0$. It is also clear that $|A_{Tt}|$ is bounded by a constant which does not depend on T. Because of assumptions (A1) and (A2) the function $\sigma^0(t) \cdot e^{\kappa_0 t}$ is L^2. Let us first suppose that $\lambda < \kappa_0/2$. Then $|A_{T4}|$ is bounded independently of T and we have

$$A_{T1} \leq c_1, \tag{8.4.16}$$

where c_1 does not depend on T. This inequality is the same as (8.4.8) and we can proceed exactly as in the proof of Theorem 8.4.1 up to the inequality (8.4.11) which is true here for all $T \in \mathbf{Z}$. Thus, we have $\vartheta^* - (N+1)\Delta \leq \sigma(t) \leq \vartheta^* + (N+1)\Delta$. So in case $\lambda < \kappa_0/2$ the theorem is proved.

In order to prove the theorem in the general case $\lambda < \kappa_0$ we note that $\dot{\eta}$ is bounded and it is possible to obtain the estimate

$$\int\limits_0^T [\eta\dot{\sigma} + \varepsilon\dot{\sigma}^2]e^{2\lambda t}\, dt \leq \int\limits_0^T [\varepsilon_2\eta^2 + \varepsilon_1\dot{\sigma}^2 + \varepsilon_4(\sigma^0)^2]e^{2\lambda t}\, dt + c_2, \tag{8.4.17}$$

where ε_1 and ε_2 are those from the proof of of Theorem 8.4.1 and ε_4 depends on ε_1, ε_2 and on the parameters of the frequency-domain condition (8.4.14). From (8.4.17) it is easy to establish the analogue of (8.4.13) in the form

$$\int\limits_0^T [\eta\dot{\sigma} + \overline{\varepsilon}\dot{\sigma}^2]e^{2\lambda t} - \int\limits_0^T \varepsilon_2\eta^2 e^{2\lambda t}\, dt \leq c_3,$$

where $\overline{\varepsilon} = \varepsilon - \varepsilon_1$ and c_3 does not depend on T. Now it remains to repeat the argumentation in the proof of Theorem 8.4.1 starting from (8.4.13). ∎

With the help of Theorems 8.2.1 and 8.2.2 the following two assertions are easily obtained.

Theorem 8.4.3 *Suppose φ is piece-wise differentiable and there exist numbers $\varepsilon > 0$, $\tau \geq 0$, $\mu_1 \leq \inf\limits_{\sigma \in [0,\Delta]} \varphi'(\sigma)$, $\mu_2 \geq \sup\limits_{\sigma \in [0,\Delta]} \varphi'(\sigma)$ and $\delta > 0$ such that for*

$$S(\omega) := \mathrm{Re}\left\{ K(i\omega) - \tau[K(i\omega) + \mu_1^{-1}i\omega]^*[K(i\omega) + \mu_2^{-1}i\omega]\right\} - \varepsilon|K(i\omega)|^2$$

we have

$$S(\omega) \geq \delta \qquad \text{for all} \qquad \omega \in \mathbf{R}.$$

Then for an arbitrary solution σ of (8.1.1) there holds

$$\begin{aligned} \dot{\sigma}(t) &\to 0 &\quad \text{as} \quad & t \to +\infty \\ \sigma(t) &\to \sigma_1 &\quad \text{as} \quad & t \to +\infty. \end{aligned} \tag{8.4.18}$$

where σ_1 is a zero of φ.

Theorem 8.4.4 *Suppose φ is C^1 and $K(0) = -\varrho + \int\limits_0^{+\infty} \gamma(t)\, dt \neq 0$. Suppose also that the assumptions of Theorem (8.4.2) are fulfilled and there exist numbers $\varepsilon > 0$, $\tau \geq 0$, $\mu_1 \leq \inf\limits_{\sigma \in [0,\Delta]} \varphi'(\sigma)$, $\mu_2 \geq \sup\limits_{\sigma \in [0,\Delta]} \varphi'(\sigma)$ such that with the function S defined in the statement of Theorem 8.4.3 the inequality $S(\omega) \geq 0$ holds for all $\omega \in \mathbf{R}$. Then any solution of (8.1.1) satisfies (8.4.18).*

In the remainder of this section we state two theorems which in contrast to the Theorems 8.4.3 and 8.4.4 include a certain algebraic assumption that guarantees the relations (8.4.18).

Theorem 8.4.5 *Consider* (8.1.1) *with* $h = 0$ *and suppose that* α *is* C^1 *and that* $t \mapsto \dot{\alpha}(t)e^{\kappa_3 t}$ *is* L^2, *where* $\kappa_3 > 0$. *Suppose also that* γ *is piece-wise* C^1, $t \mapsto \dot{\gamma}(t)e^{\kappa_4 t}$ *is* L^2 *for some* $\kappa_4 > 0$ *and the conditions of Theorem 8.4.1 are fulfilled with the positive number* $\lambda < \min_{i=1,\dots,4} \kappa_i$. *If* $\tau \mu_1^{-1} \mu_2^{-1} = 0$ *and* $\varrho(\mu_1^{-1} + \mu_2^{-1}) \le 0$, *where* τ, μ_1 *and* μ_2 *are the parameters of the frequency condition* (8.4.14), *then for an arbitrary solution* σ *of* (8.1.1) *holds* $\dot{\sigma}(t) \to 0$ *as* $t \to +\infty$.

Theorem 8.4.6 *Suppose that the conditions of Theorem 8.4.5 are satisfied. If, instead of inequality* (8.4.14), *the inequality*

$$\operatorname{Re}\left\{ K(s) - \varepsilon|K(s)|^2 - \tau[K(s) + \mu_1^{-1}s]^*[K(s) + \mu_2^{-1}s] \right\} \ge \delta > 0$$

is fulfilled for all $s = i\omega - \lambda$, $\omega \in \mathbf{R}$, *then any solution* σ *of* (8.1.1) *satisfies* (8.4.18).

8.5 Phase Synchronization Systems

In this section we investigate several concrete systems from the phase synchronization theory described by equations with delay or by integro-differential equations.

Example 8.5.1 Let us consider the equation describing the dynamics of the inertial system of a phase automatic frequency control with an integrating filter and a sinusoidal characteristic of the phase detector ([13]):

$$\ddot{\sigma} + \mu\dot{\sigma}(t) + \sin\sigma(t - h) = \beta, \tag{8.5.1}$$

where $\mu > 0$, $h > 0$ and $\beta \in [-1, +1]$ are parameters. Note that (8.5.1) coincides with the sunflower equation in biology [141, 43]. The initial condition for (8.5.1) has the form

$$\sigma(t) = \sigma_0(t) \qquad \text{for} \qquad t \in [-h, 0],$$

where $\sigma_0 \in C^1((-h, 0), \mathbf{R})$ is given. For $t > 0$ equation (8.5.1) is easily reduced to an equivalent equation of the form

$$\dot{\sigma}(t) = \alpha(t) - \int_0^t \gamma(t - \tau)\varphi(\sigma(\tau))\,d\tau, \tag{8.5.2}$$

where

$$\varphi(\sigma) = \sin\sigma - \gamma, \qquad \gamma(t) = \begin{cases} 0 & \text{for} & t < h \\ e^{-\mu(t-h)} & \text{for} & t \ge h \end{cases}$$

and the continuous function $\alpha(\cdot)$ satisfies the estimate

$$|\alpha(t)| \le Be^{-\mu t} \qquad \text{for} \qquad t > 0,$$

where $B = B(\sigma_0(\cdot))$ is a constant. We apply Theorem 8.3.1 with condition (ii) to equation (8.5.2). In Figure 8.5.1 we construct the boundaries of the stability domains for it (and, consequently, for equation (8.5.1) also) in the space of the parameters $\{\mu, \beta\}$ ("stability" means here that the limit relations (8.3.2) and (8.3.3) are fulfilled for the solutions).

The stability domains lie below the corresponding boundaries. The thick line in Figure 8.5.1 shows the actual boundary of the stability domain for the case $h = 0$ obtained by numerical methods [138]. The stability domain computed in [138] for the case $h = 0$ by means of the Bakaev-Guzh procedure is situated below the dotted line in Figure 8.5.1. This domain is narrower than the domain obtained with the help of Theorem 8.3.1. In [13] stability of stationary solutions of equation (8.5.1) with $h > 0$ has been studied using Krasovskij functionals. The sufficient stability condition obtained there is more restrictive than the assumptions of Theorem 8.3.1 (with $\beta = 0$).

186

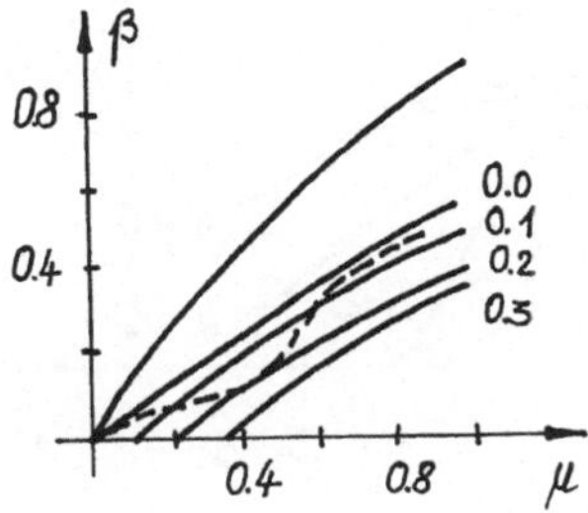

Figure 8.5.1

Example 8.5.2 Let us consider a system of phase automatic frequency control with proportional integrating filters and delay in the feedback loop [138, 13, 35]. The equations of such a system have the form

$$\begin{aligned}
\dot{x}(t) &= -x(t) - \Omega_0(1-m)\varphi(\sigma(t-h)) \\
\dot{\sigma}(t) &= F_0 x(t) - F_0\Omega_0 m\varphi(\sigma(t-h)),
\end{aligned} \qquad (8.5.3)$$

where $h \geq 0$, F_0, $\Omega_0 > 0$, $m \in (0,1)$ are parameters and $\varphi(\sigma) = \sin\sigma - \gamma_0$, $\gamma_0 \in (0,1)$. The values of the function $\sigma(t)$ for $t \in [-h,0]$ and $x(0)$ are assumed to be known. We introduce the notation $k_0^2 := F_0\Omega_0$. System (8.5.3) easily reduces, for $t \geq 0$, to an equation of the form

$$\dot{\sigma}(t) = \alpha(t) + \varrho\varphi(\sigma(t-h)) - \int_0^t \gamma(t-\tau)\varphi(\sigma(\tau))\,d\tau,$$

where

$$\gamma(t) = \left\{ \begin{array}{ll} 0 & \text{if } t < h, \\ k_0^2(1-m)e^{-(t-h)} & \text{if } t \geq h, \end{array} \right.$$

$$\alpha(t) = \left\{ \begin{array}{lll} F_0 x(0)e^{-t} - k_0^2(1-m)e^{-t}\displaystyle\int_{-h}^{t-h} e^{\tau+h}\varphi(\sigma(\tau))\,d\tau & \text{if} & t < h, \\[3ex] F_0 x(0)e^{-t} - k_0^2(1-m)e^{-t}\displaystyle\int_{-h}^{0} e^{\tau+h}\varphi(\sigma(\tau))\,d\tau & \text{if} & t \geq h. \end{array} \right.$$

In [35] system (8.5.3) was studied by the approximate method of harmonic balance. As a result, for various values of h with $m = 0.2$ there were determined the region of stability (lock-in ranges) of system (8.5.3) in the space of parameters $\{\gamma_0, k_0^2\}$. The boundaries of these regions are marked by broken lines in Figure 8.5.2. .

The regions themselves lie below the corresponding boundaries. The continuous lines in this figure indicate the boundaries of the regions of stability obtained in the present book by Theorems 8.4.2 and 8.4.3 (the regions of stability are situated below their boundaries). The verification of the conditions of Theorem 8.4.3 for each value of γ_0 was carried out as follows. The values of the parameters μ_1, μ_2 were chosen to be 1. The parameters λ and τ were varied from 0 to 1/2 and 1, respectively, with a step of 0.01. To study the reduced equation

$$\ddot{\vartheta} + 2\sqrt{\lambda\varepsilon}\dot{\vartheta} + \varphi(\vartheta) = 0$$

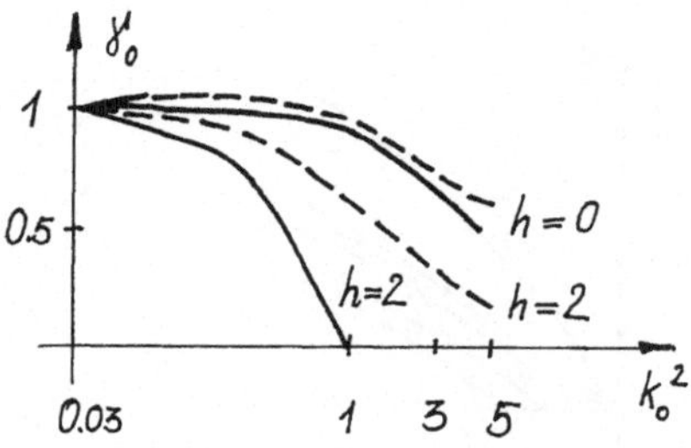

Figure 8.5.2

we used the region of stability of equation $\ddot{\vartheta} + a\dot{\vartheta} + \varphi(\vartheta) = 0$ given in [33] in the plane of the parameters $\{a, \gamma_0\}$. For the considered value of γ_0 the value of the parameter a was "fastened" onto the boundary of the region of stability of $\ddot{\vartheta} + a\dot{\vartheta} + \varphi(\vartheta) = 0$. The choice of the parameter ε for each value of λ was uniquely determined by the formula $\varepsilon = a^2/4\lambda$.

Testing of the frequency condition

$$\xi\mu_1^{-1}\mu_2^{-1}(\omega^2 + \lambda^2) - (\varepsilon + \xi)|K(i\omega - \lambda)|^2 +$$
$$+\mathrm{Re}\left\{[1 + \tau(\mu_2^{-1} - \mu_1^{-1}) + (\lambda + i\omega)]K(i\omega - \lambda)\right\} \geq 0,$$

depending, as is clear from what is written above, on two parameters, is effected for each pair $\{\lambda, \tau\}$ by a numerical calculation on the ODRA-1204 computer. First, the interval $[0, \omega_0(\lambda, \xi)]$ on which the frequency-domain inequality could fail and the variational step of the parameter ω guaranteeing a high degree of precision were analytically determined. The frequency-domain inequality $S(\omega) \geq \delta_0$ was also tested on the computer with $\mu_1 = \mu_2 = 1$. The parameters ε_0, ξ_0, δ_0 were varied from 0 to 1 with step 0.01.

Chapter 9

Cycle Slipping in Phase-Controlled Systems

In synchronization theory one is interested in conditions which ensure, that the phase error $\sigma(t)$ increases at least by $k\Delta$, where k is an integer and Δ is the period of the nonlinearity, when t tends to infinity. This property, which is called in communication systems "cycle slipping" [146], and similar problems are considered in [142, 100, 107, 143, 145, 137, 158]. We start our analysis with the ODE case.

9.1 Frequency-Domain Conditions for Cycle Slipping in ODE's

Let us consider the second canonical form of a phase-controlled system

$$\begin{aligned} \dot{z} &= Az + b\varphi(\sigma), \\ \dot{\sigma} &= c^*z + \varrho\varphi(\sigma), \end{aligned} \tag{9.1.1}$$

where A is an $n \times n$ matrix, b and c are n-vectors, ϱ is a number. The Δ-periodic function $\varphi : \mathbf{R} \to \mathbf{R}$ is assumed to be differentiable and has exactly two zeros on $[0, \Delta)$. We suppose also that $[\varphi'(\sigma)]^2 + \varphi(\sigma)^2 \neq 0$ on $\mathbf{R}$.

Let us define the transfer function $K(s) = c^*(A - sI)^{-1}b - \varrho$ and suppose that K is non-degenerate.

Definition 9.1.1 We say that (z, σ) *is a solution of* (9.1.1) *with k slipped cycles* if there exists a $T > 0$ such that

$$|\sigma(T) - \sigma(0)| = k\Delta \quad \text{and} \quad |\sigma(t) - \sigma(0)| < (k+1)\Delta$$

for all $t \geq 0$.

In order to formulate the following theorems, taken from [158], we introduce some notations. Let B denote a constant $n \times n$ matrix and G denote a constant symmetric $(n + 1) \times (n + 1)$ matrix. Let us also introduce the functions

$$\begin{aligned} d(s) &= \det(sI - B), & Q(s) &= d(s)(sI - B)^{-1}b, \\ F(w) &= -w^*Gw & (w &\in \mathbf{R}^{n+1}) \end{aligned} \tag{9.1.2}$$

and

$$\Xi(s) = \left[\begin{array}{c} Q(-s^*) \\ d(-s^*) \end{array} \right]^* G \left[\begin{array}{c} Q(s) \\ d(s) \end{array} \right]. \tag{9.1.3}$$

We assume that $F(0, \xi) =: -\Gamma\xi^2$ $(\xi \in \mathbf{R})$.

Theorem 9.1.1 *Suppose that the matrix A is Hurwitzian and that for certain numbers $\varepsilon > 0$, $\delta > 0$ and a positive integer k the following relations are true:*

(i) $\operatorname{Re} K(i\omega) - \varepsilon |K(i\omega)|^2 \geq \delta$ *for all $\omega \in \mathbf{R}$;*

(ii) $4\varepsilon\delta \geq [\nu_j(k, z(0)^* H z(0))]^2$ *for $j = 1, 2$, where $\nu_j : \mathbf{N} \times \mathbf{R} \to \mathbf{R}$ are defined by*

$$\nu_j(m, \xi) = \left[\int_0^\Delta \varphi(\sigma)\, d\sigma + (-1)^j m^{-1}\xi \right] \bigg/ \int_0^\Delta |\varphi(\sigma)|\, d\sigma. \tag{9.1.4}$$

Then for any solution (z, σ) of (9.1.1) starting in $z(0)$, $\sigma(0)$ with $\sigma(0) \in [0, 2\pi)$ it follows that

$$|\sigma(t) - \sigma(0)| < k\Delta$$

for all $t \geq 0$.

The symmetric $n \times n$ matrix $H = H^*$ mentioned in (ii) may be determined as follows:
Step 1 *Define the polynomial Ξ by (9.1.3) with $B = A$ and*

$$G = \begin{bmatrix} -\varepsilon cc^* & -(\tfrac{1}{2} + \varepsilon\varrho)c \\ -(\tfrac{1}{2} + \varepsilon\varrho)c^* & -\varrho - \varepsilon\varrho^2 - \delta \end{bmatrix}.$$

Step 2 *Find a polynomial Ψ of degree n which satisfies the factorization equation*

$$\Xi(s) = \Psi(-s)\Psi(s).$$

Such a polynomial exists and is unique.
Step 3 *Find an n-vector h and a number τ such that*

$$\Psi(s) = h^* Q(s) + \tau d(s),$$

which is equivalent to the linear problem

$$\tau = \sqrt{\Gamma}, \qquad h^* q_i = \Psi_i - \tau d_i \qquad for\, i = 0, 1, \ldots, n-1,$$

where Ψ_i, d_i resp. q_i are defined by the representations

$$\begin{aligned} \Psi(s) &= \sqrt{\Gamma} s^n + \Psi_{n-1} s^{n-1} + \cdots + \Psi_0, \\ d(s) &= s^n + d_{n-1} s^{n-1} + \cdots + d_0 \qquad resp. \\ Q(s) &= q_{n-1} s^{n-1} + \cdots + q_0. \end{aligned}$$

The vector h and the number τ can be discovered in a unique way.
Step 4 *Find a matrix $H = H^*$ satisfying $HA + A^* H = D$, where D is defined by $z^* D z := -F(z, 0) - |h^* z|^2$. This matrix H exists and is unique.*

Remark 9.1.1 The algorithm described in the statement of Theorem 9.1.1 is the well-known [49] procedure for constructing a matrix H which satisfies the inequality

$$2\operatorname{Re} z^* H(Az + b\xi) - F(z, \xi) \leq 0$$

for all $(z, \xi) \in \mathbf{C}^n \times \mathbf{C}$, where for the Hermitian form F, it is supposed that

$$F[(i\omega I - A)^{-1} b\xi, \xi] \geq 0$$

for all $\xi \in \mathbf{C}$ and $\omega \in \mathbf{R}$.

An explicit result for defining H we will give in the case $n = 2$. For $n > 2$ numerical calculations are needed.

The following lemma is useful in the proof of Theorem 9.1.1.

Lemma 9.1.1 *Let $\varphi : \mathbf{R} \to \mathbf{R}$ be a continuous Δ-periodic function, w, $\sigma : \mathbf{R}_+ \to \mathbf{R}$ functions of C^1 and $\varepsilon > 0$, $\delta > 0$ numbers such that the following conditions are satisfied:*

(i) $\dot{w}(t) + \varphi(\sigma(t))\dot{\sigma}(t) + \varepsilon\dot{\sigma}(t)^2 + \delta\varphi(\sigma(t))^2 \leq 0$ *on* $\mathbf{R}_+$;

(ii) $4\varepsilon\delta > \nu_j^2(k, w(0))$ *for* $j = 1, 2$, *where* $\nu_j(k, w(0))$ *is defined by* (9.1.4);

(iii) $w(t) \geq 0$ *on* $\mathbf{R}_+$.

Then $|\sigma(t) - \sigma(0)| < k\Delta$ *for all* $t \geq 0$.

Proof Because of (ii) we can choose an $\varepsilon_0 > 0$ with

$$4\varepsilon\delta \geq \nu_j^2(k, w(0) + \varepsilon_0) \qquad (j = 1, 2). \tag{9.1.5}$$

Let us introduce the functions $F_j : \mathbf{R} \to \mathbf{R}$ and $v_j : \mathbf{R}_+ \to \mathbf{R}$ by

$$F_j(\sigma) = \varphi(\sigma) - \nu_j(k, w(0) + \varepsilon_0)|\varphi(\sigma)|$$

and

$$v_j(t) = w(t) + \int_{\sigma(0)}^{\sigma(t)} F_j(\sigma)\, d\sigma \qquad (j = 1, 2).$$

A direct computation shows that

$$\begin{aligned}
\dot{v}_j(t) &= \dot{w}(t) + F_j(\sigma(t))\dot{\sigma}(t) \\
&= \dot{w}(t) + \varphi(\sigma(t))\dot{\sigma}(t) - \nu_j(k, w(0) + \varepsilon_0)|\varphi(\sigma(t))|\dot{\sigma}(t).
\end{aligned}$$

From this, the condition (i) and (9.1.5) it follows that

$$\dot{v}_j(t) \leq -\left[\delta - \frac{1}{4\varepsilon}\nu_j^2(k, w(0) + \varepsilon_0)\right]\varphi(\sigma(t))^2 \leq 0 \tag{9.1.6}$$

for $t \geq 0$.

Suppose now that there exists a $T > 0$ such that $\sigma(T) = \sigma(0) + k\Delta$. It follows that

$$\begin{aligned}
v_j(T) - v_j(0) &= w(T) - w(0) + \int_{\sigma(0)}^{\sigma(0)+k\Delta} F_j(\sigma)\, d\sigma \\
&= w(T) - w(0) + \int_{0}^{k\Delta} F_j(\sigma)\, d\sigma \\
&= w(T) - w(0) + w(0) + \varepsilon_0.
\end{aligned}$$

Using condition (iii), we see that $v_j(T) - v_j(0) > 0$ which contradicts (9.1.6). Thus we have shown that $\sigma(t) - \sigma(0) < k\Delta$ for $t \geq 0$.

In a similar way one proves $\sigma(t) - \sigma(0) > -k\Delta$ for $t \geq 0$. $\blacksquare$

Proof of Theorem 9.1.1 Consider the Hermitian form

$$F(z, \xi) = -\left[\operatorname{Re}\xi^*(c^*z + \varrho\xi) + \varepsilon|c^*z + \varrho\xi|^2 + \delta|\xi|^2\right]$$

on $\mathbf{C}^n \times \mathbf{C}$.

From the frequency-domain condition (i) and the fact that H is Hurwitzian it follows by Theorem 1.4.1, p. 9, that there exists an $n \times n$ matrix $H = H^* > 0$ such that

$$2z^*H(Az + b\xi) + (c^*z + \varrho\xi)\xi + \varepsilon|c^*z + \varrho\xi|^2 + \delta\xi^2 \le 0 \qquad (9.1.7)$$

for all $z \in \mathbf{R}^n$, $\xi \in \mathbf{R}$. Suppose now $z(\cdot)$, $\sigma(\cdot)$ is a non-constant solution of (9.1.1) and define the function $w : \mathbf{R}_+ \to \mathbf{R}_+$ through $w(t) = z(t)^*Hz(t)$. From (9.1.7) we get

$$\dot{w}(t) + \varphi(\sigma(t))\dot{\sigma}(t) + \varepsilon\dot{\sigma}(t)^2 + \delta\varphi(\sigma(t))^2 \le 0$$

for all $t \ge 0$.

Thus conditions (i) and (iii) of Lemma 9.1.1 are satisfied with respect to these w, σ and φ. The remaining condition (ii) of this lemma coincides with condition (ii) of the theorem. By Lemma 9.1.1 it follows the assertions of the theorem. $\blacksquare$

In the next two theorems we use the reduction method in order to extend the results concerning cycle slipping in equation $\ddot{\vartheta} + a\dot{\vartheta} + \varphi(\vartheta) = 0$ to system (9.1.1).

Theorem 9.1.2 *Suppose that there exist numbers $\varepsilon > 0$ and $\lambda > 0$ such that the following conditions are fulfilled:*

(i) $\operatorname{Re} K(i\omega - \lambda) - \varepsilon|K(i\omega - \lambda)|^2 \ge 0$ *for all $\omega \in \mathbf{R}$;*

(ii) $A + \lambda I$ *is a Hurwitz matrix;*

(iii) *the equation*

$$\ddot{\vartheta} + 2\sqrt{\lambda\varepsilon}\dot{\vartheta} + \varphi(\vartheta) = 0 \qquad (9.1.8)$$

   is Lagrange stable;

(iv) *for any solution ϑ of (9.1.8) starting in the points $(\vartheta_0, \pm\dot{\vartheta}_0)$ the inequality*

$$|\vartheta(t) - \vartheta_0| < k\Delta \qquad (9.1.9)$$

   is true for all $t \ge 0$.

Then any solution $z(\cdot)$, $\sigma(\cdot)$ of (9.1.1) with $\sigma(0) = \vartheta_0$ and $z(0)^*Hz(0) < \frac{1}{2}\dot{\vartheta}_0^2$ satisfies the inequality

$$|\sigma(t) - \sigma(0)| < k\Delta$$

for all $t \ge 0$.

The $n \times n$ matrix $H = H^*$ is determined by the procedure described in the statement of Theorem 9.1.1 with

$$G = \begin{bmatrix} -\varepsilon cc^* & -(\frac{1}{2} + \varrho\varepsilon)c \\ -(\frac{1}{2} + \varrho\varepsilon)c^* & -\varrho - \varepsilon\varrho^2 \end{bmatrix} \quad and \quad B = A + \lambda I. \qquad (9.1.10)$$

The next lemma will be needed in the proof of Theorem 9.1.2.

Lemma 9.1.2 *Let $\varphi : \mathbf{R} \to \mathbf{R}$ be a smooth Δ-periodic function having exactly two zeros on $[0, \Delta)$ and satisfying $\varphi(\vartheta)^2 + \varphi'(\vartheta)^2 \ne 0$ on $\mathbf{R}$ and let w, $\sigma : \mathbf{R}_+ \to \mathbf{R}$ be C^1-functions. Suppose that for some parameters $\lambda > 0$ and $\varepsilon > 0$ the following conditions are satisfied:*

(i) $w(t) \ge 0$ *for all $t \ge 0$;*

192

(ii) $\dot{w}(t) + 2\lambda w(t) + \varphi(\sigma(t))\dot{\sigma}(t) + \varepsilon\dot{\sigma}(t)^2 \leq 0$ *for all* $t \geq 0;$

(iii) *equation* (9.1.8) *is Lagrange stable;*

(iv) *for the solutions* ϑ *of* (9.1.8) *with* $\vartheta(0) = \sigma(0)$, *and* $\pm\dot{\vartheta}(0)$ *the inequality* (9.1.9) *is true;*

(v) $w(0) < \frac{1}{2}\dot{\vartheta}(0)^2.$

Then $|\sigma(t) - \sigma(0)| < k\Delta$ *for all* $t \geq 0.$

Proof Let us convert equation (9.1.8) into the system

$$\begin{aligned}
\dot{\vartheta} &= \eta, \\
\dot{\eta} &= -2\sqrt{\lambda\varepsilon}\eta - \varphi(\vartheta)
\end{aligned} \tag{9.1.11}$$

and consider the first order equation

$$\Phi(\vartheta)\frac{d\Phi}{d\vartheta} + 2\sqrt{\lambda\varepsilon}\Phi + \varphi(\vartheta) = 0. \tag{9.1.12}$$

The phase portrait of (9.1.11) under condition (iii) is shown in Figure 2.2.5.

Consider the two solutions Φ_1 and Φ_2 of (9.1.12) satisfying

$$\Phi_1(\sigma(0)) = -|\dot{\vartheta}(0)| \qquad \text{resp.} \qquad \Phi_2(\sigma(0)) = |\dot{\vartheta}(0)|.$$

By assumption (iv) there exist values $\sigma_1,\ \sigma_2 \in (0, k\Delta)$ with

$$\Phi_1(\sigma(0) - \sigma_1) = 0 \qquad \text{and} \qquad \Phi_2(\sigma(0) + \sigma_2) = 0. \tag{9.1.13}$$

For the functions $v_j : \mathbf{R}_+ \to \mathbf{R}$ defined by

$$v_j(t) = w(t) - \frac{1}{2}\Phi_j(\sigma(t))^2 \qquad (j = 1, 2)$$

a direct computation shows that

$$\dot{v}_j(t) = \dot{w}(t) - \Phi_j(\sigma(t))\Phi_j'(\sigma(t))\dot{\sigma}(t)$$

on $\mathbf{R}_+$. Then repeating the arguments of Theorem 5.2.1, p. 92, we obtain, (omitting the variables t and $\sigma(t)$),

$$\begin{aligned}
\dot{v}_j + 2\lambda v_j &= \dot{w} + 2\lambda w - \Phi_j\Phi_j'\dot{\sigma} - \lambda\Phi_j^2 \\
&= [\dot{w} + 2\lambda w + \varphi\dot{\sigma} - \varepsilon\dot{\sigma}^2] - [\lambda\Phi_j^2 + \Phi_j\Phi_j'\dot{\sigma} + \varphi\dot{\sigma} + \varepsilon\dot{\sigma}^2] \\
&\leq -\lambda\Phi_j^2 - \Phi_j\Phi_j'\dot{\sigma} - \varphi\dot{\sigma} - \varepsilon\dot{\sigma}^2 \\
&\leq -\lambda\Phi_j^2 + \frac{1}{4\varepsilon}[\varphi + \Phi_j\Phi_j'] \\
&= 0.
\end{aligned}$$

From Lemma 3.1.1, p. 48, it follows that for $j = 1, 2$

$$v_j(t) \leq v_j(0)$$

on $\mathbf{R}_+$. Using condition (v) of the present lemma we see that

$$v_j(0) = w(0) - \frac{1}{2}\Phi_j(\sigma(0))^2 < \frac{1}{2}\dot{\vartheta}(0)^2 - \frac{1}{2}\dot{\vartheta}(0)^2 = 0.$$

Thus

$$v_j(t) < 0 \qquad\qquad (9.1.14)$$

for all $t \geq 0$ and $j = 1, 2$.

Next we show that

$$-\sigma_1 < \sigma(t) - \sigma(0) < \sigma_2 \qquad\qquad (9.1.15)$$

for all $t \geq 0$. Suppose to the contrary that there exists a time $T > 0$ such that

$$\sigma(T) = \sigma(0) + \sigma_2 \qquad \text{or} \qquad \sigma(T) = \sigma(0) - \sigma_1.$$

Then according to (9.1.13)

$$\Phi_1(\sigma(T))\Phi_2(\sigma(T)) = 0.$$

It follows then that either $v_1(T) = w(T)$ or $v_2(T) = w(T)$. Because of (9.1.14), the last equations imply that $w(T) < 0$, a contradiction to condition (i). Thus (9.1.15) is true and Lemma 9.1.2 is proved. ∎

Proof of Theorem 9.1.2 Define the Hermitian form $F : \mathbf{C}^n \times \mathbf{C} \to \mathbf{R}$ by

$$F(z,\xi) = -\left[\operatorname{Re}\xi^*(c^*z + \varrho\xi) + \varepsilon|c^*z + \varrho\xi|^2\right].$$

Then from the frequency-domain condition (i) and the fact that $A + \lambda I$ is a Hurwitz matrix it follows by Theorem 1.4.1, p. 9, that there exists a matrix $H = H^* > 0$ such that

$$2z^*H[(A + \lambda I)z + b\xi] + (c^*z + \varrho\xi)\xi \leq -\varepsilon|c^*z + \varrho\xi|^2 \qquad\qquad (9.1.16)$$

for all $(z,\xi) \in \mathbf{R}^n \times \mathbf{R}$. This matrix may be found by means of the procedure given in Theorem 9.1.1, defining the matrices G and B by (9.1.10).

Let $(z(\cdot),\ \sigma(\cdot)$ be a solution of (9.1.1) with $\sigma(0) = \vartheta(0)$ and $z(0)^*Hz(0) < \frac{1}{2}\dot\vartheta(0)^2$. Taking H from (9.1.16) we introduce the function $w : \mathbf{R}_+ \to \mathbf{R}$ by $w(t) = z(t)^*Hz(t)$. From (9.1.16) it follows that

$$\dot w(t) + 2\lambda w(t) + \varphi(\sigma(t))\dot\sigma(t) + \varepsilon\dot\sigma(t)^2 \leq 0$$

for all $t \geq 0$. Thus conditions (i) and (iii) of Lemma 9.1.2 are fulfilled. Conditions (iii) and (iv) of Lemma 9.1.2 coincide with conditions (iii) and (iv) of Theorem 9.1.2 respectively. Because of the choice of the considered solution, condition (v) of Lemma 9.1.2 also holds. Thus, from this lemma it follows the assertion of the theorem. ∎

On the basis of Theorems 9.1.1 and 9.1.2 one obtains an estimate from above of the number of cycles which slips a solution of (9.1.1), by changing t to infinity. The next theorem deals with the estimate from below for this number of cycles.

Theorem 9.1.3 *Let be $\varrho = 0$ and $c^*b < 0$. Suppose that for certain $\varepsilon > 0$ and $\lambda > 0$ the following conditions are fulfilled:*

(i) *$\operatorname{Re} K(i\omega - \lambda) + \varepsilon|K(i\omega - \lambda)|^2 \leq 0$ for all $\omega \in \mathbf{R}$;*

(ii) *the matrix $A + \lambda I$ has one positive and $n - 1$ negative eigenvalues with negative real parts;*

(iii) *for a solution ϑ of*

$$\ddot\vartheta + \lambda\sqrt{-[c^*b]^{-1}}\dot\vartheta + \varphi(\vartheta) = 0$$

starting in $\vartheta(0)$, $\dot\vartheta(0) \neq 0$ there exists a time t_1 such that $|\vartheta(t_1) - \vartheta(0)| > k\Delta$, where k id an integer.

194

Then for any solution $z(\cdot), \sigma(\cdot)$ of (9.1.1) with initial data satisfying

$$\sigma(0) = \vartheta(0) \qquad and \qquad z(0)^* H z(0) < -\frac{1}{2}\dot{\vartheta}(0)^2$$

there exists a time t_2 such that $|\sigma(t) - \sigma(0)| > k\Delta$. The positive definite $n \times n$ matrix H can be found as in Theorem 9.1.1 putting $B = A + \lambda I$ and $G = \begin{bmatrix} -\varepsilon cc^* & \frac{1}{2}c \\ \frac{1}{2}c^* & 0 \end{bmatrix}$.

The proof of this theorem is based on the next auxiliary proposition.

Lemma 9.1.3 *Suppose that w, σ, $\psi : \mathbf{R}_+ \to \mathbf{R}$ are C^1 functions, $\varphi : \mathbf{R} \to \mathbf{R}$ is Δ-periodic and continuous and $\lambda > 0$, $\nu > 0$ are certain parameters satisfying the following conditions:*

(i) $w(t) + \nu\sigma(t)^2 \geq 0$ *for* $t \geq 0$;

(ii) $\psi(t) < \varphi(\sigma(t))$ *for all* $t \geq 0$ *with* $\dot{\sigma}(t) > 0$ *and* $\psi(t) > \varphi(\sigma(t))$ *for all* $t \geq 0$ *with* $\dot{\sigma}(t) < 0$;

(iii) *the scalar function*

$$t \mapsto w(t) + \int_0^t [2\lambda w(\tau) - \psi(\tau)\dot{\sigma}(\tau)]d\tau$$

is non-increasing on $\mathbf{R}_+$;

(iv) *for a solution ϑ of*

$$\ddot{\vartheta} + \lambda\sqrt{2\nu}\dot{\vartheta} + \varphi(\vartheta) = 0 \tag{9.1.17}$$

having the initial data $\vartheta(0)$, $\dot{\vartheta}(0) \neq 0$ there exists a time $t_1 > 0$ such that $|\vartheta(t_1) - \vartheta(0)| > k\Delta$;

(v) $\sigma(0) = \vartheta(0)$ *and* $2w(0) + \dot{\vartheta}(0)^2 < 0$.

Then there exists a time $t_2 > 0$ such that $|\sigma(t_2) - \sigma(0)| > k\Delta$.

Proof Let us consider the first-order equation

$$\Phi(\vartheta)\frac{d\Phi}{d\vartheta} + \lambda\sqrt{2\nu}\Phi(\vartheta) + \varphi(\vartheta) = 0, \tag{9.1.18}$$

corresponding to (9.1.17). Suppose $\Phi_0(\cdot)$ is the solution of (9.1.18) satisfying $\Phi_0(\vartheta(0)) = \dot{\vartheta}(0)$. Take $\bar{\vartheta}$ with $\Phi_0(\bar{\vartheta}) = 0$. Then it is clear (see Chapter 2) that $\vartheta(t_1)$ lies between $\vartheta(0)$ and $\bar{\vartheta}$.

Consider now the function $v : \mathbf{R}_+ \to \mathbf{R}$ defined by $v(t) = w(t) + \frac{1}{2}\Phi_0(\sigma(t))^2$. From condition (iv) it follows that $v(0) < 0$. Hence there exists a certain neighborhood of $t = 0$ in $\mathbf{R}_+$ in which v is defined and non-positive. Introduce the set $\Omega = \{t : t \geq 0, \sigma(t) \neq \bar{\vartheta}\}$. Since by condition (v) we have either $\sigma(0) < \bar{\vartheta}$ or $\sigma(0) > \bar{\vartheta}$ it follows that Ω contains a certain interval $[0, t_3]$ with $t_3 > 0$.

Suppose that $v(t) \leq 0$ for $t \in [0, t_3] \subset \Omega$. Then for all $t \in [0, T]$, $[0, T] \subset \Omega$, we have

$$w(t) + \nu\dot{\sigma}(t)^2 \leq -\frac{1}{2}\Phi_0(\sigma(t))^2 + \nu\dot{\sigma}(t)^2.$$

From this and condition (i) we get

$$\Phi_0(\sigma(t))^2 \leq 2\nu\dot{\sigma}(t)^2$$

195

for all $t \in [0, T]$. This inequality combined with condition (v) implies that for each $t \in [0, T]$ one of the following pairs of inequalities is true:

$$\begin{aligned}
\dot{\sigma}(t) &\geq (2\nu)^{-1/2}\Phi_0(\sigma(t)) \quad \text{and} \quad \dot{\sigma}(t) > 0, \\
\dot{\sigma}(t) &\leq (2\nu)^{-1/2}\Phi_0(\sigma(t)) \quad \text{and} \quad \dot{\sigma}(t) < 0.
\end{aligned} \tag{9.1.19}$$

Now using condition (ii) we obtain

$$\psi(t)\dot{\sigma}(t) \leq \varphi(\sigma(t))\dot{\sigma}(t) - \delta(t) \quad \text{on} \quad [0, T], \tag{9.1.20}$$

where $\delta : \mathbf{R}_+ \to \mathbf{R}_+$ is a continuous function. From (9.1.18) and (9.1.19) it follows that

$$\begin{aligned}
\lambda\Phi_0^2 + \left[\varphi + \frac{1}{2}(\Phi_0^2)'\right]\dot{\sigma} &\leq \lambda\Phi_0^2 + (2\nu)^{-1/2}\Phi_0\left[\varphi + \frac{1}{2}(\Phi_0^2)'\right] \\
&= (2\nu)^{-1/2}\Phi_0\left[\frac{1}{2}(\Phi_0^2)' + \lambda\sqrt{2\nu}\Phi_0 + \varphi\right] \\
&= 0
\end{aligned}$$

on $[0, T]$. Hence from (9.1.17) and condition (iii) it follows that the function

$$t \mapsto v(t) + \int_0^t [2\lambda v(\tau) + \delta(\tau)]\, d\tau \tag{9.1.21}$$

does not increase on $[0, T]$. Indeed,

$$\begin{aligned}
\dot{v}(t) + 2\lambda v(t) + \delta(t) &= \dot{w}(t) + 2\lambda w(t) + \frac{1}{2}(\Phi_0^2)'(\sigma(t))\dot{\sigma}(t) + \lambda\Phi_0^2(\sigma(t)) + \delta(t) \\
&\leq \frac{1}{2}(\Phi_0^2)'(\sigma(t))\dot{\sigma}(t) + \lambda\Phi_0^2(\sigma(t)) + \delta(t) + \psi(t)\dot{\sigma}(t) \\
&\leq \frac{1}{2}(\Phi_0^2)'(\sigma(t))\dot{\sigma}(t) + \lambda\Phi_0^2(\sigma(t)) + \varphi(\sigma(t))\dot{\sigma}(t) \\
&\leq 0
\end{aligned}$$

for all $t \in [0, T]$.

Suppose now that $v(t) < 0$ for $t \in [0, T)$ and

$$v(T) = 0. \tag{9.1.22}$$

Let us choose a $T_1 < T$ near T such that

$$\delta(t) > -2\lambda v(t) \tag{9.1.23}$$

for $t \in (T_1, T)$. From the fact that the function (9.1.21) is non-increasing on $[0, T]$ it follows that

$$v(T) - v(t) + \int_t^T [2\lambda v(\tau) + \delta(\tau)]\, d\tau \leq 0$$

on $[0, T]$.

From this and (9.1.21) we find that $v(T) < v(t)$ for all $t \in (t_3, T)$. Thus we have $v(T) < 0$. So if $[0, T] \subset \Omega$ the equation (9.1.22) cannot be true. Hence (9.1.19) is true for all $t \in [0, T]$

assuming that $[0, T] \subset \Omega$. Suppose now $\overline{\vartheta} > \sigma(0)$ and choose a $\overline{\overline{\vartheta}} \in (\sigma(0) + k\Delta, \overline{\overline{c}})$. We show that there exists a $t_2 > 0$ such that

$$\sigma(t_2) = \overline{\overline{\vartheta}}. \tag{9.1.24}$$

Assuming the opposite we see that for all $t \geq 0$ the inequalities (9.1.19) are true and, consequently, $|\dot{\sigma}(t)| \geq \varepsilon_0 > 0$ for all $t \geq 0$. (Notice that it follows from (i) and (v) that $\dot{\sigma}(0) \neq 0$.) Thus (9.1.24) is fulfilled for certain t_2, what implies that $\sigma(t_2) - \sigma(0) > k\Delta$. In an analogous way one considers the case $\overline{\vartheta} < \sigma(0)$. Thus Lemma 9.1.3 is proved. ∎

Proof of Theorem 9.1.3 Let us consider the Hermitian form $F : \mathbf{C}^n \times \mathbf{C} \to \mathbf{R}$ defined by

$$F(z, \xi) = -\mathrm{Re}\,(\xi^* c^* z) + \varepsilon |c^* z|^2.$$

Under our conditions there exists by Theorem 1.4.2, p. 9, a matrix $H = H^*$ satisfying

$$2z^* H[(A + \lambda I)z + b\xi] - c^* z\xi \leq -\varepsilon(c^* z)^2 \tag{9.1.25}$$

for all $(z, \xi) \in \mathbf{R}^n \times \mathbf{R}$. This matrix can be found by the procedure introduced in the statement of Theorem 9.1.1, defining the matrices G and B as in Theorem 9.1.3.

Putting $\xi = 0$ in (9.1.25) we get

$$2z^* H(A + \lambda I)z \leq -\varepsilon(c^* z)^2$$

for all $z \in \mathbf{R}^n$. By Lemma 1.2.1, p. 6, it follows that H has one negative and $n - 1$ positive eigenvalues. From (9.1.25) we see that

$$2Hb = c. \tag{9.1.26}$$

Now using the fact that for arbitrary n-vectors p, q $\det(I + pq^*) = 1 + p^* q$, we see that for arbitrary $\xi \in \mathbf{R}$

$$\det(H + \xi cc^*) = \det H \det(I + \xi H^{-1} cc^*) = \det H(1 + \xi c^* H^{-1} c).$$

From this and (9.1.26) it follows that

$$\det(H + \xi cc^*) = \det H(1 + 2\xi c^* b)$$

for all $\xi \in \mathbf{R}$. Since $\det H < 0$ and $c^* b < 0$ it follows that $\det(H + \xi cc^*) > 0$ for all $\xi > -(2c^* b)^{-1}$ and $\det[H - (2c^* b)^{-1} cc^*] = 0$. Since H has only one negative eigenvalue and $c^* b < 0$ the last relations imply that

$$H - (2c^* b)^{-1} cc^* \geq 0 \quad \text{and} \quad H + \xi cc^* > 0 \quad \text{for} \quad \xi > -(2c^* b)^{-1}. \tag{9.1.27}$$

Consider now an arbitrary solution $z(\cdot)$, $\sigma(\cdot)$ of (9.1.1) having initial points $\sigma(0) = \vartheta(0)$ and $z(0)^* Hz(0) < -\frac{1}{2}\dot{\vartheta}(0)^2$. In order to use Lemma 9.1.3 here we introduce the functions $w, \psi : \mathbf{R}_+ \to \mathbf{R}$ defined by $w(t) = z(t)^* Hz(t)$ and $\psi(t) = \varphi(\sigma(t)) - \varepsilon\dot{\sigma}(t)$ and the number $\nu = -(2c^* b)^{-1}$.

The first relation of (9.1.27) guarantees condition (i) of Lemma 9.1.3. Conditions (ii) and (v) of this lemma are fulfilled automatically; condition (iii) follows from (9.1.25). Because condition (iv) of Lemma 9.1.3 coincides with condition (iii) of the present theorem we can use Lemma 9.1.3 with respect to the functions $\sigma(\cdot)$, $w(\cdot)$, $\varphi(\cdot)$, $\psi(\cdot)$ and the number λ and $-(2c^* b)^{-1}$ to conclude that there exists a number t_2 such that $|\sigma(t_2) - \sigma(0)| > k\Delta$. ∎

Let us now consider the case when φ belongs to $\mathbf{C}^1$. Define the values $\mu_1 = \inf\limits_{\sigma \in [0, \Delta]} \varphi'(\sigma)$ and $\mu_2 = \sup\limits_{\sigma \in [0, \Delta]} \varphi'(\sigma)$. Using this additional information about φ we can weaken the frequency-domain conditions in the above theorems.

Suppose that $z(\cdot)$, $\sigma(\cdot)$ is an arbitrary solution of (9.1.1) with $\sigma_0 - \Delta \leq \sigma(0) \leq \sigma_0$, where $\varphi(\sigma_0) = 0$.

Introduce the functions $r_i : \mathbf{N} \times \mathbf{R} \to \mathbf{R}$ $(i = 1, 2)$ by

$$r_i(k, x) = \left[\int\limits_{\sigma(0)}^{\sigma_0} \varphi(\sigma)\, d\sigma + k \int\limits_0^{\Delta} \varphi(\sigma)\, d\sigma + (-1)^i x \right] \cdot \left[\int\limits_{\sigma(0)}^{\sigma_0} |\varphi(\sigma)|\, d\sigma + k \int\limits_0^{\Delta} |\varphi(\sigma)|\, d\sigma \right]^{-1} \quad (i = 1, 2).$$

Introduce also the function $y : \mathbf{R}_+ \to \mathbf{R}^{n+1}$ by

$$y(t) = \left[\begin{array}{c} z(t) \\ \varphi(\sigma(t)) \end{array} \right]$$

and the matrices of order $(n + 1) \times (n + 1)$, $(n + 1) \times 1$ resp. $(n + 1) \times 1$

$$Q = \left[\begin{array}{cc} A & b \\ 0 & 0 \end{array} \right], \qquad l = \left[\begin{array}{c} 0 \\ 1 \end{array} \right], \qquad \text{and} \qquad d = \left[\begin{array}{c} 0 \\ \varrho \end{array} \right].$$

The following two assertations are true.

Theorem 9.1.4 *Suppose A is a Hurwitz matrix and for certain numbers $\varepsilon > 0$, $\delta > 0$, $\tau \geq 0$ and a positive integer k the following conditions are valid:*

(i) $\mathrm{Re}\left\{ K(i\omega) - \varepsilon |K(i\omega)|^2 - \tau[K(i\omega) + \mu_1^{-1} i\omega]^*[K(i\omega) + \mu_2^{-1} i\omega] \right\} \geq \delta$ *for all $\omega \in \mathbf{R}$;*

(ii) $4\varepsilon\delta > [r_i(k, y(0)^* H y(0))]^2$ $(i = 1, 2)$, *where the symmetric $(n + 1) \times (n + 1)$ matrix H is defined as in Theorem 9.1.1 with $B = Q$ and*

$$F(y, \xi) = \varepsilon(d^* y)^2 + \delta(l^* y)^2 - \tau(d^* y - \mu_1^{-1}\xi)(\mu_2^{-1}\xi - d^* y) + y^* l d^* y.$$

Then for the given solution $z(\cdot)$, $\sigma(\cdot)$ the estimate

$$|\sigma(t) - \sigma(0)| < (k + 1)\Delta \tag{9.1.28}$$

is true for all $t \geq 0$.

Theorem 9.1.5 *Suppose that for certain numbers $\varepsilon > 0$, $\lambda > 0$, $\tau \geq 0$ the following conditions are valid:*

(i)

$$\begin{aligned} \mathrm{Re}\left\{ K(i\omega - \lambda) \right. & - \varepsilon |K(i\omega - \lambda)|^2 - \\ & - \left. \tau[K(i\omega - \lambda) + \mu_1^{-1}(i\omega - \lambda)]^*[K(i\omega - \lambda) + \mu_2^{-1}(i\omega - \lambda)] \right\} \geq 0 \end{aligned}$$

for all $\omega \in \mathbf{R}$;

(ii) $A + \lambda I$ *is a Hurwitz matrix;*

(iii) *the equation*

$$\ddot{\vartheta} + 2\sqrt{\lambda\varepsilon}\,\dot{\vartheta} + \varphi(\vartheta) = 0 \tag{9.1.29}$$

is Lagrange stable;

198

(iv) *for a certain solution ϑ of (9.1.29) with initial points $\vartheta(0)$, $\pm\dot\vartheta(0)$ and a positive integer k the inequality*

$$|\vartheta(t) - \vartheta(0)| < k\Delta$$

is true for all $t \geq 0$.

Then for the solution $z(\cdot)$, $\sigma(\cdot)$ of (9.1.1) satisfying $\sigma(0) = \vartheta(0)$ and $y(0)^*Hy(0) < \frac{1}{2}\dot\vartheta(0)^2$ the estimate (9.1.28) is true for all $t \geq 0$. The symmetric $(n+1) \times (n+1)$ matrix H is defined as in Theorem 9.1.1 with $B = Q + \lambda I$ and

$$F(y, \xi) = \varepsilon(d^*y)^2 + y^*ld^*y - \tau(d^*y - \mu_1^{-1}\xi)(\mu_2^{-1}\xi - d^*y).$$

We omit here the proof of these theorems. They are based on auxillary propositions of the same type as Lemmas 9.1.1 and 9.1.2.

In the next section we shall prove analogous theorems for the case of distributed parameters systems.

Example 9.1.1 Let us continue the investigation of the equation of second order PLL with proportionally-integrating filter, which has already been dealt with in Example 5.4.1, p. 100, and Example 6.3.5, p. 141. It has the transfer function

$$K(s) = T(1 + mTs)(1 + Ts)^{-1} \tag{9.1.30}$$

($T > 0$ and $m \in (0, 1)$ are constant) and the nonlinearity $\varphi(\sigma) = \sin\sigma - \gamma$, $\gamma \in (0, 1)$. System (9.1.1), (9.1.30) may be written as

$$\dot z = -T^{-1}z - (1 - m)\varphi(\sigma), \qquad \dot\sigma = z - mT\varphi(\sigma). \tag{9.1.31}$$

Let us apply Theorem 9.1.1 to this system. As it was shown in Chapter 6 the frequency-domain condition is fulfilled if we choose $\varepsilon = [T(1 + m)]^{-1}$ and $\delta = T(1 - \varepsilon T)$.

Now put $B = A$ and define the following objects by the procedure given in the statement of Theorem 9.1.1:

$$d(s) = s + T^{-1}, \qquad Q(s) = m - 1,$$

$$G = \begin{bmatrix} -\varepsilon & \frac{1}{2}\frac{m-1}{m+1} \\ \frac{1}{2}\frac{m-1}{m+1} & 0 \end{bmatrix}, \qquad \Xi(s) = \begin{bmatrix} m-1 \\ -s^* + T^{-1} \end{bmatrix}^* G \begin{bmatrix} m-1 \\ s + T^{-1} \end{bmatrix} \equiv 0.$$

It follows that we can take $\Psi(s) \equiv 0$, $\Gamma = 0$, $\tau = 0$, $h = 0$, $D = -\varepsilon$, $H = \frac{1}{2}T\varepsilon = \frac{1}{2(1+m)}$. Since

$$\nu_j(k, z(0)^*Hz(0)) = \frac{-2\pi\gamma + (-1)^{j+1}[2(1+m)k]^{-1}z(0)^2}{4(\gamma\arcsin\gamma + \sqrt{1-\gamma^2})}$$

for integer k and real $z(0)$, the condition (ii) is satisfied if

$$\frac{2\sqrt{m}}{1+m} > \frac{-2\pi\gamma + [2(1+m)k]^{-1}z(0)^2}{4(\gamma\arcsin\gamma + \sqrt{1-\gamma^2})}.$$

Thus all conditions of Theorem 9.1.1 are verified.

In order to do this for Theorem 9.1.2 we put

$$\lambda(1 + \sqrt{1-m})^{-1}T^{-1} \qquad \text{and} \qquad \varepsilon = (1 - T\lambda)(1 - mT\lambda)^{-1}T^{-1}, \tag{9.1.32}$$

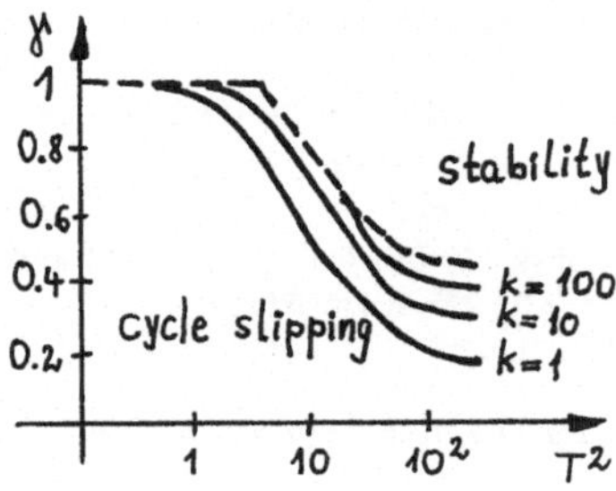

Figure 9.1.1

which guarantees that conditions (i) and (ii) of Theorem 9.1.2 are valid. The required number H we can determine by defining the following objects:

$$B = -T^{-1} + \lambda, \qquad d(s) = s - \lambda + T^{-1}, \qquad Q(s) = m - 1,$$

$$G = \begin{bmatrix} -\varepsilon & \varepsilon m T - \frac{1}{2} \\ \varepsilon m T - \frac{1}{2} & m T - \varepsilon m^2 T^2 \end{bmatrix}, \qquad \Xi(s) = -(mT - \varepsilon m^2 T^2)s^2,$$

$$\Gamma = mT - \varepsilon m^2 T^2, \quad \tau = (mT - \varepsilon m^2 T^2)^{\frac{1}{2}}, \quad h^2 = \frac{m}{T(1 + \sqrt{1-m})^2 \sqrt{1-m}}, \quad D = -(\varepsilon + h^2)$$

and, consequently

$$H = [2(1 - m)]^{-1}. \tag{9.1.33}$$

Thus if for the solutions of the reduction-type equation

$$\ddot{\vartheta} + 2\left[T(1 + \sqrt{1-m})\right]^{-1}\dot{\vartheta} + \varphi(\vartheta) = 0 \tag{9.1.34}$$

with initial data $\vartheta(0), \pm\dot{\vartheta}(0)$ the inequality $\mid \vartheta(t) - \vartheta(0) \mid < 2\pi k$ is true for $t \geq 0$ then for the solution (z, σ) of (9.1.31) with initial points $\sigma(0) = \vartheta(0)$ and $[2(1-m)]^{-1}z(0)^2 < \dot{\vartheta}(0)^2$ Theorem 9.1.2 guarantees that $\mid \sigma(t) - \sigma(0) \mid < 2\pi k$ for $t \geq 0$. In order to calculate the number of slipped cycles of the solution (9.1.31) starting in $\vartheta(0), \pm\dot{\vartheta}(0)$ one has to consider the corresponding first-order equation

$$\Phi\frac{d\Phi}{d\vartheta} + 2\left[T(1 + \sqrt{1-m})\right]^{-1}\Phi + \varphi(\vartheta) = 0 \tag{9.1.35}$$

with respect to solutions passing through the points $(\sigma_0 + 2k\pi, 0)$ in the (ϑ, Φ)-plane. Here σ_0 is such that $\varphi(\sigma_0) = 0, \varphi'(\sigma_0) > 0$; the initial point satisfies $\vartheta(0) \in [\sigma_0 - 2\pi, \sigma_0)$. Let us continue each of these solutions to the left until the integral curve intersects the line $\vartheta = \vartheta(0)$ and let us denote the ordinates of the intersection point by $v_j, j \in \mathbf{Z}$. If $\dot{\vartheta}(0) \in (v_j, v_{j+1}]$ then the number of cycles slipped by the solution of (9.1.34) starting in $\vartheta(0), \dot{\vartheta}(0)$ is $j + 1$. As an example we consider the solution of (9.1.31) with $m = 0.1$ satisfying

$$\sigma(0) = \varphi^{-1}(0), \quad z(0) = K(0)\gamma. \tag{9.1.36}$$

The curves in the (γ, T)-plane which approximate from below the boundaries of regions, where the number of cycles slipped by the solution (z, σ) of (9.1.31) with initial conditions (9.1.36) is k, are shown in Figure 9.1.1.

These approximations are obtained by using (9.1.32) and the behavior of (9.1.34). The dotted line in Figure 9.1.1 demonstrates the boundary of the stability region obtained in [33] by qualitatively-numerical methods.

200

9.2 Distributed Parameter Systems

Let us consider again the integro-differential equation

$$\dot{\sigma}(t) = \alpha(t) + \varrho\varphi(\sigma(t)) - \int_0^\sigma \gamma(t - \tau)\varphi(\sigma(\tau))\,d\tau, \qquad (9.2.1)$$

where $\alpha : [0, +\infty) \to \mathbf{R}$ is continuous, $\gamma : [0, +\infty) \to \mathbf{R}$ is $L^1(0, +\infty)$, $\varphi : \mathbf{R} \to \mathbf{R}$ is C^1 and Δ-periodic, ϱ is a constant. Suppose that there exist constants $M > 0$ and $\kappa > 0$ such that $\mid \alpha(t) \mid\, \leq M$ on $\mathbf{R}_+, t \mapsto \alpha(t)e^{\kappa t}$ and $t \mapsto \gamma(t)e^{\kappa t}$ are $L^2(0, +\infty)$. Suppose also that φ has exactly two zeros on $[0, \Delta)$. As was done in Chapter 8 we introduce the notations

$$\mu_1 = \inf_{\sigma \in [0,\Delta)} \varphi'(\sigma) \quad \text{and} \quad \mu_2 = \sup_{\sigma \in [0,\Delta)} \varphi'(\sigma).$$

Clearly, $\mu_1 < 0$ and $\mu_2 > 0$. Denote the transfer function by

$$K(s) = -\varrho + \int_0^{+\infty} \gamma(t)e^{-st}\,dt.$$

It is convenient for us to use some notations, earlier introduced, in the proof of Lemma 8.2.2. So for an arbitrary solution $\sigma(\cdot)$ of (9.2.1) and arbitrary $T > 1$ we define the following functions:

$$\eta(t) = \begin{cases} 0, & t < 0 \\ \varphi(\sigma(t)), & t \geq 0 \end{cases}, \quad \mu(t) = \begin{cases} 0, & t \leq 0 \\ t, & 0 < t \leq 1 \\ 1, & t > 1 \end{cases}, \quad \eta_T(t) = \begin{cases} \mu(t)\eta(t) & t \in [0,T] \\ 0, & t \notin [0,T], \end{cases}$$

$$\zeta_T(t) = \varrho\eta_T(t) - \int_0^t \gamma(t - \tau)\eta_T(\tau)\,d\tau,$$

$$\sigma_\mu(t) = \alpha(t) + (1 - \mu(t))\varrho\eta(t) - \int_0^t \gamma(t - \tau)(1 - \mu(\tau))\eta(\tau)\,d\tau.$$

Note that for $t \in [0, T]$

$$\dot{\sigma}(t) = \sigma_\mu(t) + \zeta_T(t). \qquad (9.2.2)$$

We introduce the constants $\overline{\varphi} := \sup_{\sigma \in [0,\Delta]} \varphi(\sigma), \Gamma := \int_0^{+\infty} \mid \gamma(t) \mid\, dt, \overline{\sigma} := M + \varrho\overline{\varphi} + \Gamma\overline{\varphi}$ and $\overline{\eta} := \overline{\sigma\varphi}$. Clearly, that $\mid \dot{\sigma}(t) \mid\, \leq \overline{\sigma}$ and $\mid \dot{\eta}(t) \mid\, \leq \overline{\eta}$ on $\mathbf{R}_+$. Furthermore let us define the functions $f_1 : [0,1] \times \mathbf{R} \times \mathbf{R}_+ \times [0, \frac{\kappa}{2}] \to \mathbf{R}, f_j : \mathbf{R}_+ \times \mathbf{R}_+ \times \mathbf{R}_+ \times [0, \frac{\kappa}{2}] \to \mathbf{R}, (j = 2, 3)$ by

$$f_1(t, \xi_1, \xi_2, \xi_3) = \int_0^t \{\dot{\sigma}(\tau)(\mu(\tau) - 1)\eta(\tau) + \xi_1(\mu(\tau)^2 - 1)\eta(\tau)^2$$

$$+ \xi_2[\mu_1^{-1}\mu_2^{-1}[\dot{\mu\eta}]^2 - \eta(\tau)^2 - (\mu_1^{-1} + \mu_2^{-1})(\dot{\mu\eta} - \dot{\eta}(\tau))]\}e^{2\xi_3\tau}\,d\tau,$$

$$f_2(t, \xi_1, \xi_2, \xi_3) = \int_0^t \{-\sigma_\mu(\tau)\eta(\tau)\mu(\tau) - 2(\xi_1 + \xi_2)\sigma_\mu(\tau)\dot{\sigma}(\tau) +$$

$$\xi_3(\mu_1^{-1} + \mu_2^{-1})\sigma_\mu(\tau)\dot{\mu\eta}\}e^{2\xi_3\tau}\,d\tau,$$

$$f_3(t, \xi_1, \xi_2, \xi_3) = \int_t^\infty (\xi_1 + \xi_2)\sigma(\tau)^2 e^{2\xi_3\tau}\,d\tau + \int_0^t (\xi_1 + \xi_2)\sigma_\mu(\tau)^2 e^{2\xi_3\tau}\,d\tau.$$

Note that $f_3 \geq 0$ and $f_1(\cdot, \xi_1, \xi_2, \xi_3)$, $f_2(\cdot, \xi_1, \xi_2, \xi_3)$ are bounded, say by the constants

$$\mid f_1(t, \xi_1, \xi_2, \xi_3) \mid \leq q_1(\xi_1, \xi_2, \xi_3), \quad t \leq 1$$

and

$$\mid f_2(t, \xi_1, \xi_2, \xi_3) \mid \leq q_2(\xi_1, \xi_2, \xi_3), \quad t \geq 0.$$

Note also that q_1 and q_2 may be calculated by means of $\overline{\sigma}, \overline{\varphi}, \overline{\eta}, \Gamma, M$ and ϱ. The first theorem in this section concerns the application of the Bakaev-Guzh technique to the cycle-slipping problem. The material of this section is taken from [66].

Theorem 9.2.1 *Consider a solution $\sigma(\cdot)$ of (9.2.1) satisfying $\sigma(0) \in (\sigma_0 - \Delta, \sigma_0)$, where σ_0 is a zero of φ. Suppose that there exist parameters $\varepsilon > 0, \delta > 0$ and $\tau \geq 0$ such that the following conditions are satisfied:*

(i) *$4\varepsilon\delta > \nu_j(k, q_0)^2$ for $j = 1, 2$, where $q_0 := q_1(\delta, \tau, 0) + q_2(\varepsilon, \tau, 0)$ is calculated for $\sigma(\cdot)$ by the previous procedure, k is a positive integer and the functions ν_j are defined in the statement of Theorem 9.1.1;*

(ii) *$\mathrm{Re}\,\{K(i\omega) - \tau[K(i\omega) + i\mu_1^{-1}\omega]^*[K(i\omega) + i\mu_2^{-1}] - \varepsilon \mid K(i\omega) \mid^2\} \geq \delta$ for all $\omega \in \mathbf{R}$.*

Then the considered solution $\sigma(\cdot)$ of (9.2.1) satisfies

$$\mid \sigma(t) - \sigma(0) \mid < (k+1)\Delta \quad t \geq 0. \tag{9.2.3}$$

Proof Define the set $Z = \{T : T \geq 1, \varphi(\sigma(T)) = 0\}$. For $T \in Z$ the function $\eta_T(\cdot)$ is continuous and C^1, maybe with the exception of $z = T$. Consider the integral

$$\sigma_T := \int\limits_0^{+\infty} \{\zeta_T(t)\eta_T(t) + \varepsilon\zeta_T(t)^2 + \delta\eta_T(t)^2 + \tau[\zeta_T(t) - \mu_1^{-1}\dot{\eta}_T(t)][\zeta_T(t) - \mu_2^{-1}\dot{\eta}_T(t)]\} \, dt.$$

Proceeding as in the proof of Lemma 8.2.1, we may establish by using assumption (ii) that

$$\sigma_T \leq 0. \tag{9.2.4}$$

Let us write σ_T in the form

$$\sigma_T = I_0(T, \delta, \varepsilon, 0) + \tau I_1(T, 0) + f_1(T, \delta, \varepsilon, 0) + f_2(T, \varepsilon, \tau, 0) + f_3(T, \varepsilon, \tau, 0),$$

where

$$I_0(t, \xi_1, \xi_2, \xi_3) := \int\limits_0^t \{\dot{\sigma}(\tau)\varphi(\sigma(\tau)) + \xi_1\varphi(\sigma(\tau))^2 + \xi_2\dot{\sigma}(\tau)^2\}e^{2\xi_3\tau} \, d\tau$$

$(\xi_1, \xi_2, \xi_3$ - parameters) and

$$I_1(t, \xi) := \int\limits_0^t [-\mu_1^{-1}\dot{\eta}(\tau) + \dot{\sigma}(\tau)][-\mu_2^{-1}\dot{\eta}(\tau) + \sigma(\tau)]e^{2\xi\tau} \, d\tau +,$$

ξ is also a parameter. Since I_1 and f_3 are non-negative and $\mid f_1 + f_2 \mid$ is bounded by q_0 it follows from (9.2.4) that

$$I_0(T, \delta, \varepsilon, 0) \leq q_0. \tag{9.2.5}$$

According to assumption (i) of the theorem we can choose a number ε_0 such that

$$4\varepsilon\delta \geq \nu_j(k, q_0 + \varepsilon_0)^2, \quad j = 1, 2. \tag{9.2.6}$$

Introducing the function $F_j : \mathbf{R} \to \mathbf{R}, j = 1, 2$ defined by

$$F_j(\sigma) = \varphi(\sigma) - \nu_j(k, q_0 + \varepsilon) \mid \varphi(\sigma) \mid,$$

we can use for $j = 1, 2$ the representation

$$I_0(T, \delta, \varepsilon, 0) = \int\limits_{\sigma(0)}^{\sigma(T)} F_j(\sigma) \, d\sigma + \int\limits_0^T \left[\dot{\sigma}(t)\eta(t) + \delta\eta(t)^2 + \varepsilon\dot{\sigma}(t)^2 - F_j(\sigma(t))\dot{\sigma}(t) \right] \, dt.$$

From (9.2.6) it follows that for $t \geq 0$ and $j = 1, 2$

$$\varepsilon\dot{\sigma}(t)^2 + \delta\varphi(\sigma(t))^2 + \dot{\sigma}(t)[\varphi(\sigma(t)) - F_j(\sigma(t))] \geq \delta\varphi(\sigma(t))^2 - \frac{1}{4\varepsilon}[F_j(\sigma(t)) - \varphi(\sigma(t))]^2. \tag{9.2.7}$$

Now suppose to the contrary that there exists a time t_1 such that $\sigma(t_1) = \sigma_0 + k\Delta$. Then we have

$$\int\limits_{\sigma(0)}^{\sigma(t_1)} F_1(\sigma) \, d\sigma = \int\limits_{\sigma(0)}^{\sigma_0} \varphi(\sigma) \, d\sigma + k \int\limits_0^\Delta \varphi(\sigma) \, d\sigma - \nu_1 \int\limits_{\sigma(0)}^{\sigma_0} \mid \varphi(\sigma) \mid \, d\sigma - k\nu_1 \int\limits_0^\Delta \mid \varphi(\sigma) \mid \, d\sigma \tag{9.2.8}$$

or

$$\int\limits_{\sigma(0)}^{\sigma(t_1)} F_1(\sigma) \, d\sigma = q_0 + \varepsilon_0. \tag{9.2.9}$$

From (9.2.7) and (9.2.9) it follows that $I_0(t_1, \delta, \varepsilon, 0) \geq q_0 + \varepsilon_0$, which contradicts the estimation (9.2.5). Therefore for all $t \geq 0, \sigma(t) < \sigma_0 + k\Delta$. By the same type of argument as above, using the integral I_0 and the function F_2, we can show that $\sigma(t) > \sigma_0 - (k + 1)\Delta$ for $t \geq 0$. This, together with the property $\sigma(0) \in (\sigma_0 - \Delta, \sigma_0)$ proves the assertion. ∎

In the next theorem the non-local reduction method will be used to get an analogous result as in Theorem 9.2.1.

Theorem 9.2.2 *Consider a solution $\sigma(\cdot)$ of (9.2.1) and suppose that there exist parameters $\lambda \in (0, \kappa/2), \tau \geq 0$ and $\varepsilon > 0$ such that the following conditions are satisfied;*

(i) $\operatorname{Re}\{K(i\omega - \lambda) - \varepsilon \mid K(i\omega - \lambda) \mid^2 - \tau[K(i\omega - \lambda) + \mu_1^{-1}(i\omega - \lambda)]^*[K(i\omega - \lambda) + \mu_2^{-1}(i\omega - \lambda)]\} \geq 0$ *for all $\omega \in \mathbf{R}$;*

(ii) *any solution of the reduction-type equation*

$$\ddot{\vartheta} + 2\sqrt{\lambda\varepsilon}\dot{\vartheta} + \varphi(\vartheta) = 0$$

is bounded on $\mathbf{R}_+$;

(iii) *there exist two solutions ϑ_1, ϑ_2 of the equation in condition (ii) with initial conditions $\vartheta_j(0) = \vartheta(0)$ and $\dot{\vartheta}_j(0) = (-1)^{j+1}\dot{\vartheta}(0), j = 1, 2$ satisfying*

$$\frac{1}{2}\dot{\vartheta}(0)^2 > q_0, \tag{9.2.10}$$

where $q_0 = q_1(0, \tau, \lambda) + q_2(\varepsilon, \tau, \lambda)$ is calculated for $\sigma(\cdot)$ by the above procedure, and

$$\mid \vartheta_j(t) - \vartheta(0) \mid < k\Delta \tag{9.2.11}$$

for all $t \geq 0$ and $j = 1, 2$, where k is a natural number.

Then the considered solution $\sigma(\cdot)$ of (9.2.1) satisfies

$$| \sigma(t) - \sigma(0) | < (k+1)\Delta \tag{9.2.12}$$

for $t \geq 0$ provided that $\sigma(0) = \vartheta(0)$.

Proof Consider again the set $Z = \{T \geq 1 : \varphi(\sigma(T)) = 0\}$ and define for $T \in Z$ the integral

$$\varrho_{T,\lambda} = \int\limits_0^{+\infty} \left\{ [\eta_T]^\lambda [\zeta_T]^\lambda + \varepsilon([\zeta_T]^\lambda)^2 + \tau([\zeta_T]^\lambda - \mu_1^{-1}[\dot\eta_T]^\lambda)([\zeta_T]^\lambda - \mu_2^{-1}[\dot\eta_T]^\lambda) \right\} \, dt,$$

where the notation $[f]^\lambda(t) := f(t)e^{\lambda t}$ is used. Because of (i) we can use the Parseval equation to conclude that for $T \in Z$

$$\varrho_{T,\lambda} < 0. \tag{9.2.13}$$

By means of the representation

$$\sigma_{T,\lambda} = I_0(T, 0, \varepsilon, \lambda) + \tau I_1(T, \lambda) + f_1(T, 0, \varepsilon, \lambda) + f_2(T, \varepsilon, \tau, \lambda) + f_3(T, \varepsilon, \tau, \lambda),$$

and the estimate (9.2.13) we establish that

$$I_0(T, 0, \varepsilon, \lambda) < q_0 \qquad \text{if} \qquad T \in Z. \tag{9.2.14}$$

Consider now the first-order equation

$$y(\vartheta)\frac{dy(\vartheta)}{d\vartheta} + 2\sqrt{\lambda\varepsilon}y(\vartheta) + \varphi(\vartheta) = 0 \tag{9.2.15}$$

which corresponds to the equation in (ii).

Let σ_0 be a zero of φ on $[0, \Delta)$ and Φ be a solution of (9.2.15) such that $\Phi(\sigma_0) = 0, \Phi(\vartheta) \neq 0$ for $\vartheta \neq \sigma_0$ and $| \Phi(\vartheta) | \to +\infty$ for $| \vartheta | \to +\infty$. Let us consider the set $\Sigma := \{\sigma \ : \ \sigma = \sigma_0 + j\Delta, j \in \mathbf{Z}\}$ and choose a $\vartheta_0 \in \Sigma$ in such a way that for the considered solution

$$\vartheta_0 - \Delta < \sigma(0) \leq \vartheta_0. \tag{9.2.16}$$

Suppose that $\vartheta_0 = \sigma_0 + j_0\Delta$ and consider the functions Φ_1 and Φ_2 defined by $\Phi_1(\vartheta) = \Phi(\vartheta - (k + j_0)\Delta)$ and $\Phi_2(\vartheta) = \Phi(\vartheta - (j_0 - k - 1)\Delta)$, where k is given in the statement of the theorem. Clearly Φ_1 and Φ_2 are solutions of (9.2.15). By the uniqueness theorem and inequility (9.2.11) we have

$$| \dot\vartheta(0) | \leq | \Phi_j(\sigma(0)) |, \quad j = 1, 2. \tag{9.2.17}$$

Let us consider for $N = 1, 2$ the integrals

$$I_{T,N} = I_0(T, 0, \varepsilon, \lambda) + \frac{1}{2}\Phi_N^2(\sigma(T))e^{2\lambda T} - \frac{1}{2}\Phi_N^2(\sigma(0)) = \int\limits_0^T \{\varepsilon\dot\sigma^2 + \dot\sigma\eta + \lambda\Phi_N^2 + \Phi_N\Phi_N'\dot\sigma\}e^{2\lambda t} \, dt.$$

$$\tag{9.2.18}$$

In a similar way as in the proof of Theorem 8.4.1, one shows that

$$I_{T,N} \geq 0 \quad \text{for} \quad N = 1, 2. \tag{9.2.19}$$

Suppose now that (9.2.12) is not true and there exists a time t_1 such that $\sigma(t_1) = \sigma_0 + k\Delta$. Then from (9.2.19) and (9.2.18) it follows that

$$\frac{1}{2}\Phi_1^2(\sigma(t_1))e^{2\lambda t_1} \geq \frac{1}{2}\Phi_1^2(\sigma(0)) - I_0(t_1, 0, \varepsilon, \lambda).$$

Using now the properties (9.2.14), (9.2.10) and (9.2.17) we obtain the inequality

$$\frac{1}{2}\Phi_1^2(\sigma(t_1))e^{2\lambda t_1} > \frac{1}{2}\Phi_1^2(\sigma(0)) - q_0 > \frac{1}{2}(\Phi_1^2(\sigma(0)) - \dot{\vartheta}(0)^2) \geq 0.$$

Thus $\Phi_1(\sigma(t_1)) \neq 0$ and therefore $\sigma(t_1) \neq \sigma_0 + k\Delta$. By the same type of argument as above we can show with the help of Φ_2 that $\sigma(t) \neq \sigma_0 - (k+1)\Delta$ for $t \in Z$. It follows that $\sigma_0 - (k+1)\Delta < \sigma(t) < \sigma_0 + k\Delta$ for all $t \geq 0$. From this and (9.2.16) we receive the assertion of Theorem 9.2.2. $\blacksquare$

Chapter 10

Discrete Systems

In the previous chapters we concentrated on the global investigation of the continuous in the time and in the states systems. Now we turn to a study of systems, which may be discrete in time and in states, considered as feedback representations.

Although continuous and discrete systems differ in several ways, we show that many of the results of the previous chapters also go through for discrete systems. This chapter is devoted to the extension of some concepts, previously developed for continuous systems (invariant cones method, Bakaev-Guzh technique and non-local reduction method) for studying the global behavior of discrete systems.

10.1   Introduction

Let us consider on the Riemannian manifold (M, G) the nonlinear discrete system

$$u_{t+1} = f(t, u_t), \quad t = \tau, \tau + 1, \ldots, \tag{10.1.1}$$

where $f : \mathbf{Z}_+ \times M \to M$ may be in general a non-continuous mapping with respect to the second argument. We denote the solution of (10.1.1) starting in $p \in M$ for $t = \tau$ by $\{\alpha_t(\tau, p)\}$. As in the continuous case we say that the solution $\{\alpha_t(\tau, p)\}$ of (10.1.1) is *bounded*, if the positive semi-orbit $\gamma^+(p, \tau) := \{\alpha_t(\tau, p) : t = \tau, \tau + 1, \ldots\}$ is relatively compact in M. System (10.1.1) is said to be *Lagrange stable* if all solutions of (10.1.1) are bounded.

Suppose now that system (10.1.1) is an autonomous one, i.e.

$$u_{t+1} = f(u_t), \quad t = 0, 1, \ldots, \tag{10.1.2}$$

with $f : M \to M$.

We say that the solution $\{\alpha_t(p)\}$ of (10.1.2) *converges* if there exists a point $q \in M$ with $\lim_{t \to +\infty} \alpha_t(p) = q$. It follows that q is an *equilibrium point* of (10.1.2), i.e. $\alpha_t(q) = q$ for $t = 0, 1, \ldots$. System (10.1.1) is said to be *gradient-like* if every solution converges. The solution $\{\alpha_t(p)\}$ of (10.1.2) is said to be *quasi-convergent* if the set $\mathcal{E}$ of equilibria attracts the orbit of $\{\alpha_t(p)\}$.

We say that system (10.1.2) is *monostable* (resp. *quasi-monostable*) if every bounded solution is convergent (resp. quasi-convergent). Proceeding as in the ODE case we define for a solution $\{\alpha_t(p)\}$ of (10.1.2) the ω-limit set $\omega(p)$ as $\omega(p) := \left\{y : \exists t_n \to +\infty \text{ with } \lim_{n \to \infty} \alpha_{t_n} = y\right\}$. It is easy to see that this set is closed and invariant for (10.1.2). If the solution $\{\alpha_t(p)\}$ is bounded on $\mathbf{Z}_+$ the limit set $\omega(p)$ is compact and non-empty, but, in contrast to the continuous time case in general, not connected. This chapter mainly deals with systems (10.1.1) or (10.1.2) on $\mathbf{R}^n$ with solutions having an *equivariance property* with respect to a discrete subgroup $\Gamma =$

$$\left\{ \sum_{j=1}^{m} k_j d_j \ : \ k_j \in \mathbf{Z}, 1 \le j \le m \right\} \text{ of } \mathbf{R}^n, \text{ where the vectors } d_j \in \mathbf{R}^n \text{ are assumed to be linearly}$$

independent. As in the ODE case equivariance means that for any solution $\{\alpha_t(\tau,p)\}$ of (10.1.1) and any $d \in \Gamma$ $\quad \alpha_t(\tau, p + d) = \alpha_t(\tau, p) + d$ for all $t = \tau, \tau + 1, \dots$. We also say in this case that (10.1.1) is *pendulum-like*. The obvious following result is a helpful criterion for the equivariance.

Proposition 10.1.1 *The solution of* (10.1.1) *satisfy the equivariance property with respect to* Γ *if and only if*

$$f(t, u + d) = f(t, u) + d \tag{10.1.3}$$

for all $t \in \mathbf{Z}_+$, $u \in \mathbf{R}^n$ *and* $d \in \Gamma$.

Proof Suppose (10.1.3) is satisfied and consider an arbitrary solution $\{\alpha_t(\tau,p)\}$ of (10.1.1). Define for an arbitrary $d \in \Gamma$ the sequence $y_t := \alpha_t(\tau, p) + d$ for $t = \tau, \tau + 1, \dots$, satisfying $y_\tau = p + d$. Furthermore we have by (10.1.3) that

$$
\begin{aligned}
y_{t+1} &= \alpha_{t+1}(\tau, p) + d = f(t, \alpha_t(\tau, p)) + d \\
&= f(t, \alpha_t(\tau, p) + d) = f(t, y_t) \quad \text{for all} \quad t \ge \tau.
\end{aligned}
$$

It follows that $\{y_t\}$ is the solution of (10.1.1) starting with $t = \tau$ in $p + d$.

Suppose now that the solutions of (10.1.1) satisfy the equivariance property. Consider an arbitrary triple $\tau \in \mathbf{Z}_+$, $p \in \mathbf{R}^n$ and $d \in \Gamma$. Using the equivariance we get

$$\alpha_{t+1}(\tau, p + d) = f(t, \alpha_t(\tau, p + d)) = f(t, \alpha_t(\tau, p)) + d.$$

Setting $t = \tau$ we establish from the last equality that $f(\tau, p + d) = f(\tau, p) + d.$ ∎

Consider now the discrete feedback control system

$$u_{t+1} = Pu_t + q\varphi(t, r^* u_t), \tag{10.1.4}$$

where P, q, r are constant real matrices of order $n \times n$, $n \times m$ and $n \times l$, respectively, and $\varphi : \mathbf{Z}_+ \times \mathbf{R}^l \to \mathbf{R}^m$. As in the continuous-time case the linear part of (10.1.4) and the pair (P, q) is called *controllable* if rank $[q, Pq, \dots, P^{n-1}q] = n$. The linear part of (10.1.4) and the pair (P, r) is *observable* if rank $[r, P^* r, \dots, (P^*)^{n-1} r] = n$.

In synchronization theory feedback control systems (10.1.4) with an equivariance property are called *discrete phase-controlled systems* [137]. In order to prove necessary and sufficient conditions for (10.1.4) to have the equivariance property with respect to the group $\Gamma = \{jd \ : \ j \in \mathbf{Z}\}$, where $d \in \mathbf{R}^n$ is a non-zero vector, we assume for simplicity in (10.1.4) $l = 1$.

Lemma 10.1.1 *Suppose that* (10.1.4) *is pendulum-like with respect to* $\Gamma = \{jd\}$, *the pair* (P, r) *is observable and* $l = 1$. *Then* $r^* d \ne 0$.

Proof Using the equivariance of (10.1.4) with respect to $\Gamma = \{jd \ : \ j \in \mathbf{Z}\}$ we get the relation (10.1.3) in the form

$$(P - I)d = q[\varphi(t, r^* u) - \varphi(t, r^* u + r^* d)] \tag{10.1.5}$$

for all $(t, u) \in \mathbf{Z}_+ \times \mathbf{R}^n$. Assume to the contrary that

$$r^* d = 0. \tag{10.1.6}$$

It follows then from (10.1.5) that

$$(P - I)d = 0. \tag{10.1.7}$$

Because (P, r) is observable it follows by Theorem 1.2.1, p. 5, that rank $[P^* - I, r] = n$. This contradicts (10.1.6) and (10.1.7), if $d \ne 0$. ∎

Theorem 10.1.1 *Suppose that system* (10.1.4) *is pendulum-like with respect to* $\Gamma = \{jd\}$, *the pair* (P, q) *is controllable, the pair* (P, r) *is observable and* $\operatorname{rank} q = m$. *Then w.l.o.g. it can be assumed that* $Pd = d$ *and* $\varphi(t, \cdot)$ *is* r^*d-*periodic.*

Proof Let us rewrite system (10.1.4) in the form

$$u_{t+1} = (P - q\kappa r^*)u_t + q\varphi_1(t, r^*u_t), \qquad (10.1.8)$$

where the m-vector κ is to be determined and

$$\varphi_1(t, \sigma) = \varphi(t, \sigma) + \kappa\sigma$$

for all $(t, \sigma) \in \mathbf{Z}_+ \times \mathbf{R}$. Because of $\operatorname{rank} q = m$ we get from (10.1.5) the representation

$$[q^*q]^{-1}q^*(P - I)d = \varphi(t, r^*u) - \varphi(t, r^*u + r^*d) \qquad (10.1.9)$$

for all $(t, u) \in \mathbf{Z}_+ \times \mathbf{R}^n$. Let us choose now κ in φ_1 in the form (with the use of Lemma 10.1.1)

$$\kappa = [r^*d]^{-1}[q^*q]^{-1}q^*(P - I)d.$$

Then from (10.1.9) it follows that

$$\varphi_1(t, \sigma + r^*d) = \varphi_1(t, \sigma)$$

for all $(t, \sigma) \in \mathbf{Z}_+ \times \mathbf{R}$. Using now the equivariance property (10.1.3) with respect to (10.1.8) we see that $(P - q\kappa r^*)d = d$. $\blacksquare$

The converse of Theorem 10.1.1 is true under less strong assumptions.

Theorem 10.1.2 *Suppose in* (10.1.4) *with* $l = 1$ *that* (P, r) *is observable,* P *has the eigenvalue* 1 *and* φ *is* Δ-*periodic in the second argument. Then* (10.1.4) *is pendulum-like.*

Proof Assume that $Ps = s$ with $s \neq 0$. By the observability of (P, r) it follows that $r^*s \neq 0$. Thus we can define the vector $d = (r^*s)^{-1}\Delta s$. It follows that

$$\varphi(t, r^*u + r^*d) = \varphi(t, r^*u + \Delta) = \varphi(t, r^*u)$$

for all $(t, u) \in \mathbf{Z}_+ \times \mathbf{R}^n$. On the other hand we have $(P - I)d = 0$. $\blacksquare$

In order to use, for boundedness and convergence investigations of system (10.1.4), the solvability of certain matrix inequalities of Lur'e type we provide the Yakubovich-Kalman-Szegö theorem [156] which may be considered as the discrete analogue of Theorem 1.4.1, p. 9.

Let χ denote the transfer function of the linear part associated with (10.1.4) defined by

$$\chi(s) = r^*(P - sI)^{-1}q \qquad (\det(P - sI) \neq 0).$$

Theorem 10.1.3 *Suppose that* P *and* q *are real matrices of order* $n \times n$ *and* $n \times m$, *respectively, and assume that the pair* (P, q) *is controllable and* P *has no eigenvalues on the unit circle. Let* $F : \mathbf{R}^n \times \mathbf{R}^m \to \mathbf{R}$ *be a quadratic form which is given. Then there exists a real matrix* $H = H^*$ *satisfying*

$$(Px + q\xi)^*H(Px + q\xi) - x^*Hx + F(x, \xi) \leq 0$$

for all $(x, \xi) \in \mathbf{R}^n \times \mathbf{R}^m$ *if and only if*

$$F_C([P - sI]^{-1}q\xi, \xi) \leq 0$$

for all $\xi \in \mathbf{C}^m$ *and* $s \in \mathbf{C}$ *with* $|s| = 1$, *where* F_C *denotes the extension of* F *to a Hermitian form.*

208

Using the following result from [155] one gets more information about the matrix H, determined by Theorem 10.1.3.

Theorem 10.1.4 *Let $H = H^*$, P and r be matrices of order $n \times n$, $n \times n$ and $n \times l$ respectively, and let the pair (P, r) be observable. Suppose that the inequality*

$$x^* P^* H P x - x^* H x \leq - \mid r^* x \mid^2$$

is satisfied for all $x \in \mathbf{R}^n$. Then P has no eigenvalues on the unit circle, $\det H \neq 0$ and the number of negative (resp. positive) eigenvalues of H is equal to the number of eigenvalues of P, which lie outside (resp. inside) the unit circle.

Theorem 10.1.3 and 10.1.4 are used to determine Lyapunov functions for the investigation of global properties of the discrete-time system (10.1.4). As in the continuous time case there are certain Lyapunov-type propositions available, written for the general system (10.1.2). We restrict ourself here to the case of the discrete-time analogue of the Barbashin-Krasovskij theorem whose proof runs parallel to the continuous-time case.

Theorem 10.1.5 *Suppose that the set of equilibria of (10.1.2) - (10.1.3) consists of isolated points only and there exists a continuous function $V : \mathbf{R}^n \to \mathbf{R}$ satisfying the following properties:*

(i) *$V(u + g) = V(u)$ for all $u \in \mathbf{R}^n$ and $g \in \Gamma$;*

(ii) *$V(u) + \sum_{i=1}^{m} (d_i^* u)^2 \to \infty$ as $\mid u \mid \to +\infty$;*

(iii) *For every solution $\{\alpha_t\}$ of (10.1.2) the function $V(\alpha_t)$ is non-increasing;*

(iv) *If $\{\alpha_t\}$ is a solution of (10.1.2) and there exists a $\tau > 0$ with $V(\alpha_\tau) = V(\alpha_0)$ then $\{\alpha_t\}$ is a stationary solution.*

Then system (10.1.2) - (10.1.3) is gradient-like.

In the next section we start our analysis of global aspects of (10.1.4) with boundedness considerations.

10.2 Boundedness by Positively Invariant Cones

The main tool of this section is to extend the method of positively invariant cones for ODE's to a class of discrete pendulum-like systems described by

$$\begin{aligned} z_{t+1} &= A z_t + b\varphi(t, \sigma_t), \\ \sigma_t &= c^* z_t, \qquad\qquad t = 0, 1, \ldots, \end{aligned} \tag{10.2.1}$$

in which A is a constant $n \times n$ matrix, q and r are constant n-vectors and $\varphi : \mathbf{Z}_+ \times \mathbf{R} \to \mathbf{R}$ is a scalar function satisfying the growth condition

$$\mu_1 \sigma^2 \leq \varphi(t, \sigma)\sigma \leq \mu_2 \sigma^2 \tag{10.2.2}$$

for all $(t, \sigma) \in \mathbf{Z}_+ \times \mathbf{R}$. The constants μ_1 and μ_2 with $-\infty \leq \mu_1 \leq 0 < \mu_2 \leq +\infty$ are assumed to be given. We suppose that φ is Δ-periodic in the second argument, i.e. $\varphi(t, \sigma + \Delta) = \varphi(t, \sigma)$ for all $(t, \sigma) \in \mathbf{Z}_+ \times \mathbf{R}$. Suppose also that the matrix A has the eigenvalue 1 and $n - 1$ eigenvalues inside the open unit circle.

The pairs (A, b) and (A, c) are supposed to be controllable and observable, respectively. Under this condition there exists a vector $r \in \mathbf{R}^n$ with

$$Ar = r \quad \text{and} \quad c^*r = \Delta. \tag{10.2.3}$$

To prove this, note that for any $r \neq 0$ with $Ar = r$ it follows that $c^*r \neq 0$. Actually, assuming that $c^*r = 0$ we get the relations $c^*A^k r = c^*r = 0$ for $k = 0, 1, \ldots, n-1$, which contradicts the observability of (A, c).

Denote by $\chi(s) = c^*(A - sI)^{-1}b$ the transfer function of the linear part of (10.1.2).

The next theorem can be viewed as the discrete analogue of Theorem 3.1.1, p. 48. All results of this section follow from [127, 128].

Theorem 10.2.1 *Suppose that for some $\lambda \in (0,1)$ the matrix $\frac{1}{\lambda}A$ has one eigenvalue outside and $n-1$ eigenvalues inside the open unit circle and the inequality*

$$\mu_1^{-1}\mu_2^{-1} + (\mu_1^{-1} + \mu_2^{-1})\operatorname{Re}\chi(\lambda s) + \mid \chi(\lambda s)\mid^2 < 0 \tag{10.2.4}$$

is satisfied for all $s \in \mathbf{C}$ with $\mid s \mid = 1$. Then any solution $\{z_t\}$ of (10.2.1) is bounded on $\mathbf{Z}_+$.

The proof of this theorem requires some auxiliary propositions.

Lemma 10.2.1 *Suppose that there are given the $n \times n$ matrices Q, G and H such that $G = G^*$ is positive semi-definite and $H = H^*$ has one negative and $(n$-$1)$ positive eigenvalues. Suppose also that there are n-vectors q $(q \neq 0)$, r and scalars g and $\delta > 0$ with*

$$(Qz + q\xi)^*H(Qz + q\xi) - z^*Hz + z^*Gz + gr^*z\xi \leq -\delta \mid z \mid^2 \tag{10.2.5}$$

for all $(z, \sigma) \in \mathbf{R}^n \times \mathbf{R}$. Then

$$\{z \; : \; z^*Hz \leq 0\} \cap \{z \; : \; r^*z = 0\} = \{0\}. \tag{10.2.6}$$

Proof As it was shown in the proof of Theorem 3.1.1, p. 48, under our conditions there exists a n-vector h such that

$$\{z \; : \; z^*Hz \leq 0\} \cap \{z \; : \; h^*z = 0\} = \{0\}. \tag{10.2.7}$$

Taking in (10.2.5) $z = 0$ we get $q^*Hq \leq 0$. From this and (10.2.7) it follows that $h^*q \neq 0$. Suppose to the contrary that (10.2.6) is not satisfied, i.e. that there exists a vector $u \neq 0$ with $r^*u = 0$ and $u^*Hu \leq 0$. Because of $u^*Gu \geq 0$ we discover from (10.2.5) the inequality

$$(Qu + q\xi)^*H(Qu + q\xi) \leq -\delta \mid u \mid^2 \tag{10.2.8}$$

for all $\xi \in \mathbf{R}$. Since $h^*q \neq 0$ we can choose $\xi_0 \in \mathbf{R}$ such that

$$h^*[Qu + q\xi_0] = 0. \tag{10.2.9}$$

By (10.2.7), (10.2.8) and (10.2.9) we obtain that $Qu + q\xi_0 = 0$, which contradicts (10.2.8). $\blacksquare$

In the next lemma we consider the non-autonomous system

$$z_{t+1} = f(t, z_t), \qquad t = 0, 1, \ldots \tag{10.2.10}$$

with $f : \mathbf{Z}_+ \times \mathbf{R}^n \to \mathbf{R}^n$.

Lemma 10.2.2 *Let be $\lambda \in (0,1]$, $\delta > 0$, $V : \mathbf{R}^n \to \mathbf{R}$ continuous and*

$$\frac{1}{\lambda} V(z_{t+1}) - V(z_t) \leq -\delta \mid z_t \mid^2 \tag{10.2.11}$$

for any solution $\{z_t\}$ of $(10.2.10)$ and for all $t \geq t_0$, where t_0 depends on $\{z_t\}$. Then any solution of $(10.2.10)$ converges to zero for $t \to \infty$ or enters the positively invariant set $\{z \ : \ V(z) < 0\}$.

Proof Suppose $\{z_t\}$ is given and write the inequality $(10.2.11)$ in the form

$$V(z_{t+1}) - V(z_t) + (1 - \lambda)V(z_t) \leq -\lambda\delta \mid z_t \mid^2 \tag{10.2.12}$$

for $t \geq t_0$. If $\{z_t\}$ does not converge to zero we have $\sum\limits_{t=t_0}^{\infty} \mid z_t \mid^2 = +\infty$. Suppose to the contrary, that $V(z_t) \geq 0$ for all $t \geq t_0$. From $(10.2.12)$ it follows then that for arbitrary $T \geq t_0$

$$V(z_{t+1}) - V(z_{t_0}) \leq -\lambda\delta \sum\limits_{t=t_0}^{T} \mid z_t \mid^2$$

and, consequently, $V(z_{T+1}) \to -\infty$ for $T \to +\infty$, a contradiction. $\blacksquare$

In the next proposition *cone* means, in contrast to the definition, given in Chapter 3, a topological cone [75] i.e. a set $K \subset \mathbf{R}^n$ which is convex, closed and satisfies $K \cap (-K) = \{0\}$ and $\lambda K \subset K$ for all $\lambda \geq 0$.

Lemma 10.2.3 *Let $H = H^*$ be an $n \times n$ matrix, h an n-vector and*

$$K := \{z \ : \ z^* H z \leq 0\} \cap \{z \ : \ h^* z \geq 0\}.$$

Then it holds that:

  (i) *The set K is a cone if and only if $(10.2.7)$ is satisfied.*

  (ii) *If $(10.2.7)$ is satisfied then the matrix H has at most one non-positive eigenvalue.*

Proof Suppose K is a cone and $u \in \{z \ : \ z^* H z \leq 0\} \cap \{z \ : \ h^* z = 0\}$. Because of $(-u)^* H(-u) \leq 0$ and $h^*(-u) = 0$ it follows that $(-u) \in K$. Using the cone property we get $u = 0$. Suppose now that $(10.2.7)$ is satisfied. Since the set $\{z \ : \ h^* z = 0\}$ has at least the dimension $n - 1$ it follows from $(10.2.7)$ that $\dim \{z \ : \ z^* H z \leq 0\} \leq 1$. There are two possibilities:

1. $H > 0$. Thus $\{z \ : \ z^* H z \leq 0\} = \{0\}$ and K is a cone.

2. The matrix H has one non-positive and $n-1$ positive eigenvalues. W.l.o.g. we may assume that $H = \begin{bmatrix} \lambda_1 & 0 \\ 0 & H_0 \end{bmatrix}$, where $\lambda_1 \leq 0$ is a scalar and $H_0 > 0$ an $(n-1) \times (n-1)$ matrix.

The direct computation shows that

$$\begin{aligned}
\{z \ : \ z^* H z \leq 0\} &= \left\{z = (z_1, z_2) \in \mathbf{R}^n \ : \ \lambda_1 z_1^2 + z_2^* H_0 z_2 \leq 0\right\} \\
&= \left\{z \ : \ \sqrt{z_2^* H_0 z_2} \leq \sqrt{-\lambda_1} z_1\right\} \cup \left\{z \ : \ \sqrt{z_2^* H_0 z_2} \leq -\sqrt{-\lambda_1} z_1\right\} \\
&=: \ M \cup (-M).
\end{aligned}$$

Note that $q : \mathbf{R}^n \to \mathbf{R}$, defined by $q(z) = \sqrt{z_2^* H_0 z_2} - \sqrt{-\lambda_1} z_1$, $(z = (z_1, z_2) \in \mathbf{R} \times \mathbf{R}^{n-1})$ is a sublinear functional. Using the results of [75] the set $M = \{z \ : \ q(z) \leq 0\}$ is a cone if $u \in M$ and $-u \in M$ implies that $u = 0$. It follows that M and $(-M)$ are cones and because of $(10.2.7)$ the set $\{z \ : \ h^* z \geq 0\}$ contains M or $(-M)$. $\blacksquare$

Lemma 10.2.4 *Suppose that for the $n \times n$ matrix $H = H^*$ and the n-vector h the relation (10.2.7) is satisfied. Then for arbitrary numbers α_1, α_2 with $\alpha_1 \leq \alpha_2$ and $\alpha_1 \alpha_2 \geq 0$ the set*

$$M := \{z \,:\, z^* H z \leq 0\} \cap \{z \,:\, \alpha_1 \leq h^* z \leq \alpha_2\}$$

is bounded.

Proof By Lemma 10.2.3 the set M is convex. If we assume that M is unbounded then there exists a ray $\{a + \lambda d, \lambda \geq 0\} \subset M$ and $d \neq 0$. By construction of M this is possible only in the case $h^* d = 0$. Because of (10.2.7) this implies $d = 0$. ∎

Proof of Theorem 10.2.1 The pair $(\frac{1}{\lambda} A, b)$ is controllable, the matrix $\frac{1}{\lambda} A$ has no eigenvalues on the unit circle and the frequency-domain condition (10.2.4) is satisfied. Since, by the Yakubovich-Kalman-Szegö theorem 10.1.3, p. 208, there exists an $n \times n$ matrix $H = H^*$ and a number $\delta > 0$ such that

$$\frac{1}{\lambda^2}(Az + b\xi)^* H(Az + b\xi) - z^* H z + (\mu_2^{-1}\xi - c^* z)(\mu_1^{-1}\xi - c^* z) \leq -\delta \mid z \mid^2 \qquad (10.2.13)$$

for all $(z, \xi) \in \mathbf{R}^n \times \mathbf{R}$. Putting in (10.2.13) $\xi = 0$ we get

$$\frac{1}{\lambda^2} z^* A^* H A z - z^* H z \leq -\delta \mid z \mid^2 \qquad (10.2.14)$$

for all $z \in \mathbf{R}^n$. Since the matrix $\frac{1}{\lambda} A$ has one eigenvalue outside and $n - 1$ eigenvalues inside the open unit circle we establish from (10.2.14) by Theorem 10.1.4 that the matrix H has one positive and $n - 1$ negative eigenvalues. Furthermore there exists an n-vector h such that (10.2.7) is satisfied.

Define the functions $V_j : \mathbf{R}^n \to \mathbf{R}$ by

$$V_j(z) = (z - jr)^* H(z - jr),$$

where r is from (10.2.3) and j is an arbitrary integer. Let us also for $j \in \mathbf{Z}$ define the sets

$$\begin{aligned}
\Gamma_j &:= \{z \,:\, V_j(z) < 0\}, \\
\Gamma_j^+ &:= \Gamma_j \cap \{z \,:\, h^*(z - jr) \geq 0\} \quad \text{and} \\
\Gamma_j^- &:= \Gamma_j \cap \{z \,:\, h^*(z - jr) \leq 0\}.
\end{aligned}$$

Taking $z = r$ we receive from (10.2.14) that

$$r^* H r < 0. \qquad (10.2.15)$$

Consider now an arbitrary solution $\{z_t(t_0, z_0)\}$ of (10.2.1). Two situations are possible:

1. There exists an integer j such that $z_t(t_0, z_0) \to jr$ for $t \to +\infty$. It follows that $\{z_t(t_0, z_0)\}$ is bounded.

2. The solution does not converge to any jr. Let us consider the term

$$V_j(z_0) = (z_0 - jr)^* H(z_0 - jr) = z_0^* H z_0 - 2jr^* H z_0 + j^2 r^* H r.$$

212

It follows from (10.2.15) that one can choose an integer j_0 such that $V_j(z_0) < 0$ for all $\mid j \mid \geq j_0$.

Assume now that j is an integer from $(-j_0, j_0)$. Since $z_t(t_0, z_0) \not\to jr$ for $t \to +\infty$ and all j, we get by the equivariance property that

$$z_t(t_0, z_0) - jr = z_t(t_0, z_0 - jr) \not\to 0 \qquad \text{for} \qquad t \to +\infty \tag{10.2.16}$$

The inequality (10.2.13) implies that

$$\frac{1}{\lambda^2} V_0(z_{t+1}(t_0, z_0 - jr)) - V_0(z_t(t_0, z_0 - jr)) \leq -\delta \mid z_t(t_0, z_0 - jr) \mid^2$$

for all $t \geq t_0$. By Lemma 10.2.2 there exists an integer t_1 with $z_t(t_0, z_0 - jr) \in \Gamma_0$, i.e. $z_t(t_0, z_0) \in \Gamma_j$ for all $t \geq t_1$. The last inclusion holds for integer j from $(-j_0, j_0)$ if t_1 is chosen sufficiently large.

As a consequence we can assume w.l.o.g. that for the given solution $\{z_t(t_0, z_0)\}$ the initial point z_0 lies in Γ_j for $j = 0, \pm 1, \ldots$. Using (10.2.7) and (10.2.15) we see that $h^* r \neq 0$ and we can take an integer m with $m h^* r \leq h^* z_0 \leq (m+1) h^* r$. It follows that

$$z_0 \in F_m := \Gamma_m \cap \Gamma_{m+1} \cap \{z \; : \; m h^* r \leq h^* z \leq (m+1) h^* r\} = \Gamma_m^+ \cap \Gamma_{m+1}^-$$

Lemma 10.2.4 guarantees that the set F_m is bounded. Next we prove the positive invariance of F_m with respect to (10.2.1), i.e. we show that

$$z_t(t_0, z_0) \in F_m \tag{10.2.17}$$

for all $t \geq t_0$. Because of $z_0 \in \Gamma_m^+$ one has $z_0 - mr \in \Gamma_0^+$. Suppose that κ/λ denotes the eigenvalue of $\frac{1}{\lambda} A$ which lies outside the unit circle and u is a corresponding eigenvector. From (10.2.14)

$$\left[\left(\frac{\kappa}{\lambda}\right)^2 - 1\right] u^* H u \leq -\delta \mid u \mid^2$$

and therefore

$$u^* H u < 0. \tag{10.2.18}$$

We infer from (10.2.7) and (10.2.18) that $h^* u \neq 0$. Thus, we can assume w.l.o.g. that

$$h^* u > 0. \tag{10.2.19}$$

Since (A, c) is observable, we obtain that

$$c^* u \neq 0. \tag{10.2.20}$$

In order to prove (10.2.17) we show first that $z_0 \in \Gamma_0^+$ implies $A z_0 + b\varphi(t, c^* z_0) \in \Gamma_0^+$ for arbitrary $t \geq 0$. Suppose to the contrary that there are $z_0 \in \Gamma_0^+$ and $t_0 \geq 0$ with

$$A z_0 + b\varphi(t_0, c^* z_0) \notin \Gamma_0^+. \tag{10.2.21}$$

It follows from (10.2.2) that $G(c^* z_0, \varphi(t_0, c^* z_0)) \geq 0$, where $G(\sigma, \xi) = (\mu_2^{-1} \xi - \tau)(\mu_1^{-1} \xi - \sigma)$. Therefore we have from (10.2.13) that $A z_0 + b\varphi(t_0, c^* z_0) \in \Gamma_0$. Consequently, (10.2.21) implies that

$$h^*[A z_0 + b\varphi(t_0, c^* z_0)] < 0. \tag{10.2.22}$$

Because of (10.2.20) we can choose a number $\varrho > 0$ such that $z_1 := \varrho u$ satisfies $c^* z_1 \neq c^* z_0$. Let us construct a continuous function $\widetilde{\varphi} : \mathbf{R} \to \mathbf{R}$ with $\widetilde{\varphi}(c^* z_1) = 0$, $\widetilde{\varphi}(c^* z_0) = \varphi(t_0, c^* z_0)$ and

$G(\sigma, \widetilde{\varphi}(\sigma)) \geq 0$ for all $\sigma \in \mathbf{R}$. This is possible since the set $\{(\sigma, \xi) \, : \, G(\sigma, \xi) \geq 0\}$ is connected and the points $(\sigma, 0)$ belong to this set. By Lemma 10.2.3 the set Γ_0^+ is convex. Consequently,

$$y(s) := sz_0 + (1-s)z_1 \in \Gamma_0^+ \tag{10.2.23}$$

for $s \in [0,1]$. Using the fact that $h^*y(s) > 0$ on $[0,1]$, we get by (10.2.7) and (10.2.23) that

$$y(s) \neq 0 \qquad \text{on} \qquad [0,1]. \tag{10.2.24}$$

Employing (10.2.13), (10.2.23) and the inequality

$$G(c^*y(s), \widetilde{\varphi}(c^*y(s))) \geq 0 \qquad (s \in [0,1])$$

we conclude that

$$[Ay(s) + b\widetilde{\varphi}(c^*y(s))]^*H[Ay(s) + b\widetilde{\varphi}(c^*y(s))] \leq -\delta \mid y(s) \mid^2 \tag{10.2.25}$$

for $s \in [0,1]$. The inequalities (10.2.23) and (10.2.25) imply that

$$Ay(s) + b\widetilde{\varphi}(c^*y(s)) \neq 0 \tag{10.2.26}$$

for $s \in [0,1]$. If we take now the continuous function $f : [0,1] \to \mathbf{R}$ defined by $f(s) = h^*[Ay(s) + b\widetilde{\varphi}(c^*y(s))]$ we get from (10.2.19) that $f(0) = h^*Az_1 = \kappa \varrho h^* u > 0$ and from (10.2.22) that $f(1) = h^*[Az_0 + b\varphi(t_0, c_0^*z_0)] < 0$. Using the continuity of f and a mean value theorem we see that $f(\bar{s}) = 0$ for some $\bar{s} \in (0,1)$. From this and (10.2.7), (10.2.24) it follows that

$$Ay(\bar{s}) + b\widetilde{\varphi}(c^*y(\bar{s})) = 0. \tag{10.2.27}$$

It is elementary to see, that (10.2.27) contradicts (10.2.25). Thus we have shown the positive invariance of Γ_0^+. In a similar way one shows the positive invariance of Γ_0^- and, therefore, (10.2.17). ∎

Parallel to the continuous-time case we introduce the following definition.

Definition 10.2.1 The pendulum-like system (10.2.1) is called Δ-*stable* or *Bakaev-stable* if for any solution $\{\sigma_t\}$ there exists a $T > 0$ such that $\mid \sigma_k - \sigma_m \mid < \Delta$ for all $k, m > T$.

The next theorem may be viewed as the discrete analogue of Theorem 3.4.1, p. 59.

Theorem 10.2.2 *Suppose that the assumptions of Theorem* 10.2.1 *are satisfied with* $\mu_1 = -\infty$ *and* $\mu_2 < +\infty$ *or* $\mu_1 > -\infty$ *and* $\mu_2 = +\infty$. *Then system* (10.2.1) *is Bakaev-stable.*

Proof We can proceed exactly as in the proof of Theorem 10.2.1. In contrast to the situation in Theorem 10.2.1 however we can use by Lemma 10.2.1 the relation

$$\{z \, : \, z^*Hz \leq 0\} \cap \{z \, : \, c^*z = 0\} = \{0\}.$$

It follows that c can play the role of h. By the same type of argument as in the proof of Theorem 10.2.1 one shows that the sets

$$\Gamma_j := F_j \cap F_{j+1} \cap \{z \, : \, j\Delta \leq c^*z \leq (j+1)\Delta\}$$

are positively invariant for system (10.2.1) and any non-convergent solution enters such a set. ∎

The next theorem demonstrates a further realization of the boundedness-concept based on the use of invariant cones.

Convention 10.2.1 *In the following theorem inequality $B > 0$ (resp. $B \geq 0$), where B is an arbitrary matrix, means that all the components of B are positive (resp. non-negative).*

Theorem 10.2.3 *Assume that (10.2.2) is satisfied with $\mu_2 = +\infty$. Assume also that $A > 0$, $b \geq 0$, $c \geq 0$ and $A + \mu_1 bc^* \geq 0$. Then any solution of (10.2.1) is bounded on $\mathbf{Z}_+$.*

Proof Since $A > 0$ and A has the eigenvalue 1, one can assume that the vector r in (10.2.3) satisfies $r > 0$. Let us define for arbitrary integer $j < m$ the convex sets

$$K_j^+ = \{z \ : \ z \geq jr\}, \qquad K_m^+ = \{z \ : \ -z \geq -mr\}$$

and $\Gamma_{j,m} = K_j^+ \cap K_m^-$. Then the set $\Gamma_{j,m}$ is bounded.

Indeed, $\Gamma_{j,m}$ is convex and supposing that it is unbounded, we get the existence of vectors $a, q \in \mathbf{R}^n, q \neq 0$, which satisfy $a + sq \in \Gamma_{j,m}$ for all $s \geq 0$. But from the last fact it follows that $q = 0$. Consider now an arbitrary solution $\{z_t(\tau, p)\}$ of (10.2.1). Because of $r > 0$ there exist integer j, m with $j < m$ and $p \in \Gamma_{j,m}$. Let us show that $z_t(\tau, p) \in \Gamma_{j,m}$ for $t = \tau, \tau - 1, \dots$. Using the equivariance property it sufficies to show that

$$z_t(\tau, p - jr) \in K_0^+ \quad \text{and} \quad z_t(\tau, p - mr) \in K_0^-$$

for all $t \geq \tau$. But this is equivalent to the fact that from $z \in K_0^+$ and $t \geq 0$ it follows that $Az + b\varphi(t, c^*z) \in K_0^+$ and from $z \in K_0^-$ and $t \geq 0$ it follows that $Az + b\varphi(t, c^*z) \in K_0^-$.

Let us write

$$z_1 := Az + b\varphi(t, c^*z) = (A + \mu_1 bc^*)z + b[\varphi(t, c^*z) - \mu_1 c^*z].$$

From (10.2.2) it follows that $\varphi(t, c^*z) - \mu_1 c^*z \geq 0$ if $z \in K_0^+$ and, consequently, $z_1 \geq 0$. If $z \in K_0^-$ we have $\varphi(t, c^*z) - \mu_1 c^*z \leq 0$. By the same argument as above if follows that $z_1 \leq 0$. Thus we have shown the positive invariance of $\Gamma_{j,m}$. Since $\Gamma_{j,m}$ is bounded the assertion of Theorem 10.2.3 follows. ∎

Example 10.2.1 Consider the discrete PLL system [137]

$$\vartheta_{t+1} = \vartheta_t + \Omega_H T_p - \Omega_y T_p F(\vartheta_t), \quad t = 0, 1, \dots, \tag{10.2.28}$$

where Ω_H, Ω_y, T_p are constant and the 2π-periodic function $F : \mathbf{R} \to \mathbf{R}$ is defined by

$$F(\vartheta) = \begin{cases} +1 & if \quad 0 \leq \vartheta < \pi \\ -1 & if \quad -\pi \leq \vartheta < 0. \end{cases}$$

Putting $\gamma_H = \Omega_H / \Omega_y$ we introduce the 2π-periodic function

$$F_1(\vartheta) = \begin{cases} \Omega_y T_p(\gamma_H - 1) & \text{if} & 0 \leq \vartheta < \pi \\ \Omega_y T_p(\gamma_H - 1) & \text{if} & -\pi \leq \vartheta < 0. \end{cases}$$

The change of variables $\vartheta = \sigma + \pi$ converts (10.2.28) into the system

$$\sigma_{t+1} = \sigma_t + F_2(\sigma_t), \quad t = 0, 1, \dots, \tag{10.2.29}$$

where $F_2(\sigma) \equiv F_1(\sigma + \pi)$. The transfer function of the linear part of (10.2.29) is

$$\chi(s) = \frac{1}{1 - s}.$$

Suppose now that $|\gamma_H| < 1$. Defining μ_1 by

$$\mu_1 = \min\left\{\frac{\Omega_y T_p}{\pi}(\gamma_H - 1),\, \frac{-\Omega_y T_p}{\pi}(\gamma_H + 1)\right\}$$

we get the inequality

$$F_2(\sigma)\sigma \geq \mu_1\sigma^2 \quad\text{for all}\quad \sigma \in \mathbf{R}.$$

A direct computation shows that for arbitrary $\alpha \in [-\pi, \pi)$ and $\lambda \in (0,1)$

$$\operatorname{Re}\chi(\lambda e^{i\alpha}) = \frac{1 - \lambda\cos\alpha}{(1 - \lambda\cos\alpha)^2 + \lambda^2\sin^2\alpha}$$

and

$$|\chi(\lambda e^{i\alpha})|^2 = \frac{1}{(1 - \lambda\cos\alpha)^2 + \lambda^2\sin^2\alpha}.$$

Thus the frequency-domain condition of Theorem 10.2.2 is satisfied if $1 - \lambda\cos\alpha + \mu_1 > 0$ for $\alpha \in [-\pi, \pi)$. Since we can choose a suitable $\lambda \in (0,1)$ the last inequality is valid if $1 + \mu_1 > 0$, i.e. if

$$\beta := \max\left\{\frac{-\Omega_y T_p}{\pi}(\gamma_H - 1),\, \frac{\Omega_y T_p}{\pi}(\gamma_H + 1)\right\} < 1. \tag{10.2.30}$$

Under these conditions Theorem 10.2.2 guarantees that system (10.2.29) is Bakaev stable. Note that Theorem 10.2.3 is also applicable. In the notation of this theorem we have $A = 1 > 0$, $c = b = 1 > 0$ and $A + \mu_1 bc^* = 1 + \mu_1$.

We see that for β, defined by (10.2.30), with $\beta \leq 1$ the conditions of Theorem 10.2.3 are satisfied.

Example 10.2.2 Suppose that an impulse PLL system with a sinusoidal impulse-phase detector [137] is given by

$$\vartheta_{t+1} - \vartheta_t + \Omega_c[\sin(\vartheta_t + \vartheta_0) - \sin\vartheta_0] = 0, \tag{10.2.31}$$

where $\vartheta_0 = \arcsin\dfrac{\Omega_H}{\Omega_y} \in [0, \pi/2]$ and $\Omega_c = \Omega_y T_p$ is the magnitude of the synchronization band. Introducing in (10.2.31) the new variable $\sigma = \vartheta - \pi + 2\vartheta_0$, we get the system

$$\sigma_{t+1} - \sigma_t + \Omega_c[\sin(\vartheta_0 - \sigma_t) - \sin\vartheta_0] = 0. \tag{10.2.32}$$

It is easy to see that the inequality

$$[\sin(\vartheta_0 - \sigma) - \sin\vartheta_0]\sigma \leq \frac{1 + \sin\vartheta_0}{\sqrt{2} - 1 + 1.25\pi - \vartheta_0}\sigma^2$$

takes place for all $\sigma \in \mathbf{R}$. The transfer function of the linear part of (10.2.32) with the nonlinearity $\varphi(\sigma) = \Omega_c[\sin(\vartheta_0 - \sigma) - \sin\vartheta_0]$ is given by $\chi(s) = \dfrac{1}{s - 1}$. According to Theorem 10.2.1 we get that any solution $\{\sigma_t\}$ of (10.2.32) is bounded on $\mathbf{Z}_+$ provided that

$$\Omega_c < \min\left\{2,\, \frac{\sqrt{2} - 1 + 1.25\pi - \vartheta_0}{1 + \sin\vartheta_0}\right\}.$$

Example 10.2.3 Consider the discrete phase-controlled system with proportionally-integrating filter, described by the second-order equation [140]

$$\sigma_{t+2} - (1 + b)\sigma_{t+1} + b\sigma_t + d\varphi(\sigma_{t+1}) + (1 - d - b)\varphi(\sigma_t) = 0 \tag{10.2.33}$$

where $\varphi(\sigma) = \Omega_c F(\sigma) - \Omega_H T_p$ and b, d, Ω_c, Ω_H and T_p are positive parameters with $0 < b < 1$ and $1 - d - b \neq 0$. A direct computation gives

$$\chi(s) = \frac{ds + (1 - d - b)}{(s - b)(s - 1)}. \tag{10.2.34}$$

The 2π-periodic function F is given by

$$F(\sigma) = \begin{cases} -1 & \sigma \in [0, \pi), \\ 1 & \sigma \in [-\pi, 0). \end{cases}$$

Clearly, that

$$\varphi(\sigma)\sigma \leq \frac{\Omega_c + \Omega_H T_p}{\pi}\sigma^2, \qquad \sigma \in \mathbf{R}. \tag{10.2.35}$$

One can show that the frequency condition (10.2.4) for χ given by (10.2.34) and $\mu_1 = -\infty$ is satisfied if

$$\mu_2 \leq \frac{(1 - b)(\sqrt{1 - d} - 1)^2}{d^2}. \tag{10.2.36}$$

It follows then by (10.2.35) and (10.2.36) that condition (10.2.4) of Theorem 10.2.2 is satisfied if

$$\frac{\Omega_c + \Omega_H T_p}{\pi} < \frac{(1 - b)(\sqrt{1 - d} - 1)^2}{d^2}.$$

According to Theorem 10.2.2 the considered system (10.2.33) is 2π-stable in this case.

10.3 Bakaev-Guzh Technique for Discrete Systems

In this section we provide sufficient conditions for the monostability of system (10.1.2). The technique which is used here is similar to the constructions for continuous time systems on Riemannian manifolds used in Chapter 4. On the basis of this abstract result, frequency-domain conditions for gradient-like behavior of discrete pendulum-type systems are derived

We start our disposition with a result taken from [60], which may be considered as the analogon of Theorem 4.3.1, p. 68.

Theorem 10.3.1 *Suppose that for system (10.1.2) there exists a C^1-function $V : M \to \mathbf{R}$, a closed vector field h on M and a continuous function $\vartheta : M \to \mathbf{R}_+$ such that the following conditions hold:*

(i) *For any $u \in M$ there is a piece-wise C^1-path β_u from u to $f(u)$ satisfying*

$$\int\limits_0^1 \langle \operatorname{grad} V(\beta_u(\tau)) + h(\beta_u(\tau)), \dot{\beta}_u(\tau) \rangle \, d\tau \leq -\vartheta(u);$$

(ii) *If $\sum\limits_{t=0}^{\infty} \vartheta(\alpha_t(p)) < +\infty$ for some bounded on $\mathbf{Z}_+$ solution $\{\alpha_t(p)\}$ of (10.1.2) then this solution converges for $t \to +\infty$ to an equilibrium;*

(iii) *For an arbitrary solution $\{\alpha_t(p)\}$ of (10.1.2) and arbitrary integer $k < l$ let $\alpha[k, l, p]$ denote the closed piecewise C^1-path, constructed by the curves β from (i) between α_t and α_{t+1} ($t = k, k+1, \ldots, l-1$) and the geodesic between α_k and α_l. Then for any $\alpha[k, l, p]$ which is not* *contractible in M, there exist numbers $\varepsilon_1 > 0$ and $\varepsilon_2 > 0$, depending only on the homotopy class of $\alpha[k, l, p]$, such that*

$$\sum_{t=k}^{l-1} \vartheta(\alpha_t(p)) \geq \varepsilon_1 + \int\limits_{\alpha[k,l,p]} \langle h, du \rangle,$$

proposed that $d(\alpha_t, \alpha_l) < \varepsilon_2$.

Then system (10.1.2) is monostable.

Proof Assume that $\alpha_t \equiv \alpha_t(p)$ is an arbitrary bounded on $\mathbf{Z}_+$ solution of (10.1.2). Along this solution we have for arbitrary $k < l$

$$
\begin{aligned}
V(\alpha_l) - V(\alpha_k) &= \sum_{t=k}^{l-1} V(\alpha_{t+1}) - V(\alpha_t) \\
&= \sum_{t=k}^{l-1} \int_0^1 \langle \operatorname{grad} V(\beta_{\alpha_t}(\tau)) + h(\beta_{\alpha_t}(\tau)), \dot\beta_{\alpha_t}(\tau) \rangle \, d\tau \\
&\quad - \sum_{t=k}^{l-1} \int_0^1 \langle h(\beta_{\alpha_t}(\tau)), \dot\beta_{\alpha_t}(\tau) \rangle \, d\tau,
\end{aligned}
\tag{10.3.1}
$$

where β_{α_t} is the path beween α_t and α_{t+1} defined in assumption (i). Because $\{\alpha_t\}$ is bounded the ω-limit set $\omega(p)$ is not empty. Let us choose a point $q \in \omega(p)$ and two sequences $k_n \to +\infty$ and $l_n \to +\infty$ with $\alpha_{k_n} \to q, \alpha_{l_n} \to q$ and $l_n - k_n \to +\infty$ as $n \to +\infty$.

For every n we consider the closed piece-wise smooth path $\alpha[k_n, l_n, p]$. It follows that for every n

$$\left| \int\limits_{\gamma[\alpha_{k_n}, \alpha_{l_n}]} \langle h, du \rangle \right| < \lambda_n,$$

where $\gamma[u, v]$ denotes the geodesic from u to v, and $\lambda_n \to 0$ for $n \to +\infty$. The following two situations are possible.

First, every cycle $\alpha[k_n, l_n, p]$ for $n \geq n_0$, n_0 sufficiently large, is contractible and $d(\alpha_{k_n}, \alpha_{l_n}) < \varepsilon_2$. Then by (10.3.1) we conclude that for $n \geq n_0$

$$V(\alpha_{l_n}) - V(\alpha_{k_n}) \leq -\sum_{t=k_n}^{l_n-1} \vartheta(\alpha_t) + \lambda_n.$$

Second, for any $n_0 \geq 0$ there exists a cycle $\alpha[k_n, l_n, p]$ with $n \geq n_0$ and $d(\alpha_{l_n}, \alpha_{k_n}) < \varepsilon_2$, which is not contractible in M. Because of (10.3.1), condition (iii) and the properties of $\{\alpha_{k_n}\}$ and $\{\alpha_{l_n}\}$ this implies that there exists a sequence of positive numbers $\{\delta_n\}$ satisfying $\delta_n \to 0$ for $n \to \infty$ and

$$\delta_n \geq V(\alpha_{k_n}) - V(\alpha_{l_n}) \geq \sum_{t=k_n}^{l_n-1} \vartheta(\alpha_t) - \lambda_n + \int\limits_{\alpha[k_n,l_n,p]} \langle h, du \rangle \geq \varepsilon_1 - \lambda_n.$$

It is clear that for large n the inequality $\delta_n \geq \varepsilon_1 - \lambda_n$ is impossible. From the above it follows that one can take a fixed k_{n_0} sufficiently large and $l_n \to \infty$ such that the closed paths $\alpha[k_{n_0}, l_n, p]$ are all contractible for sufficiently large n. Thus, there exists a sequence $l_n \to \infty$ with

$$\sum_{t=k_{n_0}}^{l_n-1} \vartheta(\alpha_t) \leq c,$$

where c does not depend on n. Using now assumption (ii) of the theorem we get the stated assertion. ∎

Let us now investigate the discrete-time pendulum-like system

$$\begin{aligned} z_{t+1} &= Az_t + b\varphi(\sigma_t), \\ \sigma_{t+1} &= \sigma_t + c^*z_t + \varrho\varphi(\sigma_t), \quad t = 0, 1, \ldots, \end{aligned} \tag{10.3.2}$$

where A is an $n \times n$ matrix with all eigenvalues inside the open unit circle, b and c are n-vectors and ϱ is a number.

Suppose that the pair (A, b) is controllable and $\varrho \neq c^*(A - I)^{-1}b$. Let us assume further that $\varphi : \mathbf{R} \to \mathbf{R}$ is Δ-periodic, Lipschitz continuous on $\mathbf{R}$ with a constant L and has a finite number of zeros on $[0, \Delta)$. Denote by $K(s) = c^*(A - sI)^{-1}b - \varrho$ the transfer function of the linear part of (10.3.2). Under the conditions described for system (10.3.2) we can interpret it on the flat cylinder $\mathbf{R}^{n+1}/\Gamma$, where $\Gamma = \{kd \ : \ k \in \mathbf{Z}\}$ and $d = (0, \ldots, 0, \Delta)$. We want to establish a convergence result for system (10.3.2) by means of Theorem 10.3.1, which can be viewed as the discrete-time analogue of Theorem 4.3.2, p. 70.

Theorem 10.3.2 *Suppose that there exist parameters $\varepsilon > 0$, $\delta > 0$ and*

$$\nu > \frac{\left| \int_0^{\Delta} \varphi(\sigma)\, d\sigma \right|}{\int_0^{\Delta} |\varphi(\sigma)|\, d\sigma}$$

such that the following conditions hold:

(i) $\operatorname{Re} K(s) - \varepsilon \mid K(s) \mid^2 - \delta \geq 0$ *for all $s \in \mathbf{C}$ with $\mid s \mid = 1$;*

(ii) $4[\varepsilon - (1 + \nu)L]\delta > \nu^2$.

Then system (10.3.2) is gradient-like.

Proof It may be argued exactly the same way as in the proof of Theorem 4.3.2, p. 70. Since the pair (A, b) is controllable, and the frequency-domain condition (i) is satisfied, the Yakubovich-Kalman-Szegö theorem (Theorem 10.1.3) guarantees the existence of an $n \times n$ matrix $H = H^*$ such that

$$(Az + b\xi)^*H(Az + b\xi) - z^*Hz + \xi(c^*z + \varrho\xi) + \varepsilon(c^*z + \varrho\xi)^2 + \delta\xi^2 \leq 0 \tag{10.3.3}$$

for all $(z, \sigma) \in \mathbf{R}^n \times \mathbf{R}$. Using this matrix H we define on $\mathbf{R}^{n+1}/\Gamma$ the C^1-function $V(z, \sigma) := z^*Hz$.

Besides, we determine the closed vector field $h(z, \sigma) := [0, \ldots, 0, \varphi(\sigma)]$ and consider for an arbitrary point $(z, \sigma) \in \mathbf{R}^{n+1}/\Gamma$ the path

$$\beta(\tau) = (z, \sigma) + \tau[(\overline{z}, \overline{\sigma}) - (z, \sigma)], \ \tau \in [0, 1],$$

which connects (z, σ) with the image $(\overline{z}, \overline{\sigma})$ under (10.3.2). According to (10.3.3) we obtain for

$$\int_0^1 \langle \operatorname{grad} V(\beta(\tau)) + h(\beta(\tau)), \dot{\beta}(\tau) \rangle\, d\tau = V(\overline{z}, \overline{\sigma}) - V(z, \sigma) + \int_{\sigma}^{\overline{\sigma}} \varphi(\tau)\, d\tau$$

the inequality

$$\int_0^1 \langle grad\, V + h, \dot\beta(\tau)\rangle\, d\tau - \int_\sigma^{\overline\sigma} \varphi(\tau)\, d\tau + \varphi(\sigma)(\overline\sigma - \sigma) + \varepsilon(\overline\sigma - \sigma)^2 + \delta\varphi(\sigma)^2 \le 0. \qquad (10.3.4)$$

Applying a mean value theorem to the integral and using the Lipschitz continuity of φ we get for $(z,\sigma) \in \mathbf{R}^{n+1}/\Gamma$ the relation

$$\begin{aligned}
\int_\sigma^{\overline\sigma} \varphi(\tau)\, d\tau &= \varphi(\widetilde\sigma)(\overline\sigma - \sigma) \\[2mm]
&= \varphi(\sigma)(\overline\sigma - \sigma) + (\varphi(\widetilde\sigma) - \varphi(\sigma))(\overline\sigma - \sigma) \\
&\le \varphi(\sigma)(\overline\sigma - \sigma) + L(\overline\sigma - \sigma)^2,
\end{aligned}$$

with $\widetilde\sigma$ meaning a value between σ and $\widetilde\sigma$. Combining the last inequality and (10.3.4) we receive

$$\int_0^1 \langle grad\, V + h, \dot\beta(\tau)\rangle\, d\tau \le -(\varepsilon - L)(\overline\sigma - \sigma)^2 - \delta\varphi(\sigma)^2 \qquad (10.3.5)$$

for all $(z,\sigma) \in \mathbf{R}^{n+1}/\Gamma$. Thus conditions (i) and (ii) of Theorem 10.3.1 are satisfied if $L < \varepsilon$ and ϑ is given by $\vartheta(z,\sigma) := (\varepsilon - L)(c^*z + \varrho\varphi(\sigma))^2 + \delta\varphi(\sigma^2)$. Indeed, from $\sum_{t=0}^\infty \vartheta(z_t, \sigma_t) < \infty$ for a solution $\{z_t, \sigma_t\}$ it follows that $\varphi(\sigma_t) \to 0$ and $\sigma_{t+1} - \sigma_t \to 0$ for $t \to +\infty$. Using this, we conclude by (10.3.2) and the discreteness of the set of equilibria of (10.3.2) that σ_t converges to an equilibrium. Furthermore, $\mid z_t \mid\to 0$ for $t \to +\infty$.

In order to verify condition (iii) of Theorem 10.3.1 we consider an arbitrary non-contractible path γ of the type $\alpha[k,l,p]$. W.l.o.g. we can assume γ is homotopic to the generator of the flat fundamental group $\pi_1(S^1)$, and consequently $\int_\gamma \langle h, du\rangle = \int_0^\Delta \varphi(\tau)\, d\tau$. To show that

$$\sum_{t=k}^{l-1}(\varepsilon - L)(c^*z_t + \varrho\varphi(\sigma_t))^2 + \delta\varphi(\sigma_t)^2 > \varepsilon_1 + \int_0^\Delta \varphi(r)\, dr$$

for some $\varepsilon_1 > 0$, it is sufficient to guarantee that

$$\sum_{t=k}^{l-1}(\varepsilon - L)(c^*z_t + \varrho\varphi(\sigma_t))^2 + \delta\varphi(\sigma_t)^2 \ge \nu \int_0^\Delta \mid \varphi(r) \mid\, dr,$$

where

$$\nu > \left| \int_0^\Delta \varphi(r)\, dr \right| \Big/ \int_0^\Delta \mid \varphi(r) \mid\, dr.$$

Obviously, if $d((z_k, \sigma_k), (z_l, \sigma_l)) < \varepsilon_3$, $\varepsilon_3 > 0$ is sufficiently small, there is a $\varepsilon_4 = \varepsilon_4(\varepsilon_3) > 0$, converging to zero with $\varepsilon_3 \to 0$ such that

$$\int_0^\Delta \mid \varphi(r) \mid\, dr - \varepsilon_4 \le \sum_{t=k}^{l-1} \int_{T[\sigma_t, \sigma_{t+1}]} \mid \varphi(\sigma) \mid\, d\sigma = \sum_{t=k}^{l-1} \mid \varphi(\widetilde\sigma_t) \mid \mid \sigma_{t+1} - \sigma_t \mid.$$

Thus we must guarantee that

$$\sum_{t=k}^{l-1}(\varepsilon - L)(\sigma_{t+1} - \sigma_t)^2 + \delta\varphi(\sigma_t)^2 \geq \nu \sum_{t=k}^{l-1} |\varphi(\widetilde{\sigma}_t)| \, |\sigma_{t+1} - \sigma_t| \,.$$

Note that

$$\varphi(\widetilde{\sigma}_t) = \varphi(\sigma_t) + \varphi'(\widetilde{\widetilde{\sigma}}_t)(\sigma_{t+1} - \sigma_t)$$

and, consequently,

$$|\varphi(\widetilde{\sigma}_t)| \leq |\varphi(\sigma_t)| + L\,|\sigma_{t+1} - \sigma_t|\,.$$

It follows that the required inequality is fulfilled if

$$\sum_{t=k}^{l-1}(\varepsilon - L - \nu L)(\sigma_{t+1} - \sigma_t)^2 - \nu\,|\varphi(\sigma_t)|\,|\sigma_{t+1} - \sigma_t| + \delta\varphi(\sigma_t)^2 \geq 0.$$

A sufficient condition for this, according to the Sylvester criterion, is $(\varepsilon - L - \nu L)\delta - \frac{\nu^2}{4} > 0$, which is satisfied by assumption (ii). $\blacksquare$

Let us continue the investigation of system (10.3.2). Suppose now the more general form

$$\begin{aligned}
z_{t+1} &= Az_t + B\varphi(\sigma_t), \\
\sigma_{t+1} &= \sigma_t + C^*z_t + R\varphi(\sigma_t), \quad t = 0, 1, \ldots,
\end{aligned} \tag{10.3.6}$$

where A, B, C and R are real matrices of order $n \times n$, $n \times m$, $n \times m$, $m \times m$, respectively.

Suppose that the pair (A, B) is controllable and the eigenvalues of A lie inside the open unit circle. Suppose also that $\varphi : \mathbf{R}^m \to \mathbf{R}^m$ is a vector-valued function having the property $\varphi(\sigma) = (\varphi_1(\sigma_1), \ldots, \varphi_m(\sigma_m))$ for $\sigma = (\sigma_1, \ldots, \sigma_m) \in \mathbf{R}^m$. We asume that every component φ_j is Δ_j-periodic, belongs to C^1 and has a finite number of zeros on $\mathbf{R}^m/\Gamma'$ with $\Gamma' = \{j\Delta, j \in \mathbf{Z}\}$ and $\Delta = (\Delta_1, \Delta_2, \ldots, \Delta_m)$. Assume also that for the components φ_j of φ we have $\varphi_j(\sigma) \not\equiv 0$ and

$$\mu_{1j} \leq \frac{d}{d\sigma}\varphi_j(\sigma) \leq \mu_{2j} \tag{10.3.7}$$

for all $\sigma \in \mathbf{R}$, where $\mu_{1j} < 0 < \mu_{2j}$ $(j = 1, 2, \ldots, m)$ are certain numbers.

Introduce the vectors $d_j = [0, \ldots, 0, \Delta_j, 0, \ldots, 0]$ where Δ_j is the $(n + j)$-th component of d_j. Our aim is to weaken the frequency-domain condition (i) of Theorem 10.3.2 by employing the information from the derivative of φ, given by (10.3.7). This may be done on the base of Theorem 10.3.1. We work however in the covering space $\mathbf{R}^n \times \mathbf{R}^m$ in order to demonstrate the classical Bakaev-Guzh procedure.

Let us determine for each $j = 1, 2, \ldots, m$ the value

$$\nu_j = \int\limits_0^{\Delta_j} \varphi_j(\sigma)\,d\sigma \;\bigg/\; \int\limits_0^{\Delta_j} |\varphi_j(\sigma)|\,d\sigma$$

and define the $m \times m$ matrix $\nu = \mathrm{diag}\,(\nu_1, \ldots, \nu_m)$. In the statement of the next lemma the symbol $(\overline{z}, \overline{\sigma}) \in \mathbf{R}^n \times \mathbf{R}^m$ denotes the image of $(z, \sigma) \in \mathbf{R}^n \times \mathbf{R}^m$ under system (10.3.6).

Lemma 10.3.1 *Suppose there exist diagonal $m \times m$ matrices $\kappa = \mathrm{diag}\,(\kappa_1, \ldots, \kappa_m)$, $\delta = \mathrm{diag}\,(\delta_1, \ldots, \delta_m) > 0$, $\varepsilon = \mathrm{diag}\,(\varepsilon_1, \ldots, \varepsilon_m) > 0$ and a function $W : \mathbf{R}^n \times \mathbf{R}^m \to \mathbf{R}$ such that the following properties hold:*

1. $W(\overline{z},\overline{\sigma})-W(z,\sigma) \leq -\varphi(\sigma)^*\kappa(\overline{\sigma}-\sigma)-(\overline{\sigma}-\sigma)^*\varepsilon(\overline{\sigma}-\sigma)-\varphi(\sigma)^*\delta\varphi(\sigma)$ *for all* $(z,\sigma) \in \mathbf{R}^n\times\mathbf{R}^m$;

2. $4\delta(\varepsilon - \mu(I+\nu)) > (\kappa\nu)^2$, *where* $\mu = \mathrm{diag}\,(\mu_1,\ldots,\mu_m)$ *and* $\mu_j = \begin{cases} \mu_{2j} & \text{if} \quad \kappa_j \geq 0 \\ \mu_{1j} & \text{if} \quad \kappa_j < 0. \end{cases}$

Then every solution $\{z_t,\sigma_t\}$ *of* (10.3.6) *with* $\{W(z_t,\sigma_t)\}$ *bounded below satisfies* $\lim\limits_{t\to\infty} \varphi(\sigma_t) = 0$.

Proof Define the function $V : \mathbf{R}^n \times \mathbf{R}^m \to \mathbf{R}$ by

$$V(z,\sigma) = W(z,\sigma) + \int_0^\sigma [\varphi(\tau) - \nu \mid \varphi \mid (\tau)]^*\kappa\, d\tau$$

with $\mid \varphi \mid (\tau) := (\mid \varphi_1(\tau_1) \mid,\ldots,\mid \varphi_m(\tau_m) \mid)$ for $\tau \in \mathbf{R}^m$. We then get by assumption (i) that

$$\begin{aligned} V(\overline{z},\overline{\sigma}) - V(z,\sigma) \;\leq\; & -\varphi(\sigma)^*\kappa(\overline{\sigma}-\sigma) - (\overline{\sigma}-\sigma)^*\varepsilon(\overline{\sigma}-\sigma) \\ & -\varphi(\sigma)^*\delta\varphi(\sigma) + \int_\sigma^{\overline{\sigma}}[\varphi(\tau) - \nu \mid \varphi \mid (\tau)]^*\kappa\, d\tau \end{aligned} \tag{10.3.8}$$

for all $(z,\sigma) \in \mathbf{R}^n \times \mathbf{R}^m$. Using the property (10.3.7) and the mean value theorem we conclude that

$$\begin{aligned} \int_\sigma^{\overline{\sigma}} [\varphi(\tau) - \nu \mid \varphi \mid (\tau)]^*\kappa\, d\tau \;=\; & [\varphi(\widetilde{\sigma}) - \nu \mid \varphi \mid (\widetilde{\sigma})]^*\kappa(\overline{\sigma}-\sigma) \\[4pt] =\; & [\varphi(\sigma) - \nu \mid \varphi \mid (\sigma)]^*\kappa(\overline{\sigma}-\sigma) + \\ & +[(\varphi(\widetilde{\sigma}) - \nu \mid \varphi \mid (\widetilde{\sigma})) - (\varphi(\sigma) - \nu \mid \varphi \mid (\sigma))]^*\kappa(\overline{\sigma}-\sigma) \\ \leq\; & [\varphi(\sigma) - \nu \mid \varphi \mid (\sigma)]^*\kappa(\overline{\sigma}-\sigma) + (\overline{\sigma}-\sigma)^*\mu(I+\nu)(\overline{\sigma}-\sigma), \end{aligned}$$

where $\widetilde{\sigma}$ is some vector with components $(\widetilde{\sigma})_j$ between $(\sigma)_j$ and $(\widetilde{\sigma})_j$ for $j = 1,2,\ldots,m$.

Combining (10.3.8) and the last inequality we see that

$$V(\overline{z},\overline{\sigma}) - V(z,\sigma) \leq -(\overline{\sigma}-\sigma)^*[\varepsilon - \mu(I+\nu)](\overline{\sigma}-\sigma) - \nu \mid \varphi \mid (\sigma)^*\kappa(\overline{\sigma}-\sigma) - \varphi(\sigma)^*\delta\varphi(\sigma) \tag{10.3.9}$$

for all $(z,\sigma) \in \mathbf{R}^n \times \mathbf{R}^m$. Because of assumption (ii) there exist positive definite diagonal matrices ε_0 and δ_0 such that

$$\sigma^*[\varepsilon - \mu(I+\nu)]\sigma + \nu \mid \xi \mid^* \kappa\sigma + \xi^*\delta\xi \geq \sigma^*\varepsilon_0\sigma + \xi^*\delta_0\xi \tag{10.3.10}$$

for all $(\sigma,\xi) \in \mathbf{R}^m \times \mathbf{R}^m$ and $\mid \xi \mid := (\mid \xi_1 \mid,\mid \xi_2 \mid,\ldots,\mid \xi_m \mid)$. Suppose now that $\{z_t,\sigma_t\}$ is a solution of (10.3.6) with $\{W(z_t,\sigma_t)\}$ bounded below. Because of

$$\int_0^\Delta [\varphi(\sigma) - \nu \mid \varphi \mid (\tau)]^*\kappa\, d\sigma = 0$$

the integral

$$\int_0^{\sigma_t} [\varphi(\sigma) - \nu \mid \varphi \mid (\tau)]^*\kappa\, d\sigma$$

is bounded on $\mathbf{Z}_+$. Thus for the given solution $\{V(z_t,\sigma_t)\}$ is bounded below.

From (10.3.9) and (10.3.10) we get

$$\sum_{t=0}^{T} [\sigma_t^* \varepsilon_0 \sigma_t + \varphi(\sigma_t)^* \delta_0 \varphi(\sigma_t)] \leq V(z_0, \sigma_0) - V(z_{T+1}, \sigma_{T+1}) \leq c,$$

independently of T.

It follows that $\sum_{t=0}^{\infty} \varphi(\sigma_t)^* \delta_0 \varphi(\sigma_t) \leq c$ and $\varphi(\sigma_t) \to 0$ for $t \to +\infty$. $\blacksquare$

In the following theorem we use the notation

$$\mu_1 = \operatorname{diag}(\mu_{11}, \mu_{12}, \ldots, \mu_{1m}) \qquad \text{and} \qquad \mu_2 = \operatorname{diag}(\mu_{21}, \mu_{22}, \ldots, \mu_{2m})$$

with the numbers μ_{1j} and μ_{2j} from (10.3.7).

Theorem 10.3.3 *Suppose that there exist diagonal $m \times m$ matrices $\kappa = \operatorname{diag}(\kappa_1, \kappa_2, \ldots, \kappa_m)$, $\delta > 0$, $\varepsilon > 0$ and $\tau \geq 0$ such that condition 2) of Lemma 10.3.1 holds and*

$$\kappa \operatorname{Re} K(\lambda) - K^*(\lambda)\varepsilon K(\lambda) - \delta + \operatorname{Re}[\mu_1 K(\lambda) + (\lambda - 1)I]^* \tau [(\lambda - 1)I + \mu_2 K(\lambda)] \geq 0 \quad (10.3.11)$$

for all complex λ with $|\lambda| = 1$. Then system (10.3.6) is gradient-like.

Proof Let us introduce the notations $P = \begin{bmatrix} A & B \\ 0 & I \end{bmatrix}$, $Q = \begin{bmatrix} 0 \\ I \end{bmatrix}$, $D = \begin{bmatrix} C \\ R \end{bmatrix}$, $y = \begin{bmatrix} z \\ \varphi(\sigma) \end{bmatrix}$.
Thus P, Q, D and y are matrices of order $(n+m) \times (n+m)$, $(n+m) \times m$, $(n+m) \times m$, $(n+m) \times 1$, respectively. Then system (10.3.6) may be written as

$$\begin{aligned} y_{t+1} &= P y_t + Q \xi_t, & \xi_t &= \varphi(\sigma_{t+1}) - \varphi(\sigma_t) \\ \sigma_{t+1} &= \sigma_t + D^* y_t, & t &= 0, 1, \ldots. \end{aligned} \qquad (10.3.12)$$

Consider the quadratic form on $\mathbf{R}^{n+m} \times \mathbf{R}^m$

$$\begin{aligned} G(y, \xi) &= [Py + Q\xi]^* H[Py + Q\xi] - y^* Hy + y^* Q\nu D^* y + y^* D\varepsilon D^* y \\ &\quad + y^* Q^* \delta Q y + (\mu_1 D^* y - \xi)^* \tau(\xi - \mu_2 D^* y), \end{aligned}$$

where the $(n+m) \times (n+m)$ matrix $H = H^*$ is to be determined. From the controllability of (A, B) it follows that the pair (P, Q) is also controllable. Using this fact and the frequency-domain inequality (10.3.11) we get, by the Yakubovich-Kalman-Szegö theorem (Theorem 10.1.3), the existence of an $(n+m) \times (n+m)$ matrix $H = H^*$ such

$$G(y, \xi) \leq 0 \quad \text{for all} \quad (y, \xi) \in \mathbf{R}^{n+m} \times \mathbf{R}^m. \qquad (10.3.13)$$

Define the function $W : \mathbf{R}^{n+m} \to \mathbf{R}$ by $W(y) = y^* Hy$ and consider an arbitrary solution $\{y_t, \sigma_t\}$ of (10.3.12). From (10.3.7), the discrete stability of A and the boundedness of φ it follows that $\{y_t\}$ is bounded on $\mathbf{Z}_+$. Thus $\{W(y_t)\}$ is also bounded. Furthermore we have by (10.3.7) and (10.3.13) that

$$W(\overline{y}) - W(y) + \varphi(\sigma)^* \nu(\overline{\sigma} - \sigma) + (\overline{\sigma} - \sigma)^* \varepsilon(\overline{\sigma} - \sigma) + \varphi(\sigma)^* \delta \varphi(\sigma)$$

$$\equiv G(y, \varphi(\overline{\sigma}) - \varphi(\sigma)) - (\mu_1 D^* y - \varphi(\overline{\sigma}) + \varphi(\sigma))^* \tau(\varphi(\overline{\sigma}) - \varphi(\sigma) - \mu_2 D^* y) \leq 0.$$

Thus the assumptions of Lemma 10.3.1 are fulfilled. We conclude that $\varphi(\sigma_t) \to 0$ for $t \to +\infty$ and, consequently, $\sigma_t \to \text{const}$ for $t \to +\infty$. From the first equation of (10.3.6) we get that $z_t \to 0$ for $t \to \infty$. $\blacksquare$

Example 10.3.1 Consider an impulse PLL with one proportionally-integrating filter [140], which may be described by (10.3.6) with $m = 1$, $\varphi(\sigma) = T_p\Omega_y[\sin(\sigma + \sigma^0) - \sigma^0]$ and the transfer function $K(s) = (ds + 1 - b - d)/(s - b)$. We suppose that T_p, Ω_y, $\sigma^0 \in (0, \pi/2)$, d and $b \in (0, 1)$ are positive constants.

The transfer function K may be rewritten in the form $K(s) = d + c/(s - b)$ with $c := db + 1 - b - d$. A direct and rough estimation shows that the frequency condition (i) of Theorem 10.3.2 is satisfied if for some $\varepsilon > 0$ and $\delta > 0$

$$-c(1 + b)[(1 + b)^2 + 1] + d(1 - b)^2 - \varepsilon c^2 - \varepsilon \left\{ c(1 + b) + d[(1 + b)^2 + 1]^2 \right\} > \delta(1 - b)^2, \quad (10.3.14)$$

proposed that $(1 - b)^2 \geq c(1 + b)$.

Also in a direct way one shows that $|\varphi'(\sigma)| \leq T_p\Omega_y =: L$ and

$$\left| \int_0^{2\pi} \varphi(\sigma)\, d\sigma \right| \Big/ \int_0^{2\pi} |\varphi(\sigma)|\, d\sigma < \frac{2\pi \sin \sigma^0}{4 - 2\pi \sin \sigma^0} =: \nu.$$

Thus condition (ii) of Theorem 10.3.2 is fulfilled if

$$4[\varepsilon - (1 + \nu)L]\delta > \nu^2. \quad (10.3.15)$$

It follows that the considered discrete-time system is gradient-like if for some $\varepsilon > 0$ and $\delta > 0$ conditions (10.3.14) and (10.3.15) are satisfied.

10.4 The Method of Non-Local Reduction

In this section the method of non-local reduction is extended to a class of discrete-time systems in order to get conditions for boundedness, global convergence and the existence of circular solutions, respectively. The present material is taken from [65].

Consider the discrete-time pendulum-like system

$$\begin{aligned} z_{t+1} &= Az_t + b\varphi(\sigma_t), \\ \sigma_{t+1} &= \sigma_t + c^*z_t + \varrho\varphi(\sigma_t), \quad t = 0, 1, \ldots, \end{aligned} \quad (10.4.1)$$

where A is a real $n \times n$ matrix, b and c are real n-vectors and ϱ is scalar. Suppose that the function $\varphi : \mathbf{R} \to \mathbf{R}$ is of class C^1, is Δ-periodic and has exactly the two zeros 0 and σ^0 on $[0, \Delta)$ with $\varphi'(0) > 0$ and $\varphi'(\sigma) < 0$. Suppose also that φ' is bounded with the constant $\mu_1 < 0$ and $\mu_2 > 0$:

$$\mu_1 \leq \varphi'(\sigma) \leq \mu_2 \text{ for all } \sigma \in \mathbf{R}. \quad (10.4.2)$$

Define for (10.4.1) the transfer function

$$K(p) = c^*(A - pI)^{-1}b - \varrho$$

and suppose that K is non-degenerate. From Section 2.2 it follows, that under the assumption, that any solution of the second-order equation $\ddot{\sigma} + \alpha\dot{\sigma} + \kappa\varphi(\sigma) = 0$ with constants $\alpha > 0$ and $\kappa > 0$, is bounded on $\mathbf{R}_+$, there exists a solution $F(\cdot)$ of the Cauchy problem

$$F'F + \alpha F + \kappa\varphi(\sigma) = 0, \qquad F(\sigma^0) = 0 \quad (10.4.3)$$

which is defined on $(-\infty, +\infty)$ and satisfies $F(\sigma)^2 \to \infty$ for $|\sigma| \to +\infty$.

Assume that F is such a solution of (10.4.3). It is clear that there exists the value

$$\beta(\alpha, \kappa) := \max_{\sigma \in \mathbf{R}} \frac{d}{d\sigma}[-F'(\sigma)F(\sigma)], \quad (10.4.4)$$

which will be needed in the statement of the following theorem.

224

Theorem 10.4.1 *Suppose that there exist numbers $\lambda \in (0,1]$, $\varepsilon > 0$, $\tau \geq 0$ and $\kappa > 0$ such that the following conditions are satisfied:*

(1) the eigenvalues of $\frac{1}{\lambda}A$ have moduli less than one;

(2) any solution of $\ddot{\sigma} + \sqrt{2\varepsilon(1-\lambda^2)}\,\dot{\sigma} + \kappa\varphi(\sigma) = 0$ is bounded on $\mathbf{R}_+$;

(3) for all complex s with $\mid s \mid = 1$ we have

$$\kappa\,\mathrm{Re}\,K(\lambda s) - (\varepsilon + \frac{1}{2}\beta) \mid K(\lambda s) \mid^2 > \tau\,\mathrm{Re}\,[\mu_2^{-1}(1-\lambda s) + K(\lambda s)]^* \cdot [\mu_1^{-1}(1-\lambda s) + K(\lambda \varepsilon)], \quad (10.4.5)$$

where the number $\beta = \beta\left(\sqrt{2\varepsilon(1-\lambda^2)}, \kappa\right)$ is defined by (10.4.4).

Then every solution $\{z_t, \sigma_t\}$ of (10.4.1) is bounded on $\mathbf{Z}_+$. If in addition to the above conditions the inequalities
$$(-1)^m[\mathrm{Re}\,K(s) - \frac{1}{2}\mu_m \mid K(s) \mid^2] > 0$$
take place for $m = 1, 2$ and all $s \in \mathbf{C}$ with $\mid s \mid = 1$, the system (10.4.4) is gradient-like.

In preparing the proof of Theorem 10.4.1 we state and prove the following lemma.

Lemma 10.4.1 *Suppose that there exist numbers $\varepsilon > 0$, $\kappa > 0$ and $\lambda \in (0,1]$ and a continuous function $W : \mathbf{R}^n \times \mathbf{R} \to \mathbf{R}$ with the following properties:*

(1) $W(z, 0) \geq z^*Gz$ for all $z \in \mathbf{R}^n$, where $G = G^*$ is a positive definite $n \times n$ matrix;

(2) any solution σ of the second-order equation

$$\ddot{\sigma} + \sqrt{2\varepsilon(1-\lambda^2)}\,\dot{\sigma} + \kappa\varphi(\sigma) = 0$$

is bounded on $\mathbf{R}_+$;

(3) the inequality

$$\frac{1}{\lambda^2}W(\overline{z}, \varphi(\overline{\sigma})) - W(z, \varphi(\sigma)) + \beta_1(\overline{\sigma} - \sigma)^2 + \kappa(\overline{\sigma} - \sigma)\varphi(\sigma) \leq 0 \qquad (10.4.6)$$

is satisfied for all $(z, \sigma) \in \mathbf{R}^n \times \mathbf{R}$, where $\beta_1 = \varepsilon + \frac{1}{2}\beta$, $\beta = \beta\left(\sqrt{2\varepsilon(1-\lambda^2)}, \kappa\right)$ is defined by (10.4.4) and $\overline{z}, \overline{\sigma}$ denotes the image of (z, σ) under (10.4.1).

Then any solution $\{z_t, \sigma_t\}$ of (10.4.1) with respect to σ_t is bounded on $\mathbf{Z}_+$.

Proof Suppose that F is a solution of the Cauchy problem (10.4.3) with $\alpha = \sqrt{2\varepsilon(1-\lambda^2)}$. Under condition (2) Section 2.2 guarantees that there exist functions $\{F_k\}_{k \in \mathbf{Z}_+}$ with

$$F_k(\sigma) = F(\sigma - k\Delta) \qquad \text{for} \qquad k = 0, 1, \ldots \quad \text{and} \quad \sigma \in \mathbf{R},$$

which are defined on $\mathbf{R}$ and satisfy the following conditions:

$$F_k'(\sigma)F_k(\sigma) + \sqrt{2\varepsilon(1-\lambda^2)}F_k(\sigma) + \kappa\varphi(\sigma) = 0 \qquad (10.4.7)$$

on $\mathbf{R}$;

$$F_k(\sigma^0 + k\Delta) = 0 \qquad \text{for} \qquad k = 0, 1, \ldots; \qquad (10.4.8)$$

$$\lim_{|\sigma|\to\infty} F_k^2(\sigma) = +\infty. \tag{10.4.9}$$

Let us now define for any $k = 0, 1, \dots$ the function $V_k : \mathbf{R}^n \times \mathbf{R} \to \mathbf{R}$ by

$$V_k(z, \sigma) := W(z, \varphi(\sigma)) - \frac{1}{2}\lambda^2 F_k^2(\sigma).$$

We show at first that for any $k = 0, 1, \dots$ and $z \in \mathbf{R}^n$, $\sigma \in \mathbf{R}$ the inequality

$$\frac{1}{\lambda^2}V_k(\overline{z}, \overline{\sigma}) - V_k(z, \sigma) =: g(z, \sigma) + f_k(\sigma) \leq 0 \tag{10.4.10}$$

holds. Here we have used the notations

$$g(z, \sigma) = \frac{1}{\lambda^2}W(\overline{z}, \varphi(\overline{\sigma})) - W(z, \varphi(\sigma)) + \beta_1(\overline{\sigma} - \sigma)^2 + \kappa(\overline{\sigma} - \sigma)\varphi(\sigma)$$

and

$$f_k(\sigma) = -\frac{1}{2}[F_k^2(\overline{\sigma}) - F_k^2(\sigma)] - \frac{1 - \lambda^2}{2}F_k^2(\sigma) - \beta_1(\overline{\sigma} - \sigma)^2 - \kappa(\overline{\sigma} - \sigma)\varphi(\sigma).$$

Using the Taylor expansion for $F_k^2(\overline{\sigma})$ in σ and taking account of (10.4.4) and the equation (10.4.7) we can estimate f_k for all σ

$$\begin{aligned}
f_k(\sigma) &\leq -\varepsilon(\overline{\sigma} - \sigma)^2 - (\overline{\sigma} - \sigma)[F_k'(\sigma)F_k(\sigma) + \kappa\varphi(\sigma)] - \frac{1 - \lambda^2}{2}F_k^2(\sigma) \\
&\leq \frac{1}{4}[F_k'(\sigma)F_k(\sigma) + \kappa\varphi(\sigma)]^2 - \frac{1 - \lambda^2}{2}F_k^2(\sigma) \\
&\equiv 0.
\end{aligned}$$

Employing this inequality and the assumption (10.4.6) we get the inequality (10.4.10). Since φ and W are continuous and (10.4.10) is satisfied it follows from (10.4.9) that for any $(z, \sigma) \in \mathbf{R}^n \times \mathbf{R}$ there exists a $k_0 = k_0(z, \sigma)$ such that for all $s \in [0, 1]$

$$V_k((1 - s)z + s\overline{z}, (1 - s)\sigma + s\overline{\sigma}) < 0 \tag{10.4.11}$$

whenever $|k| > k_0$.

Suppose now that there exists a solution $\{z_t, \sigma_t\}$ starting in (z_0, σ_0) such that σ_t is not bounded above on $\mathbf{Z}_+$. Then there exist integer numbers m and l with $l > k_0 = k_0(z_0, \sigma_0)$ satisfying

$$\sigma_t < \sigma^0 + l\Delta \qquad \text{for} \qquad t \leq m$$

and

$$\sigma_{m+1} \geq \sigma^0 + l\Delta. \tag{10.4.12}$$

Consider now for $s \in [0, 1]$ the continuous function $h(s) = \sigma_m(z(s), \sigma(s))$ where $z(s) = (1 - s)z_0 + sz_1$, $\sigma(s) = (1 - s)\sigma_0 + s\sigma_1$, $z_1 = \overline{z}_0$, $\sigma_1 = \overline{\sigma}_0$. Because of (10.4.12) we have

$$h(0) < \sigma^0 + l\Delta \quad \text{and} \quad h(1) \geq \sigma^0 + l\Delta.$$

It follows that there exists an $s^* \in (0, 1]$ with $h(s^*) = \sigma^0 + l\Delta$. Because of (10.4.8) and the assumptions on φ we get

$$F_l(h(s^*)) = \varphi(h(s^*)) = 0.$$

Using assumption (1) on W we conclude that

$$V_l(z_m(z(s^*), \sigma(s^*)), \varphi(\sigma_m(z(s^*), \sigma(s^*)))) \geq 0$$

which contradicts (10.4.10) and (10.4.11). In a similar way one shows that σ_t is bounded below.
∎

Proof of Theorem 10.4.1 Because of (10.4.5) and the non-degeneracy of $K(\cdot)$ there exists by the Yakubovich-Kalman-Szegö theorem an $(n+1) \times (n+1)$ matrix $H = H^*$ such that for the form

$$G(w,\xi) \;:=\; \frac{1}{\lambda^2}[Qw + q\xi]^* H[Qw + q\xi] - w^* H w + (\varepsilon + \frac{\beta}{2}) \mid r^* w \mid^2$$
$$+ \;\; \kappa w^* q r^* w + \tau(\mu_2^{-1}\xi - r^* w)^*(\mu_1^{-1}\xi - r^* w)$$

we have

$$G(w,\xi) \leq 0 \quad \text{for all} \quad w \in \mathbf{R}^{n+1}, \xi \in \mathbf{R}.$$

Here we have used the notations

$$Q = \begin{bmatrix} A & b \\ 0 & 1 \end{bmatrix}, \quad q = \begin{bmatrix} c \\ \varrho \end{bmatrix}, \quad r = \begin{bmatrix} 0 \\ \vdots \\ 0 \\ 1 \end{bmatrix}.$$

Note that for the function $W(z,\varphi(\sigma)) := [z,\varphi(\sigma)]H \begin{bmatrix} z \\ \varphi(\sigma) \end{bmatrix}$ the condition (2) of Lemma 10.4.1 is satisfied, which immediately follows from (10.4.2) and (10.4.12). Setting in (10.4.12) $w = \begin{bmatrix} z \\ 0 \end{bmatrix}$ and $\xi = 0$ we get the inequality

$$\frac{1}{\lambda^2} z^* A^* H_{11} A z - z^* H_{11} z \leq -(\varepsilon + \frac{1}{2}\beta + \tau) \mid c^* z \mid^2 \tag{10.4.13}$$

for all $z \in \mathbf{R}^n$, where H_{11} denotes the $n \times n$ matrix in the representation $H = \begin{bmatrix} H_{11} & H_{12} \\ H_{21} & H_{22} \end{bmatrix}$.

Using (10.4.13), the observability of (A,c) and the fact that the eigenvalues of $\frac{1}{\lambda}A$ lie inside the unit circle we conclude by Theorem 10.1.4 that H_{11} is positive definite. Thus for the function W defined in this way the assumptions of Lemma 10.4.1 are fulfilled and the boundness of $\{\sigma_t\}$ follows.

Condition (1) and the boundness of φ guarantee that every solution component z_t is bounded on $\mathbf{Z}_+$. ∎

Consider again system (10.4.1) and suppose for simplicity that $\varrho = 0$. In contrast to the previous assumptions on the zeros 0 and σ^0, the function φ is assumed to have arbitrary isolated zeros. Besides we assume that the assumption (10.4.2) is replaced by

$$\mid \varphi'(\sigma) \mid \leq \mu \quad \text{for all} \quad \sigma \in \mathbf{R}. \tag{10.4.14}$$

Theorem 10.4.2 *Suppose that there exist numbers $\lambda \neq 0$, $\beta > 0$ and $\alpha > 0$ such that the following conditions hold:*

(i) $\alpha > (1 - \lambda^2)(2 \mid \lambda c^* b \mid \sqrt{2\beta})^{-1}$;

(ii) *the matrix $\frac{1}{\lambda}A$ has one eigenvalue outside and $n-1$ eigenvalues inside the open unit circle;*

(iii) *the eigenvalues of $\frac{1}{\lambda}(I - \frac{1}{c^*b}bc^*)A$ lie inside the open unit circle;*

(iv) *the equation*
$$y'(\tau)y(\tau) + \alpha y(\tau) + \varphi(\tau) = 0 \tag{10.4.15}$$
has a periodic solution $y(\cdot)$ with $y(\tau) > 0$ for all $\tau \in \mathbf{R}$.

(v) *for all $s \in \mathbf{C}$, $\mid s \mid = 1$, there is*
$$\mathrm{Re}\,\chi(\lambda s) + \vartheta \mid \chi(\lambda s) \mid^2 + \beta \mid \lambda s \chi(\lambda p) - c^* b \mid^2 < 0, \tag{10.4.16}$$
where
$$\vartheta = \max\left\{0, \max_\tau[-\frac{1}{2}(y'(\tau) + \varphi'(\tau))]\right\}$$
and $\chi(s) = (s-1)K(s)$.

Then there exists for system (10.4.1) a positively invariant unbounded set Ω such, that every solution $\{z_t, \sigma_t\}$ starting in Ω does not converge to the set of equilibria of (10.4.1) and for all $t = 0, 1 \ldots$ the inequality $\sigma_{t+1} - \sigma_t > 0$ holds.

If in addition to the above conditions $\mid \lambda \mid \geq 1$, then any solution starting in Ω satisfies $\mid z_t \mid \to +\infty$ and $\mid \sigma_t \mid \to +\infty$ as $t \to +\infty$.

Proof Let us consider the function
$$V(z,\sigma) := z^* H z + \frac{1}{2} F^2(\sigma), \tag{10.4.17}$$

where $H = H^*$ is an $n \times n$ matrix and $F : \mathbf{R} \to \mathbf{R}_+$ a C^1-function to be determined. Define the sets
$$\Gamma := \{(z,\sigma) \ : \ V(z,\sigma) < 0\} \text{ and } \Omega := \{z \ : \ c^* z > 0\} \cup \Gamma. \tag{10.4.18}$$

We want to show that under the conditions of Theorem 10.4.2 there exists a matrix $H = H^*$ and a function F such that Ω is positively invariant for (10.4.1). As a first step we show that for an arbitrary point (z,σ) the inclusion $(z,\sigma) \in \Omega$ implies $(\overline{z}, \overline{\sigma}) \in \Gamma$. In order to prove that, we define the first difference of V with respect to (10.4.1) as
$$\Delta_\lambda V(z,\sigma) := \frac{1}{\lambda^2} V(\overline{z}, \overline{\sigma}) - V(z,\sigma) \tag{10.4.19}$$

for all $(z,\sigma) \in \mathbf{R}^n \times \mathbf{R}$. It is easily verified that $\Delta_\lambda V$ may be written in the form
$$\Delta_\lambda V(z,\sigma) = W(z, \varphi(\sigma)) + L(z, \sigma, \varphi(\sigma)) \tag{10.4.20}$$

where
$$\begin{aligned}
W(z,\xi) &= \frac{1}{\lambda^2}(Az + b\xi)^* H(Az + b\xi) - z^* H z + G(z,\xi), \\
G(z,\xi) &= -\frac{1}{\lambda^2}\xi c^* z + \frac{\vartheta}{\lambda^2}(c^* z)^2 + \frac{\beta}{\lambda^2}(c^* Az)^2, \\
L(z,\sigma,\xi) &= \frac{1}{2\lambda^2}F(\sigma + c^* z)^2 - \frac{1}{2}F^2(\sigma) - G(z,\xi)
\end{aligned}$$

for all $z \in \mathbf{R}^n$, $\xi \in \mathbf{R}$ and $\sigma \in \mathbf{R}$. Because of the frequency-domain condition (10.4.16) and the controllability of (A, b) Theorem 10.1.3 guarantees the existence of a matrix $H = H^*$ and a number $\delta > 0$ such that for all $(z, \xi) \in \mathbf{R}^n \times \mathbf{R}$ the inequality
$$W(z,\xi) \leq -\delta \mid z \mid^2 \tag{10.4.21}$$

228

holds.

Setting $\xi = 0$ in (10.4.21) we get that for all $z \in \mathbf{R}^n$

$$\frac{1}{\lambda^2} z^* A^* H A z - z^* H z \leq -\delta \mid z \mid^2 . \tag{10.4.22}$$

Since the matrix $\frac{1}{\lambda} A$ has one eigenvalue outside and $n - 1$ eigenvalues inside the unit circle we conclude by the last inequality that H has one negative and $n - 1$ positive eigenvalues. Because of $b \neq 0$ and $G(z, 0) \geq 0 \, (z \in \mathbf{R}^n)$ Lemma 10.2.1 says that

$$\{z \, : \, z^* H z \leq 0\} \cap \{z \, : \, c^* z = 0\} = \{0\} . \tag{10.4.23}$$

The relation (10.4.23) implies the existence of a number $\tau > 0$ such that

$$z^* H z + \tau (c^* z)^2 \geq 0 \tag{10.4.24}$$

for all $z \in \mathbf{R}^n$. Employing (10.4.21) for $z = 0$ we get

$$b^* H b \leq 0. \tag{10.4.25}$$

Using (10.4.23) and (10.4.25) together with $b \neq 0$ we get

$$c^* b \neq 0. \tag{10.4.26}$$

Let us now estimate the parameter $\tau > 0$ in (10.4.24). If we define the quadratic form G_1 by

$$G_1(z, \xi) := \frac{\tau}{\lambda^2} (Az + b\xi)^* cc^* (Az + b\xi) - \tau z^* cc^* z$$

we receive from (10.4.21) the inequality

$$\frac{1}{\lambda^2} (Az + b\xi)^* (H + \tau cc^*)(Az + b\xi) - z^* (H + \tau cc^*) z \leq -\delta \mid z \mid^2 + G_1(z, \xi) - G(z, \xi) \tag{10.4.27}$$

for all $(z, \xi) \in \mathbf{R}^n \times \mathbf{R}$. Now select for every $z \in \mathbf{R}^n$ the vector $u(z) = -\frac{1}{c^* b} c^* A z$ having the property

$$c^* [Az + bu(z)] = 0. \tag{10.4.28}$$

It follows that for every $z \in \mathbf{R}^n$

$$Az + bu(z) = (I - \frac{1}{c^* b} bc^*) Az =: Cz. \tag{10.4.29}$$

Finally combining (10.4.27) and (10.4.29) we get

$$\frac{1}{\lambda^2} z^* C^* (H + \tau cc^*) Cz - z^* (H + \tau cc^*) z \leq -\delta \mid z \mid^2 + G_1(z, u(z)) - G(z, u(z)) \tag{10.4.30}$$

for all $z \in \mathbf{R}^n$. We have for all $z \in \mathbf{R}^n$ the relation

$$G_1(z, u(z)) - G(z, u(z)) = -[\tau (c^* z)^2 - \frac{1}{\lambda^2 c^* b} (c^* Az) c^* z + \frac{\beta}{\lambda^2} (c^* z)^2] - \frac{\vartheta}{\lambda^2} (c^* z)^2. \tag{10.4.31}$$

If we take now $\tau \geq \tau_0$, where

$$\tau_0 = [4\lambda^2 (c^* b)^2 \beta]^{-1} \tag{10.4.32}$$

the term $[\ldots]$ in (10.4.31) is non-negative. Taking $\tau \geq \tau_0$ we see by (10.4.30) that

$$\frac{1}{\lambda^2} z^* C^* (H + \tau cc^*) Cz - z^* (H + \tau cc^*) z \leq -\delta \mid z \mid^2 \tag{10.4.33}$$

229

for all $z \in \mathbf{R}^n$. Since by assumption (iii) the matrix $\frac{1}{\lambda}C$ has only eigenvalues with moduli less one, Lemma 10.2.1 guarantees on the base of (10.4.33) that $H + \tau cc^* > 0$ for all $\tau \geq \tau_0$ and (10.4.24) is proved.

Consider now the function L from above. We write this function in the form

$$L(z, \sigma, \xi) = \frac{1}{2\lambda^2}[F^2(\sigma + c^*z) - F^2(\sigma)] + \frac{1}{2}(\frac{1}{\lambda^2} - 1)F^2(\sigma) - G(z, \xi). \tag{10.4.34}$$

Using the mean value theorem we get the representation

$$\frac{1}{2\lambda^2}[F^2(\sigma + c^*z) - F^2(\sigma)] = \frac{1}{\lambda^2}F'(\sigma)F(\sigma)c^*z + \frac{1}{2\lambda^2}[F''(\sigma_*)F(\sigma_*) + (F'(\sigma_*))^2](c^*z)^2 \tag{10.4.35}$$

where σ_* lies in the interval $[\sigma, \sigma + c^*z]$ for $c^*z \geq 0$ or $[\sigma + c^*z, \sigma]$ for $c^*z < 0$.

We now take the function F as a solution of

$$F'F + \alpha F + \varphi(\sigma) = 0. \tag{10.4.36}$$

By assumption (iv) such a function exists for sufficiently large α. It is clear that this function satisfies

$$\frac{1}{\lambda^2}[F'(\sigma)F(\sigma) + \varphi(\sigma)]c^*z = -\frac{\alpha}{\lambda^2}F(\sigma)c^*z. \tag{10.4.37}$$

Thus it follows that L can be written as

$$\begin{aligned} L(z, \sigma, \varphi(\sigma)) = \quad & -\frac{\alpha}{\lambda^2}F(\sigma)c^*z + \frac{1}{2}(\frac{1}{\lambda^2} - 1)F^2(\sigma) \\ & +\frac{1}{2\lambda^2}\left\{\frac{1}{2}[F''(\sigma_*)F(\sigma_*) + (F'(\sigma_*))^2] - \vartheta\right\}(c^*z)^2 - \frac{\beta}{\lambda^2}(c^*Az)^2 \end{aligned} \tag{10.4.38}$$

for arbitrary $(z, \sigma) \in \mathbf{R}^n \times \mathbf{R}$, where $\sigma_* = \sigma_*(\sigma, z)$. Because of (10.4.16) the term $\{\ldots\}$ in (10.4.38) is non-positive.

Let us estimate now the first two terms in (10.4.38). Suppose that $(z, \sigma) \in \Omega$, i.e. $z^*Hz + \frac{1}{2}F^2(\sigma) < 0$ and $c^*z > 0$. Using (10.4.23) we get for this point (z, σ)

$$\frac{1}{2}F^2(\sigma) \leq z^*Hz + \frac{1}{2}F^2(\sigma) + \tau(c^*z)^2 \leq \tau(c^*z)^2. \tag{10.4.39}$$

Since $F(\sigma) > 0$ and $c^*z > 0$ (10.4.39) implies that

$$c^*z \geq \frac{1}{\sqrt{2\tau}}F(\sigma). \tag{10.4.40}$$

Thus for a pair $(z, \sigma) \in \Omega$ it follows that

$$\bar{\sigma} - \sigma \geq \frac{1}{\sqrt{2\tau}}F(\sigma). \tag{10.4.41}$$

Using this estimate and taking (10.4.16) into account, we can estimate the form L as

$$L(z, \sigma, \varphi(\sigma)) \leq \left[\frac{1}{2}(\frac{1}{\lambda^2} - 1) - \frac{\alpha}{\lambda^2\sqrt{2\tau}}\right]F^2(\sigma) \tag{10.4.42}$$

for all $z \in \mathbf{R}^n$ and $\sigma \in \mathbf{R}$.

Now we take $\alpha > 0$ such that in (10.4.42) the term $[\ldots]$ is negative:

$$\alpha > \sqrt{\frac{\tau}{2}(1 - \lambda^2)}. \tag{10.4.43}$$

230

Because it suffices to guarantee (10.4.43) for $\tau = \tau_0$, the condition (i) for α implies the validity of (10.4.43). It follows that under the conditions of the theorem we have

$$L(z, \sigma, \varphi(\sigma)) \leq 0 \tag{10.4.44}$$

for all $(z, \sigma) \in \mathbf{R}^n \times \mathbf{R}$. Thus for an arbitrary pair $(z, \sigma) \in \Omega$ it follows that

$$\frac{1}{\lambda^2} V(\overline{z}, \overline{\sigma}) - V(z, \sigma) \leq -\delta \mid z \mid^2, \tag{10.4.45}$$

and, consequently, $(\overline{z}, \overline{\sigma}) \in \Gamma$.

Let us now show that the set Ω is positively invariant for (10.4.1). Suppose the opposite. Then there exists a pair $(z_0, \sigma_0) \in \Omega$ such that $(\overline{z}_0, \overline{\sigma}_0) \notin \Omega$ which is possible only in this case if

$$c^* \overline{z}_0 \leq 0. \tag{10.4.46}$$

Note, that there exists a point $a_0 \in \mathbf{R}^n$ with $(a_0, \sigma_0) \in \Omega$ and $(\overline{a}_0, \overline{\sigma}_0) \in \Omega$, that is

$$c^* \overline{a}_0 > 0. \tag{10.4.47}$$

To show this, note that the matrix $\frac{1}{\lambda} A$ has an (real) eigenvalue ν/λ with $\mid \nu/\lambda \mid > 1$. For the corresponding eigenvector u it follows from (10.4.22) that

$$(\mid \nu/\lambda \mid^2 - 1) u^* H u \leq -\delta \mid u \mid^2$$

and, consequently, $u^* H u < 0$. From this and (10.4.23) it follows that $c^* u \neq 0$. W.l.o.g. we can assume that

$$c^* u > 0. \tag{10.4.48}$$

Let us determine the vector a_0 as $a_0 = \omega \lambda u$, where the scalar ω is unknown. We have

$$V(a_0, \sigma_0) = \lambda^2 \omega^2 u^* H u + \frac{1}{2} F^2(\sigma_0)$$

and see that for

$$\omega > \frac{F(\sigma_0)}{\sqrt{-2\lambda^2 u^* H u}} \tag{10.4.49}$$

the inequality $V(a_0, \sigma_0) < 0$ is satisfied. Thus for this pair (a_0, σ_0) with ω satisfying (10.4.49) one has $(\overline{a}_0, \overline{\sigma}_0) \in \Gamma$.

Consider now the term

$$c^* \overline{a}_0 = c^* A a_0 + c^* b \varphi(\sigma_0) = \omega \lambda c^* A u + c^* b \varphi(\sigma_0) = \omega \lambda \nu c^* u + c^* b \varphi(\sigma_0). \tag{10.4.50}$$

By assumption $\lambda \nu > 0$. From this and (10.4.48) it follows that for

$$\omega > -c^* b \varphi(\sigma_0)(\lambda \nu c^* u)^{-1} \tag{10.4.51}$$

the inequality (10.4.47) is satisfied. Because Ω is path connected and $z \to c^* \overline{z}$ is continuous there exists an $x_0 \in \Omega$ with

$$c^* \overline{x}_0 = 0. \tag{10.4.52}$$

From $(x_0, \sigma_0) \in \Omega$ we get $(\overline{x}_0, \overline{\sigma}_0) \in \Gamma$ and in particular

$$\overline{x}_0^* H \overline{x}_0 + \frac{1}{2} F^2(\overline{\sigma}_0) < 0. \tag{10.4.53}$$

Using this, (10.4.52) and (10.4.23) we see that $\overline{x}_0 = 0$, which contradicts (10.4.53). Thus we have proved the positive invariance of Ω. $\blacksquare$

Bibliography

[1] S.M. Abramovich, Yu.A. Koryakin, G.A. Leonov, and V.Reitmann. Frequenzbedingungen für Schwingungen in diskreten Systemen. II. Schwingungen in diskreten Phasensystemen. *Wiss. Z. Techn. Univers. Dresden*, 26(1):115–122, 1977.

[2] H. Amann. *Gewöhnliche Differentialgleichungen*. Walter deGruyter, Berlin - New York, 1983.

[3] L. Amerio. Determinazione delle condizioni di stabilità per gli integrali di un'equazione interessante l'elettrotecnica. *Ann. Mat. pura ed appl.*, 30(4):75–90, 1949.

[4] L. Amerio. Studio asintotica del moto di un punto su una linea chiusa per azione di forze indipendenti dal tempo. *Ann. Scuola Norm. Sup. Pisa*, 3(3):17–57, 1950.

[5] A.A. Andoronov, A.A. Witt, and S.E. Chaikin. *Theorie der Schwingungen*. Akademie-Verlag, Berlin, 1965.

[6] A.A. Andronov, E.A. Leontevich, I.I. Gordon, and A.G. Maier. *Qualitative Theory of Second-Order Dynamical Systems*. Fizmatgiz, Moscow, 1966.

[7] A. Arapostathis, S.S. Sastry, and P. Varaiya. Global analysis of swing dynamics. *IEEE Trans. on Circuits and Systems*, 29(10):673–679, 1982.

[8] E. Arie, M. Botgros, A. Halanay, and D. Martac. Transient stability of the synchronous machine. *Rev. Roum. Sci. Techn. Serie Electrotechn. et Energy*, 19(4):611–625, 1974.

[9] Yu.N. Bakaev. Approximate integration of the differential equation of the pendulum(russian). *Priklad. Mat. i Mekh.*, 16(32):723–728, 1952.

[10] Yu.N. Bakaev. Investigation of an inertial system in the television synchronization. *Radiotekhnika i Elektronika*, 3(2):342–351, 1958.

[11] Yu.N. Bakaev. *Applied theory of phase synchronization (russian)*. PhD thesis, 1962.

[12] Yu.N. Bakaev. Stability investigation of synchronization systems with retarded arguments (russian). *Izvest. Akad. Nauk SSSR, Energetika i Avtomatika*, 6, 1962.

[13] Yu.N. Bakaev. Influence of delay on the conditions of synchronization of an automatic phase-controlled system (russian). *Izv. Akad. Nauk SSSR, Ser. Tekh. Kibernetika*, 1:139–143, 1963.

[14] Yu.N. Bakaev. Synchronizing properties in PLL systems of the third order (russian). *Radiotekhnika i Elektronika*, 16(6), 1965.

[15] Yu.N. Bakaev and A.A. Guzh. Optimal reception of frequency modulated signals under Doppler effect conditions (russian). *Radiotekhnika i Elektronika*, 10(1):175–196, 1965.

[16] A. Barbalat and A. Halanay. Evaluation of the critical value for the generalised equation of a pendulum (rumanian). *Communic. Acad. R.P.R.*, 10:385–389, 1960.

[17] E.A. Barbashin. Conditions for the existence of recurent trajectories in dynamical systems with a cylindrical phase space. *Differencial'nye Uravneniya*, 3:1632–1640, 1967.

[18] E.A. Barbashin and N.N. Krasovskij. On the stability of motion in the large (russian). *Dokl. Akad. Nauk SSSR*, 86:453–456, 1952.

[19] E.A. Barbashin and N.N. Krasovskij. On the existence of a Lyapunov function in the case of asymptotic stability in the whole (russian). *Priklad. Mat. i Mekh.*, 18(3):345–350, 1954.

[20] E.A. Barbashin and V.A. Tabueva. *Dynamical Systems with Cylindrical Phase Space (russian).* Nauka, Moscow, 1969.

[21] N.N. Bautin. Qualitative investigation of the equation of a PLL (russian). *Prikl. Mat. Mekh.*, 34(5):850–860, 1970.

[22] N.N. Bautin and E.A. Leontevich. *Methods and Techniques for the Qualitative Investigation of Dynamical Systems in the Plane (russian).* Nauka, Moscow, 1976.

[23] R. Bellman. Vector Lyapunov functions. *SIAM J. Control*, 1:32–34, 1962.

[24] V.N. Belykh. On the qualitative investigation of a non-autonomous equation of the second order (russian). *Differencial'nye Uravneniya*, 11(10):1738–1753, 1975.

[25] V.N. Belykh. *Qualitative Methods of Nonlinear Oscillations Theory for Lumped Parameter Systems.* Text-book. State University, Gorky, 1980.

[26] V.N. Belykh. On the qualitative structure and bifurcations of concrete dynamical systems. *Naukova Dumka*, pages 45–48, 1984.

[27] V.N. Belykh. *A two-dimensional comparison systems method in the qualitative theory of particular dynamical systems (russian).* Doctoral theses. State University, Gorky, 1985.

[28] V.N. Belykh and V.I. Nekorkin. Qualitative investigation of a system of three differential equations from phase synchronization theory (russian). *Prikl. Mat. Mekh.*, 39(4):642–649, 1975.

[29] L.N. Belyustina. *On an equation in the theory of electrical machines (russian)*, pages 173–186. In memory of A.A. Andronov. Izd. Akad. Nauk SSSR, Moscow, 1955.

[30] L.N. Belyustina. Investigation of a nonlinear system of PLL (russian). *Izv. Vys. Uchebn. Zaved. Radiofizika*, 2(2), 1959.

[31] L.N. Belyustina. *On the locking band and the numerical investigation of point mappings in certain synchronization problems.* In the book: Dinamika sistem. Meshvuz. sb. Vyp. 11. Gorky, 1976.

[32] L.N. Belyustina and V.N. Belykh. A qualitative investigation of dynamical systems on the cylinder (russian). *Differencial'nye Uravneniya*, 9(3):403–415, 1973.

[33] L.N. Belyustina, V.V. Bykov, K.G. Kivelyova, and V.D. Shalfeev. On the lock-in value of an AFC system with proportional integrating filter (russian). *Izv. Vyss. Uchebn. Zaved. Radiofiz.*, 13(4):561–567, 1970.

[34] R. Best. *Theorie und Anwendungen des Phase-locked loops. 4.th ed.* AT Verlag, Aarau-Stuttgart, 1987.

[35] B.N. Biswas, P. Banerjee, and A.K. Bhattacharya. Heteroclyne phase-locked loops - revisted. *IEEE Trans. on Communications*, 25(10):1164–1170, 1977.

[36] Z.U. Blyagoz, G.L. Komarova, and G.A. Leonov. *On the stability of phase systems.* In Analytical and Numerical Methods for Solving Problems in Mathematics and Mechanics. Alma-Ata, 1984.

[37] C. Böhm. Nuovi criteri di esistenza di soluzione periodiche di una nota equazione differenziale nonlineare. *Ann. Mat. Pura Appl.*, 35(4):343–352, 1953.

[38] H. Börner. *Phasenkopplungssysteme in der Nachrichten-, Mess- und Regelungstechnik.* Verlag d. Technik, Berlin, 1976.

[39] R.W. Brockett. *Finite Dimensional Linear Systems.* Wiley, New York, 1970.

[40] R.W. Brockett. On the asymptotic properties of solutions of differential equations with multiple equilibria. *J. Diff. Equations*, 44:249–262, 1982.

[41] I.M. Burkin, L.I. Burkina, and G.A. Leonov. The Barbashin problem in the theory of phase systems (russian). *Differencial'nye Uravneniya*, 17(11):1932–1944, 1981.

[42] I.M. Burkin and V.A. Yakubovich. Frequency-domain conditions for the existence of two almost periodic solutions of a nonlinear system of automatic control (russian). *Sibirskii Mat. Zh.*, 16(5), 1975.

[43] T.A. Burton. *Stability and Periodic Solutions of Ordinary and Functional Differential Equations.* Academic Press, Inc., London, 1985.

[44] T.K. Caughey. Hula-hoop: An example of heteroparametric exitation. *Am. J. Phys.*, 28(2), 1960.

[45] S.A. Chaplygin. *A New Method of Approximate Integration of Differential Equations.* GITTL, Moscow, 1950.

[46] C. Corduneanu. Applications of differential inequalities in stability theory. *An. Sti. Univ. Al. I. Cusa, Iasi Sect. I a Mat.*, 6:47–58, 1960.

[47] E. Fagiuoli and G.P. Szegö. Qualitative analysis by modern methods of a stability problem in power-systems analysis. *J. Franklin Institute*, 290(2), 1970.

[48] F.M. Gardner. *Phaselock Techiques.* J. Wiley & Sons, New York, 1966.

[49] A.Kh. Gelig, G.A. Leonov, and V.A. Yakubovich. *The Stability of Nonlinear Systems with a Non-unique Equilibrium State (Russian).* Nauka, Moscow, 1978.

[50] A. Giger. Ein Grenzproblem einer technisch wichtigen nichtlinearen Differentialgleichung. *Z.A.M. Ph.*, 7:121–129, 1956.

[51] C. Godbillon. *Dynamical Systems on Surfaces.* Springer, Berlin-Heidelberg-New York, 1983.

[52] A.A. Gorev. *Transient Processes in the Synchronous Machine (russian).* Gosenergoizdat, Leningrad-Moscow, 1950.

[53] N.A. Gubar'. Investigation of a piece-wise linear dynamical system with three parameters (russian). *Prikl. Mat. i Mekh.*, 25(6), 1961.

[54] P. Habets and K. Peiffer. Classification of stability-like concepts and their study using Lyapunov functions. *J. Math. Anal. Appl.*, 43:537–570, 1973.

[55] A. Halanay. *Stability problems for synchronous machines.* VII. Internationale Konferenz über nichtlineare Schwingungen. Abh. d. Akad. Wiss. DDR, 1975.

[56] A. Halanay, G.A. Leonov, and Vl. Răsvan. From pendulum equation to an extended analysis of synchronous machines. *Rend. Sem. Mat. Univers. Politecn. Torino*, 45(2):91–106, 1987.

[57] J.K. Hale. *Some examples of infinite dimensional dynamical systems.* Contemporary Mathematics 58, Part III. 1987.

[58] W.D. Hayes. On the equation for a damped pendulum under constant torque. *Z.A.M. Ph.*, 4(5):398–401, 1953.

[59] M.W. Hirsch. Stability and convergence in strongly monotone dynamical systems. *J. reine angew. Math*, 383:1–53, 1988.

[60] Th. Jerofsky and V. Reitmann. Bakaev-Guzh technique for discrete dynamical systems on Riemannian manifolds (russian). *Differencial'nye Uravneniya (submitted)*.

[61] R.E. Kalman. Physical and mathematical mechanisms of instability in nonlinear automatic control systems. *Trans. Amer. Soc. Mech. Eng.*, 79(3), 1957.

[62] R.E. Kalman. Lyapunov functions for the problem of Lur'e in automatic control. *Proceedings of the National Academy of Science of USA*, 49(2), 1963.

[63] E. Kamke. Zur Theorie der Systeme gewöhnlicher Differentialgleichungen. *II. Acta Matematica*, 58:57–85, 1932.

[64] M.V. Kapranov. The lock-in band in PLL systems. *Radiotekhn. i Elektron.*, 11:37–52, 1956.

[65] A.N. Karpichev, Yu.A. Koryakin, G.A. Leonov, and A.I. Shepelyavyi. *Frequency-domain criteria for stability and instability of multidimensional discrete phase synchronization systems (russian).* Voprosy Kibernetiki i vycisl. tekhn. Diskretnye sistemy, 1990.

[66] O.B. Kiseleva, G.A. Leonov, and V.B. Smirnova. *Estimation of the number of slipped cycles in PLL systems with disturbed parameters (russian).* Numerical Methods for Boundary Problems in Mathematical Physics. LISI, Leningrad, 1985.

[67] I. Klapper and T. Frankle. *Phase-locked and Frequency-Feedback Systems.* Academic Press, New York, 1972.

[68] H.W. Knobloch and F. Kappel. *Gewöhnliche Differentialgleichungen.* B.G. Teubner, Stuttgart, 1974.

[69] H.W. Knobloch and H. Kwakernaak. *Lineare Kontrolltheorie.* Akademie-Verlag, Berlin, 1986.

[70] N. Koksch. Construction of outward impermeable surface systems by means of comparison systems. *ZAMM (submitted)*.

[71] V.A. Korotkov. Estimation of stability of a synchronous motor with constantly acting perturbations (russian). In *Trudy 2.ovo seminara simpoziuma po primeneniyu funkcii Lyapunova v energetike*, pages 79–95, Nowosibirsk, 1970.

[72] Yu.A. Koryakin and G.A. Leonov. The Bakaev-Guzh technique for systems with several angular coordinates (russian). *Izvestiya Akad. Nauk Kazakhskoj SSR*, 3:41–46, 1976.

[73] Yu.A. Koryakin, G.A. Leonov, and V. Reitmann. Konvergenz im Mittel von Phasensystemen. *ZAMM*, 58(10):435–441, 1978.

[74] P.K. Kovács. *Transient Phenomena in Electrical Machines*. Akadémiai Kiadó, Budapest, 1984.

[75] M.A. Krasnosel'skij, B.Sh. Burd, and Yu.S. Kolesov. *Non-linear Almost Periodic Oscillations*. Nauka, Moscow, 1966.

[76] N.N. Krasovskij. *Some Problems of Stability of Motion (russian)*. Fismatigiz, Moscow, 1959.

[77] V. Lakshmikantham and S. Leela. *Differential and Integral Inequalities, Theory and Applications*. Academic Press, New York, London, 1969.

[78] S. Lefschetz. *Stability of Nonlinear Control Systems*. Academic Press, New York, 1965.

[79] G.A. Leonov. On the boundedness of the trajectories of phase systems (russian). *Sibirsk. Math. Zh.*, 15:687–692, 1973.

[80] G.A. Leonov. Stability and oscillations in phase-controlled systems (russian). *Sibirsk. Math. Zh.*, 16(5):1031–1052, 1975.

[81] G.A. Leonov. On a class of dynamical systems with cylindrical phase space (russian). *Sibirsk. Math. Zh.*, 17(1):91–112, 1976.

[82] G.A. Leonov. On the boundedness of solutions of phase systems. *Vestnik Leningrad Univ., Ser. Mat., Mekh., Astron.*, 1, 1976.

[83] G.A. Leonov. The second Lyapunov method in phase synchronization theory (russian). *Prikl. Math. i Mekh.*, 40(2):238–244, 1976.

[84] G.A. Leonov. A reduction theorem for time-dependent nonlinearities (russian). *Vestnik Leningrad Univ.*, 7(2):38–42, 1978.

[85] G.A. Leonov. Extension of Popov's frequency criterion for time-dependent nonlinearities (russian). *Avtomat. i Telemekh.*, 11:21–26, 1980.

[86] G.A. Leonov. Frequency-domain instability criteria for phase synchronization systems (russian). *Radiotekhnika i Elektronika*, 28(6):1101–1108, 1983.

[87] G.A. Leonov. On boundedness of solutions of non-autonomous differential equations (russian). *Vestnik Leningrad Univ., Ser. Mat., Mekh., Astron.*, 7, 1983.

[88] G.A. Leonov. *The non-local reduction method in nonlinear systems absolute stability theory. I, II. (russian)*. Avtomat. i. Telemekh. 2, 3. 1984.

[89] G.A. Leonov. On global stability of differential equations for phase synchronization systems (russian). *Differencial'nye Uravneniya*, 21(2):213–224, 1985.

[90] G.A. Leonov. Frequency conditions for the existence of limit cycles in dynamical systems with cylindrical phase space. *Differencial'nye Uravneniya*, 23(12):2047–2051, 1987.

[91] G.A. Leonov, S.M. Abramovich, L.I. Burkina, A.E. Kozyaruk, Yu.A. Koryakin, and V. Reitmann. *The reduction method for dynamical systems with cylindrical phase space and its use in the stability investigation of power systems (russian).* Stability theory and its applications. Nauka, Novosibirsk, 1979.

[92] G.A. Leonov, I.M. Burkin, and A.I. Shepelyavi. *Frequency-Domain Methods in Oscillations Theory. Multi-Dimensional analogue of the Van der Pol Equation and Dynamical Systems with Cylindrical Phase Space.* Leningrad State Univers., Leningrad, 1991.

[93] G.A. Leonov and A.N. Churilov. Frequency-domain conditions for boundedness of solutions of phase systems (russian). *Dynamics of systems, Meshvuz. Sb., Gorky,* (10):3–20, 1976.

[94] G.A. Leonov and A.N. Churilov. Frequency-domain stability criterion for systems with angular coordinates. *Vestn. Leningrad Univ., Ser. Mat., Mekh., Astron.,* 13, 1982

[95] G.A. Leonov and V. Reitmann. Lokalisierung der Lösung diskreter Systeme mit instationärer periodischer Nichtlinearität. *ZAMM,* 66(2):103–111, 1986.

[96] G.A. Leonov and V. Reitmann. *Attraktoreingrenzung für nichtlineare Systeme.* Teubner-Texte zur Mathematik. Teubner-Verlag, Leipzig, 1987.

[97] G.A. Leonov and V. Reitmann. Asymptotic behavior of solutions of differential equations on flat manifolds (russian). *Vestnik Leningr. Univ., Ser. Mat., Mekh., Astron.,* 1(1):33–38, 1991.

[98] G.A. Leonov, V. Reitmann, and T.L. Chshiyova. Eine Frequenzvariante der Vergleichsmethode von Belykh-Nekorkin in der Theorie der Phasensynchronisation. *Wiss. Z. d. Techn. Univers. Dresden,* 32(1):51–59, 1983.

[99] G.A. Leonov, V. Reitmann, and V.B. Smirnova. Convergent solutions of ordinary and functional differential pendulum-like equations. *ZAA (submitted).*

[100] G.A. Leonov and V.B. Smirnova. Asymptotic behavior of the solutions of integro-differential equations with periodic nonlinearities (russian). *Sibirsk. Mat. Zh.* 19(6):1406–1412, 1978.

[101] G.A. Leonov and V.B. Smirnova. *Certain properties of solutions of Volterra integro-differential equations with piece-wise periodic nonlinear functions.* Problems of contemporary theory of periodic motions. Izhevsk, 1980.

[102] G.A. Leonov and V.B. Smirnova. The non-local reduction method for integro-differential equations (russian). *Sibirsk. Mat. Zh.,* 21(4):112–124, 1980.

[103] G.A. Leonov and V.B. Smirnova. Non-local reduction method in differential equations theory. Series in Pure mathematics V. II. Topics in mathematical analysis, pages 658–694. World Scientific, Singapore et. al., 1989.

[104] M. Levi, F.C. Hoppensteadt, and W.L. Miranker. Dynamics of the Josephson junction. *Quarterly of Appl. Math.,* 7:157–188, 1978.

[105] A.Yu. Levin. On the stability of the solutions of a second-order equation (russian). *Dokl. Akad. Nauk SSSR,* 141:1298–1301, 1961.

[106] Z.c. Liang. The boundedness of solutions of certain nonlinear differential equations. *Chinese Math.*, 3(2):169–183, 1963.

[107] W.C. Lindsey. *Synchronization Systems in Communication and Control.* Prentice-Hall, Inc., New Jersey, 1972.

[108] A.M. Lyapunov. *General Problem of Stability of Motion (russian).* Charkov, 1892.

[109] J. Mamrilla and S. Sedsiwy. The existence of periodic solutions of a certain dynamical system in a cylindrical space. *Bollitino U.M.I.*, 4(4), 1971.

[110] J.E. Marsden and M. McCracken. *The Hopf Bifurcation and its Applications.* Springer-Verlag, Berlin, 1976.

[111] V.M. Matrosov. On stability of motion (russian). *Prikl. Mat. Mekh.*, 26(6):992–1002, 1962.

[112] J. Mawhin and M. Willem. *Critical Point Theory and Hamiltonian Systems.* Springer, New York et. al., 1989.

[113] R.K. Miller and A.N. Michel. *Ordinary Differential Equations.* Academic Press, New York et. al., 1982.

[114] Yu.A. Mitropol'skij. *The Method of Averaging in Nonlinear Mechanics (russian).* Naukova Dumka, Kiev, 1971.

[115] A. Morary. Nonlinear oscillations of synchronous machines started with a pulsating rotating torque (russian). *Trudy meshd. konf. po nelin. koleb., Kiev*, 4, 1970.

[116] I. Newton. *Philosophiae naturalis principia mathematica.* Imprimatur S. Pepys, Reg. Soc. Praeses, Julii 5, 1686, Londini anno MDCLXXXVII, 1687.

[117] E.J. Noldus. On the stability of systems having several equilibrium states. *Appl. Sci. Res.*, 21:218–233, 1969.

[118] E.J. Noldus. New direct Lyapunov-type method for studying synchronisation problems. *Automatica*, 13(2):139–151, 1977.

[119] Yu.G. Panovko and I.I. Gubanova. *Stability and Oscillations in Solid Systems (russian).* Nauka, Moscow, 1979.

[120] V.A. Pliss. Reduction principle in the theory of motion stability (russian). *Izv. AN SSSR*, 28(6), 1966.

[121] H. Poincaré. Mémoire sur les courbes définies par les équations différentiebles. *J. math. pures et appl.*, 7:375–422, 1881.

[122] H. Poincaré. *Les méthodes nouvelles de la mécanique céleste.* T. 1. Gauthier-Villars, Paris, 1892.

[123] V.M. Popov. On absolute stability of nonlinear automatic control systems (russian). *Avtomat. i Telemekh.*, 22(8):961–979, 1961.

[124] V.M. Popov. *Hyperstability of Control Systems.* Springer, Berlin, 1973.

[125] Vl. Rasvan. *Stability Theory (rumanian).* Ed. s.s. Enciclop., Bukarest, 1987.

238

[126] W.T. Reid. Anatomy of the ordinary differential equation. *Amer. Math. Monthly*, 82:971–984, 1975.

[127] V. Reitmann. Über Instabilität im ganzen von nichtlinearen diskreten Systemen. *ZAMM*, 59:652–655, 1979.

[128] V. Reitmann. Über die Beschränktheit der Lösungen nichtstationärer Phasensysteme. *ZAA*, 1:83–93, 1982.

[129] V. Reitmann. Globale Stabilität und Umlauflösungen für ein System zweier gekoppelter Josephson-Kontakt-Gleichungen. *ZAMM*, 72(2), 1992.

[130] N. Rouche, P. Habets, and M. Laloy. *Stability Theory by Lyapunov's Direct Method*. Springer, New York, Heidelberg, Berlin, 1977.

[131] V.V. Rumyancev. *The Method of Lyapunov Functions in Stability Theory of Motion (russian)*, volume 1 of *Mechanics in USSR for 50 years*. Nauka, Moscow, 1968.

[132] G. Sansone and R. Conti. *Nonlinear Differential Equations*. Pergamon Press, New York, 1964.

[133] G. Seifert. On the existence of certain solutions of nonlinear differential equations. *Z.A.M. Ph.*, 3(6):468–471, 1952.

[134] G. Seifert. On stability questions for pendulum-type equations. *Z.A.M. Ph.*, 7(3):238–247, 1956.

[135] G. Seifert. The asymptotic behavior of solutions of pendulum-type equations. *Ann. Math.*, 69(1):75–87, 1959.

[136] V.S. Serebyakova and E.A. Barbashin. On circular motions of coupled pendulums (russian). *II. Izvestiya VUZ, Matematika*, 23(4), 1961.

[137] V.V. Shakhgil'dyan, editor. *Phase Synchronization Systems with Discrete Elements (russian)*. Svyaz, Moscow, 1979.

[138] V.V. Shakhgil'dyan and L.N. Belyustina, editors. *Phase Synchronization (russian)*. Svyaz, Moscow, 1975.

[139] V.V. Shakhgil'dyan and L.N. Belyustina, editors. *Phase Synchronization Systems*. Radio i Svyaz, Moscow, 1982.

[140] V.V. Shakhgil'dyan and A.A. Lyakhovkin. *Systems of Phase-shift Automatic Frequency Control (russian)*. 2nd. ed. Svyaz, Moscow, 1972.

[141] A.S. Somolinos. Periodic solutions of the sunflower equation:
$\ddot{x} + (a/r)\dot{x} + (b/r)\sin x(t - r) = 0.$. *Quarterly of Appl. Math.*, 1:465–478, 1978.

[142] J.J. Stoker. *Nonlinear Vibrations in Mechanical and Electrical Systems*. Interscience, New York, 1950.

[143] R.A. Stratonovich. *Selected Problems in Fluctuations Theory in Radiotechnique*. Sov. Radio, Moscow, 1991.

[144] V.A. Tabueva. Evaluation of the critical value of the parameter a for the differential equations $d^2x/dt^2 + a\,dx/dt + f(x) = 0$. (russian). *Izv. Vysh. Uchebn. Zaved Matematika*, 2:227–237, 1958.

[145] V.I. Tikhonov. *Problems in Random Processes*. Nauka, Moscow, 1970.

[146] R.C. Transworthe. Cycle slipping in phase-locked loops. *IEEE Trans. on Communications*, 15(3):417–421, 1967.

[147] F. Tricomi. Sur une équation differentielle de l'électrotechnique. *C.R. Acad. Sci. Paris*, 193:635–636, 1931.

[148] F. Tricomi. Integrazione di un'equazione differenziale presentatasi in electrotecnica. *Ann. R. Scuola Norm. Sup. Pisa*, 2(2):1–20, 1933.

[149] A.J. Viterbi. *Principles of Coherent Communication*. McGraw-Hill, New York, 1966.

[150] N.P. Vlasov. *Selfoscillations of a Synchronous motor (russian)*. Uchenye zapiski GGU, vyp. 12. 1939.

[151] A.A. Voronov. *Foundations of Automatic Control Theory. Particular Linear and Nonlinear Systems (russ.)*. Energoizdat, Moscow, 1981.

[152] T. Wazewski. Systèmes des équations et des inégalités differentielles ordinaires aux deuxièmes membres monotones et leurs applications. *Ann. Soc. Polonaise Math.*, 23:112–116, 1950.

[153] V.A. Yakubovich. The solution of certain matrix inequalities in automatic control theory. *Dokl. Akad. Nauk SSSR*, 143(6):1304–1307, 1962.

[154] V.A. Yakubovich. The S-procedure in nonlinear control theory (russ.). *Vestn. Leningr. Univ.*, 1:62–67, 1971.

[155] V.A. Yakubovich. Frequency-domain criteria for auto-oscillation in nonlinear systems with one stationary nonlinear component (russ.). *Sibirsk. Mat. Zh.*, 14(5):1100–1129, 1973.

[156] V.A. Yakubovich. The frequency theorem in control theory (russ.). *Sibirsk. Mat. Zh.*, 14(2):384–420, 1973.

[157] A.A. Yanko-Trinitskij. *A New Method for Analysing Synchronous Processes in the Presence of Abrupt Changed Loads (russ.)*. Gosenergoizd., Moscow, 1958.

[158] O.B. Yershova and G.A. Leonov. Frequency estimates of the number of cycle slippings in phase control systems (russ.). *Avtomat. i Telemekh.*, 5:65–72, 1983.

[159] T. Yoshizawa. *Stability theory by Lyapunov's second method*. The Math. Soc. of Japan, Tokyo, 1966.

Index